Managing Air Quality and Scenic Resources at National Parks and Wilderness Areas

Westview Replica Editions

The concept of Westview Replica Editions is a response to the continuing crisis in academic and informational publishing. Library budgets for books have been severely curtailed. Ever larger portions of general library budgets are being diverted from the purchase of books and used for data banks, computers, micromedia, and other methods of information retrieval. Interlibrary loan structures further reduce the edition sizes required to satisfy the needs of the scholarly community. Economic pressures (particularly inflation and high interest rates) on the university presses and the few private scholarly publishing companies have severely limited the capacity of the industry to properly serve the academic and research communities. As a result, many manuscripts dealing with important subjects, often representing the highest level of scholarship, are no longer economically viable publishing projects—or, if accepted for publication, are typically subject to lead times ranging from one to three years.

Westview Replica Editions are our practical solution to the problem. We accept a manuscript in camera-ready form, typed according to our specifications, and move it immediately into the production process. As always, the selection criteria include the importance of the subject, the work's contribution to scholarship, and its insight, originality of thought, and excellence of exposition. The responsibility for editing and proofreading lies with the author or sponsoring institution. We prepare chapter headings and display pages, file for copyright, and obtain Library of Congress Cataloging in Publication Data. A detailed manual contains simple instructions for preparing the final typescript, and our editorial staff is always available to answer questions.

The end result is a book printed on acid-free paper and bound in sturdy library-quality soft covers. We manufacture these books ourselves using equipment that does not require a lengthy make-ready process and that allows us to publish first editions of 300 to 600 copies and to reprint even smaller quantities as needed. Thus, we can produce Replica Editions quickly and can keep even very specialized books in print as long as there is a demand for them.

About the Book and Editors

*Managing Air Quality
and Scenic Resources at National Parks
and Wilderness Areas*
edited by Robert D. Rowe
and Lauraine G. Chestnut

Changes in scenic resources, whether through the impairment of visibility or through degradation of the landscape, may significantly affect a visitor's enjoyment of a national park or other recreation area. This book presents the most current research on methods for determining the value to visitors and others of preventing or mitigating the aesthetically undesirable impact on visual resources that may result from human activity. The contributors discuss a broad range of research topics, including human perceptions of air pollution, the role of air quality and visual resources in the recreation experience, visual-resource management systems, and the application of social, psychological, and economic concepts to measuring the value of protecting visibility and visual resources. They offer perspectives representative of industry, as well as environmental, concerns.

Dr. Rowe and Ms. Chestnut are economists for Energy and Resource Consultants, Inc., in Boulder, Colorado, and are authors of *The Value of Visibility: Economic Theory and Applications for Air Pollution Control*.

Managing Air Quality and Scenic Resources at National Parks and Wilderness Areas

edited by Robert D. Rowe
and Lauraine G. Chestnut
foreword by Russell E. Dickenson

Westview Press / Boulder, Colorado

363.7392
M266

A Westview Replica Edition

All rights reserved. No part of this publication may be reproduced or transmitted in any form or by any means, electronic or mechanical, including photocopy, recording, or any information storage and retrieval system, without permission in writing from the publisher.

Copyright © 1983 by Westview Press, Inc.

Pubished in 1983 in the United States of America by
 Westview Press, Inc.
 5500 Central Avenue
 Boulder, Colorado 80301
 Frederick A. Praeger, President and Publisher

Library of Congress Cataloging in Publication Data
Main entry under title:
Managing air quality and visual resources at national parks and wilderness areas.
 (A Westview replica edition)
 Readings from the Visual Values Workshop held May 1982, Keystone, Colo., sponsored by U.S. National Park Service, Electric Power Research Institute, American Petroleum Institute, and Edison Electric Institute.
 1. Air quality management—Congresses. 2. Visual perception—Congresses. 3. Landscape assessment—Congresses. 4. National parks and reserves—United States—Congresses. 5. Wilderness areas—United States—Congresses. I. Rowe, Robert D., 1949–   . II. Chestnut, Lauraine G., 1953–   . III. Visual Values Workshop (1982 : Keystone, Colo.). IV. United States. National Park Service.
TD883.1.M36 1982   363.7'392   82-20078
ISBN 0-86531-941-3

Printed and bound in the United States of America

Contents

List of Tables.. x

List of Figures... xiii

Foreword: Russell E. Dickenson, Director, National Park
          Service xv

Preface xvii

Introduction xix

PART I

DEFINING THE EXPERIENCE 1

1 DEFINING THE RECREATION EXPERIENCE.
   Perry J. Brown.. 3

2 SHAPING THE VISUAL EXPERIENCE: HISTORIC ORIGINS OF
   WILDERNESS AND DESERT AESTHETIC. John Opie 13

3 INTERPRETATION AND VISITOR VALUES. Douglas Bruce
   McHenry.. 21

PART II

MEASURING VISUAL PERCEPTIONS 25

4 HUMAN PERCEPTION OF VISUAL AIR QUALITY (LAYERED
   HAZE). William Malm, Karen Kelley MacFarland, John
   Molenar and Terry Daniel 27

5 EFFECTS OF VISUAL RANGE ON THE BEAUTY OF NATIONAL
   PARKS AND WILDERNESS AREA VISTAS. Douglas A. Latimer,
   Henry Hogo, Don H. Hern and Terry C. Daniel............ 41

6 IMPLICATIONS OF NCAR'S URBAN VISUAL AIR QUALITY
   ASSESSMENT METHOD FOR PRISTINE AREAS. Paulette
   Middleton, Thomas R. Stewart, Robin L. Dennis and
   Daniel Ely.. 51

7 PSYCHOPHYSICS, VISIBILITY, AND PERCEIVED ATMOS-
   PHERIC TRANSPARENCY. Ronald C. Henry............... 64

PART III

VISUAL RESOURCE MANAGEMENT SYSTEMS 75

8   THE BUREAU OF LAND MANAGEMENT'S VISUAL RESOURCE
    MANAGEMENT SYSTEM. Stanley V. Specht. 77

9   ASSESSING THE RELIABILITY, VALIDITY AND
    GENERALIZABILITY OF OBSERVER-BASED VISUAL IMPACT
    ASSESSMENT METHODS FOR THE WESTERN UNITED STATES.
    Richard C. Smardon, Nickolaus R. Feimer, Kenneth H.
    Craik and Stephen R.J. Sheppard 84

10 OBJECTIVE EVALUATION OF VISUAL VALUES.
    Ross T. Newkirk 103

PART IV

SOCIAL AND PSYCHOLOGICAL APPROACHES
TO VALUE ASSESSMENT 115

11 A ROLE THEORETIC ANALYSIS OF SCENIC QUALITY
    JUDGMENTS. Kenneth H. Craik 117

12 VISUAL AIR QUALITY VALUES: PUBLIC INPUT AND
    INFORMED CHOICE. Thomas R. Stewart 127

13 SOCIAL RESEARCH METHODS FOR PUBLIC LAND
    MANAGERS. Glen E. Haas and David M. Ross 143

14 AN EXAMINATION OF METHODOLOGIES FOR ASSESSING
    THE VALUE OF VISIBILITY. Karen Kelley MacFarland,
    William Malm and John Molenar 151

15 TWO EXAMPLES OF PSYCHOLOGICAL ASSESSMENT OF
    VISUAL VALUES. Ross J. Loomis and Thomas C. Greene 173

16 ASSESSING THE EFFECT OF VISUAL AIR QUALITY
    DEGRADATION ON VISITOR ENJOYMENT. William C. Malm,
    David Shaver and Gerald E. McGlothin 182

17 POTENTIAL CONTRIBUTIONS OF CANONICAL ANALYSIS TO
    VISUAL VALUE RESEARCH. Thomas Buchanan, Marcia J.
    Hayter and Jacquelin P. Buchanan 195

18 ALTERING THE VISUAL QUALITY OF A RECREATION
    RESOURCE AND ACTIVITY DISPLACEMENT. Rabel J. Burdge,
    Leo McAvoy, James Absher and James H. Gramann 202

PART V

ECONOMIC APPROACHES TO VALUE ASSESSMENT ... 215

19 THE VALUE TO VISITORS OF IMPROVING VISIBILITY AT MESA VERDE AND GREAT SMOKY NATIONAL PARKS. Douglas A. Rae 217

20 ECONOMIC VALUATION OF POTENTIAL SCENIC DEGRADATION AT BRYCE CANYON NATIONAL PARK. F. Reed Johnson and Abraham E. Haspel.................... 235

21 PRIORITIES FOR ECONOMIC ANALYSIS OF VISIBILITY VALUES. Robert D. Rowe and Lauraine G. Chestnut 246

22 EXISTENCE AND BEQUEST VALUE. Kenneth E. McConnell ... 254

23 EXISTENCE VALUE IN A TOTAL VALUATION FRAMEWORK. Alan Randall and John R. Stoll 265

24 UNREVEALED EXTRAMARKET VALUES: VALUES OUTSIDE THE NORMAL RANGE OF CONSUMER CHOICES. Daniel R. Talhelm.. 275

PART VI

MANAGEMENT PERSPECTIVES AND CASE STUDIES ... 287

25 MANAGING OUR VISUAL RESOURCES. John A. Taylor 289

26 THE IMPORTANCE OF VISIBILITY PROTECTION IN THE NATIONAL PARKS AND WILDERNESS. Robert E. Yuhnke 296

27 PROTECTION OF THE VISUAL EXPERIENCE IN THE FLAT TOPS WILDERNESS. Dennis Haddow and James Blankenship ... 298

28 THE LAKE TAHOE ENVIRONMENTAL THRESHOLDS STUDY. Sheila Brady 302

PART VII

CONCLUSIONS AND FUTURE DIRECTIONS 311

Tables

Chapter 1

1.1 Highly Valued Recreation Experiences for Selected Recreation Activities 5
1.2 Common National Park Activities and Recreation Experiences Possibly Affected by Changes in Visual Resources 9

Chapter 3

3.1 Examples of Active and Passive Recreation Activities 22
3.2 Examples of Active and Passive Educational Experiences in Recreation 22

Chapter 4

4.1 Slide Characteristics 30

Chapter 5

5.1 Percentage of Scenic Beauty (SBE) Explained by Visual Range and Illumination-Cloud Conditions 48

Chapter 9

9.1 Landscape Classification of Western United States 90
9.2 Average Single Rater Reliabilities for Direct Ratings 94
9.3 Scenic Beauty Change Scores Correlated with Direct Rating Change Score 95

Chapter 10

10.1 Landscape Diversity Components 106

Chapter 14

14.1 Correlation Matrix for Grand Canyon 162
14.2 Correlation Matrix for Mesa Verde 163
14.3 Principle Component Factor Analysis (Grand Canyon) 165
14.4 Principle Component Factor Analysis (Mesa Verde) 166
14.5 Analysis of Variance (Grand Canyon) 167
14.6 Analysis of Variance (Mesa Verde) 169

Chapter 15
- 15.1 Photograph Content for Polaroid Task in Trail Instrumentation of Visibility-Related Behavior 177
- 15.2 Observed Incidents of Visibility-Related Behaviors 178

Chapter 17
- 17.1 Canonical Analysis Results 199
- 17.2 Structure Matrix Coefficients 199

Chapter 18
- 18.1 Ranking of Site Characteristics Used in Selecting Place for Recreation Activity 205
- 18.2 Percentage of Respondents Having Done Fifteen River Related Recreation Activities During Sample Year, 1981 205
- 18.3 Percentage of Recreation Facilities on the UMRS Providing Each of Fourteen River Related Activities and Ranking of Manager-Provider Outdoor Activities 206
- 18.4 Relationship Between Participation in Selected River Related Recreation Activities and Ranking of Recreation Site Characteristics (Recreationists Only) 207
- 18.5 List of Perceptions of Management Issues Related to Expanded Navigation and Their Relationship to Recreation Activity on the UMRS 209
- 18.6 Relationship Between Perceptions of the Cleanliness of the Water and Activity Displacement, if the Water Became Any Dirtier, for Selected Water Based and Water Enhanced Recreation 211
- 18.7 Relationship Between Perceptions that Water Levels are Too High and Activity Displacement, if Water Levels Fluctuate More, for Selected Water Based and Water Enhanced Recreation 212

Chapter 19
- 19.1 Information on Slides Used in Visibility Survey 221
- 19.2 Frequency of Occurrence of Visibility Conditions 222
- 19.3 Aggregate Model Parameter Estimates: Mesa Verde Deterministic Case 224
- 19.4 Aggregate Model Parameter Estimates and Benefits: Mesa Verde Probabilistic Case 226
- 19.5 Aggregate Model Parameter and Benefit Estimates: Great Smoky Deterministic Case 228
- 19.6 Aggregate Model Parameter and Benefit Estimates: Great Smoky Probabilistic Case 230
- 19.7 Benefits of Improving Visibility at Mesa Verde and Great Smoky Based on Changes in the Frequency of Occurrence of Four Visibility Conditions 232

Chapter 20

20.1	Comparision of Samples From the Three Surveys	239
20.2	Comparison of Per Vehicle and Total Willingness to Pay for BCNP by Yovimpa Point Visitors, 1980	241
20.3	Per Vehicle Willingness to Pay for Yovimpa Point, B(Y) Sample	241
20.4	Regression and Probit Results	243
20.5	Crosstabulation: Change in Stay at YP and Change in Value of YP for Worst Case Photograph	244

Chapter 22

22.1	Surplus Measure Assuming that R is a Good Resource	258

Chapter 25

25.1	Sulfur Dioxide Emissions (1976 to 2010) in 100,000 Tons for the Mountain States Based on ICF and NERA (1982) Estimates for Copper Smelters	295
25.2	Sulfur Dioxide Emissions (1976 to 2000) in 100,000 Tons for the Mountain States Based on Manging and Mead (1980) Estimates for Copper Smelters	295

Figures

Chapter 4

4.1 Average PVAQ Ratings of Navajo Mountain as a Function of Apparent Contrast of Navajo Mountain 34

4.2 Average PVAQ Ratings of the Navajo Mountain Vista with Varying Levels of Plume/Sky Contrast 35

4.3 Average PVAQ Ratings of the Navajo Mountain Vista with Dark Plumes of Varying Levels of Plume/Sky Contrast 36

4.4 Average PVAQ Ratings of the Navajo Mountain Vista with Dark and Light Layers of Haze Superimposed on the Lower Two-Thirds of the Mountain 37

4.5 Average PVAQ Ratings of the Navajo Mountain Vista with Dark Plumes with Varying Vertical Dispersion 38

Chapter 5

5.1 Example of Variability in Observer Group Ratings of Scenic Beauty (SBE) of a Baseline Slide 45

Chapter 6

6.1 VAQ Components and Models 53
6.2 Structural Models for the Three Studies 57
6.3 Data Matrix 60

Chapter 7

7.1 Schematic of Typical Appearance of Transparent Haze 66

7.2 Brightness of Mountains Calculated from Teleradiometer Readings Versus Observed Brightness Using a Munsell Gray Scale 68

7.3 Frequency of Occurrence of the Ratio of Predicted to Observed Brightness for Distant Mountains 69

7.4 Frequency of Occurrence of Subjective Transparency Index, α 70

7.5	Scatter Diagram of Perceptual Transparency Index Versus Physical Contrast	71
7.6	Scatter Diagram of Transparency and Physical Contrast with Some Points Edited Out	72

Chapter 8
8.1	Management Class Matrix	81

Chapter 9
9.1	Map of Landscape Continuity and Provinces	89
9.2	Old BLM Rating Sheet	96
9.3	VIA Detailed Procedure	97
9.4	Sample Rating Form	99

Chapter 10
10.1	General Landscape Diversity	107
10.2a	Neighborhood SCANS for Relative Viewability	109
10.2b	Results of West-East SCANS	109
10.3	Relative Viewability	110
10.4	Viewable General Diversity	111

Chapter 11
11.1	Influences in the Perception Process	120

Chapter 12
12.1	Model of Individual Choice	129

Chapter 14
14.1	Slides Used in the Willingness to Pay Methodology	153
14.2	Slides Corresponding to Air Quality Levels Used in the Allocation of Time Methodology	154
14.3	Mean Bids for Willingness to Pay Sequence for Grand Canyon National Park	155
14.4	Mean Bids for Willingness to Pay Sequence for Mesa Verde National Park	156
14.5	Mean Amount of Time Visitors Would Increase or Decrease Their Length of Stay at a Grand Canyon Vista With Varying Visual Air Quality	158
14.6	Mean Amount of Time Visitors Would Increase or Decrease Their Length of Stay in Grand Canyon NP With Varying Visual Air Quality	160
14.7	Relative Ranking of Ten Items Visitors Considered Important to Their Enjoyment of the Parks	170

Chapter 21
21.1	Total Preservation Values Related to Number of Days of Adverse Impact Prevented	249

Chapter 23
23.1	Existence Value and Relative Scarcity	269

Foreword
by Russell E. Dickenson

As spokesman for the National Park Service, I intend to set forth the essential meaning and value of our national parks as they are perceived by the American people. It is this essential meaning and value that motivates the kind of research discussed and presented in this volume. The contributors to this volume will help guide and refine the merits and means of identifying significant visual values, and help develop the tools and techniques to preserve those values, where they are appropriate within this framework.

There are, it seems to me, three levels of public perception of the national parks. First, they are special. They have been chosen for preservation in perpetuity for particular, unique values; natural, scientific, cultural, or historical. Their special meaning in that context gives a linkage with the past: a constant reminder and revered remembrance of our heritage. Secondly, the national parks provide stability and reassurance. They give purpose and meaning to our national life. From every advantage of American perception and perspective, the parks rekindle patriotism. They boost the best in all of us as citizens--pride for our land as a steady beacon for this and future generations. Third, the national parks, in the minds of the American people, are comfortingly institutionalized. They reflect and add to the quality of life for all citizens. It is a truism--and one we of the National Park Service view with great pride--that while every American may not visit a national park, knowing they are there is sufficient reward.

There is also an economic value to the positive public perception of parks. Parks, as providers of both recreation services and facilities, mean jobs and income. Economic influence is a by-product of these parks, but when park values--be they natural, historical, or cultural--are damaged, an economic impact is created. Therefore, it is just as much in the public interest to preserve parks from the economic vantage as from the aesthetics of the visual.

There is also a philosophical as well as an economic perspective. One of the strengths of the national park system is its diversity. It is, and must be, as varied as the nation which embraces it--deserts, mountains, rivers, canyons, remnants of antiquity, and samples of more modern history--all deserving a place in the national park system.

Russell E. Dickenson has been the Director of the National Park Service since 1980 and with the National Park Service since 1946.

This, of course, does not mean that all parks mean the same thing to all people. If I, personally, prefer mountainous landscapes, fine. If you prefer the rugged, rock-strewn shore of Acadia National Park, that is your right. Some will be impressed by the giant holly trees of the Congaree Swamp, while others are transfixed by the stark outline of a giant saguaro cactus against a cloudless sky or the softened images of the Blue Ridge Mountains. We must preserve samples of each of these scenes if future generations are to truly know and understand the heritage of America.

National parks also must offer outlets for a wide range of active and passive recreation, they must allow room for mountain climbers, bird watchers, campers, recreationists of every stripe, from those seeking the quiet contemplation of nature or history to those reaching out for the adventure of challenging, rugged, white-water rivers in fragile kayaks.

It has been encouraging in recent years to recognize the astounding growth of public support for the national parks. Consider the vast acreage added nationwide; the enormous increase in units, especially those close to large urban centers; and the visitation, which swelled last year to a record 320 million persons.

Parks indeed are for people. This relationship was heightened by the environmental awareness of the 1960s with the often-stated desire for clean air, clean water, and a land unimpaired by degradation.

While some people today may be fearful that many environmental gains are threatened by unwise growth created by development and industrial decisions, we in the National Park Service believe that the depth of understanding of the American people is such that they will not permit deterioration of the environment around parks or other threats that could overwhelm them. Ultimately, the protection of national parks and its environment depend on the citizens who live in the immediate vicinity as well as several miles from the site.

How great a role should the National Park Service be allowed to play in matters beyond the boundaries of the parks? This consideration is a philosophical discussion point long debated within the ranks of the service. We must, I feel, be careful to avoid isolation. To paraphrase: no park is an island unto itself alone. We must work together with our neighbors, provide them with an early warning network to the threats that concern both our mutual interests. Our desire is a place at the table with the local citizenry. We do not want veto powers over issues affecting both of us. We simply want to be cooperative and consultative.

As we work together, we must keep in mind that no one value can transcend all other considerations. We must be careful to maintain a balance, seeking answers which are workable, compatible, and fair. We must identify those places where no compromise will ever be acceptable. But we must also identify those where compromise is appropriate.

This then is the charge of the National Park Service in its management of natural resources values: to build upon the partnership and support of the American public and to bring into closer harmony the reality of our mandate with the perception of our people.

Preface

This volume contains selected papers from the Visual Values Workshop, held May 10-12, 1982, that we were fortunate enough to conduct for Abt Associates under the joint sponsorship of the National Park Service (NPS), the Electric Power Research Institute (EPRI), the American Petroleum Institute (API), and the Edison Electric Institute (EEI). We were particularly well assisted by a steering committee that included Karen Kelley MacFarland, Dave Shaver, William Malm, and Jim Carroll of the NPS; Rob Farber of Southern California Edison, who was instrumental in securing cooperation and funding from industry representatives; and Ross Loomis of Colorado State University, who provided both a link to the 1979 workshop held in Fort Collins, Colorado, and assistance in planning the social and psychological value sessions. Ron Wyzga of EPRI, Hal Dunham of EEI, Jim Nelson of API, Paul Roberts of Chevron Research, and Rabel Burdge of the University of Illinois were instrumental in securing funding and advertising the workshop. Linda Gray of Keystone Resort, Maureen Sweeney, Doug Shaw, and Lynn Stuart assisted with conference coordination; Dean Birkenkamp of Westview Press and Sheri Harms, Marianne Brown, Barry Hooten, Dianne Sales, Leah Blondeau, and Mark Willoughby of Document Control, helped direct, edit, and produce the manuscript.

We would like to acknowledge the efforts of several individuals whose presentations at the workshop are not contained here, but are available elsewhere. Mordechai Shechter of Resources for the Future and Haifi University presented a paper entitled "Evaluation of Landscape Resources for Recreation Planning," which can be found in Regional Studies 15(5): 373-390, 1981. William Desvousges of Research Triangle Institute (RTI) presented a paper entitled "A Comparative Analysis of Travel Cost and Contingent Valuation Methods for User and Non-User Benefit Estimations," co-authored with V. Kerry Smith and presented with Ann Fisher. The paper was based upon an RTI report entitled "Alternative Methods for Estimating Recreation and Related Benefits of Water Quality Improvements: A Case Study of the Monongahela River," prepared for the Office of Policy Analysis, U.S. EPA, February 1982. Edward H. Stone, Assistant Director of Recreation for the Forest Service in Washington, D.C., gave a slide presentation and discussion on the Forest Service's perspectives in managing visual resources. The verbal comments are available from the author and from the editors of this volume. Finally, we would like to thank Jack Borden of For Spacious Skies, Inc., and WCVB-TV in Boston and Barbara Brown of the National Park Service who gave enlightening luncheon presentations, and the many participants who took the extra time and effort to conduct simulations of their surveys in the evenings, to chair sessions, and to participate in the discussions throughout the workshop.

In closing, we will always remember those spring days in the mountains of Colorado that started out warm and sunny, provided good company, enlightening discussions, grand vistas, and closed with a foot and a half of snow capturing many of us for an extra day.

Robert D. Rowe
Lauraine G. Chestnut
Energy and Resource Consultants, Inc.
Boulder, Colorado

Introduction

Consider the disappointment a person might experience after traveling a great distance to view a national scenic treasure, only to find it partially obscured by air pollution, or impaired by other human activities. Changes in visual aesthetic resources, whether they be through visibility impairment or through changes in scenic content, may significantly affect the enjoyment of a visitor's recreation experience at a national park, wilderness or other recreation area. How and to what extent visits to and enjoyment of these areas are affected by such impacts are important questions for the potential visitor and the resource manager. This includes the federal land managers, who control over one-third of our nation's lands and an even larger share of our unique scenic resources; the state officials who must both implement environmental regulations and be concerned with their often conflicting impacts upon industry and tourism; and finally those in industry whose activities may result in visual impacts.

This volume presents the most recent and ongoing research on state-of-the-art techniques and applications to address the human perception of changes in visual aesthetic resources and to assign psychological, social, and economic measures of value to visitors and others of preventing or reducing undesirable man-made visual aesthetic impacts. These measures of value help determine the adverseness of a potential impact and can therefore provide useful input to the management decision concerning whether it is to be allowed. Some measures of the value of protecting against visual aesthetic impacts can also be used in comparisons with the costs of control or costs of not undertaking a proposed action in a cost-benefit framework.

BACKGROUND AND ORGANIZATION

These papers and directions for future research are from the 1982 Visual Values Workshop jointly sponsored by the National Park Service, the Electric Power Research Institute, the American Petroleum Institute, and the Edison Electric Institute. The workshop was convened to focus upon the current state of research concerning how air quality impacts are perceived and valued by national park and wilderness visitors and how these values can be measured by researchers to assist the land manager with resource management decisions.

These managers are charged with protecting and preserving our unique scenic and natural resources in national parks, wilderness areas, and wildlife refuges for the enjoyment of all present and future generations.

With increasing population growth and demand for and development of energy resources near these protected federal lands, the incidence of and potential for visibility and visual impacts within or visible from these lands has been increasing. The land managers cannot, however, simply attempt to halt <u>all</u> potential development, and thus <u>all</u> human activity, energy development, and the like, but rather must be able to assess when and how these activities will adversely affect the resources they are mandated to protect. The appropriate method to measure impacts to scenic resources is to determine how it affects the experience of the recreationist at the affected site. This is where the research in this volume attempts to assist the land managers and those in industry who must be aware of how these decisions will be made.

This workshop followed an earlier effort (Fox et al. 1979) with the belief that considerable progress had since been made, but not widely distributed or professionally reviewed, and that researchers examining related issues were using methodologies of interest to, but not known to, visibility values researchers. It was also desired to obtain perspectives from those in federal agencies charged with making visual resource management decisions, and from those in industry and with environmental groups concerning the visual resource management decision making process and the effectiveness of current research to assist in this process.

Early in the workshop planning it became apparent that the value of preventing visibility impacts from haze or plume blight could only be fully evaluated if the content of the scene that is impacted is also considered. It seemed that value assessment techniques used for visual resource management and visibility value assessment techniques address similar questions. Consequently, the workshop and this volume also examine visual resource management systems and methodologies used to examine visual impacts, or changes in scenic content, such as may be caused by mineral exploration activities like strip mines and oil and gas exploration that may occur in and around natural recreation areas.

The organizing principle of this volume follows those steps necessary to effectively implement the current Clean Air Act regulations for mandatory Class I federal areas, which include national parks and wilderness areas (U.S. EPA 1980). As discussed below, a similar process must be undertaken for other related visual impact analyses. The Clean Air Act regulations specifically require states to assure they will make reasonable progress toward the goal of preventing future and remedying existing manmade impairment of visibility in mandatory Class I federal areas where visibility is determined to be an important value. The federal land manager and the federal official responsible for a particular Class I area are the key decision makers in determining whether a proposed facility will have an effect on the <u>air quality-related values</u> (including visibility) of a Class I area. If an applicant can demonstrate that a proposed facility will not have an adverse effect on air quality-related values, even if the facility will violate the allowable Class I prevention of significant deterioration (PSD) increments, the state may issue a permit to construct.

Under the visibility regulations, implementing a visibility impact analysis requires determining whether there exists (or may exist) a visibility impairment--a humanly perceptible change in visibility--and if so, whether it is deemed to be "significant" or "adverse"--meaning it interferes with the management, protection, preservation, or enjoyment of a mandatory Class I federal area, including interference with and impairment of the visitors' visual experience. A simplified version of this process is depicted in Figure 1 in four steps. In Step 1, existing or expected future conditions with and without a proposed activity are used to determine the expected change to the visual resource due to the activity. This may require simulation modeling of emission rates, transport, dispersion, and the resultant effect on environmental conditions. In Step 2, the characteristics of the potential impacts and human perception process are examined to determine if and what characteristics of the impacts will be humanly perceptible. If the impacts are not perceptible, no control or modification on the proposed activity will be required to protect visual resources. If an impact is perceptible, a determination of whether that impact is adverse to the visitor enjoyment of an area proceeds in Step 3. If it is not adverse, no controls or modifications are required to protect visitor enjoyment of scenic resources. If it is adverse, the impacts are undesirable and may not be allowed or may be required to be modified under current regulations protecting visitor enjoyment of visual resources at Class I areas. It may be the case under special circumstances, or where costs of control far exceed benefits, that some impacts will be allowed to occur with fewer facility modifications than would be required to prevent all adverse impacts.

The same analytical logic is also required if one is making a suitability determination under the Surface Mining Control and Reclamation Act regulations, which consider the visual aesthetic impacts of scenic alterations as one of several criteria for determining suitability. Specifically, Section 522 establishes several such criteria for determining whether an activity's impacts are unsuitable, including whether mining operations will "affect fragile or historic lands in which such operations could result in significant damage to important historic, cultural, scientific and aesthetic values and natural systems." A specific application of this concern for a proposed strip mine near Bryce Canyon National Park is discussed in Johnson and Haspel's paper in Chapter 20.

Decision making under other federal mandates, such as the Organic Act, which established the National Park Service, and similar Forest Service and Bureau of Land Management mandates, requires much the same analytic process of determining if an impact is perceptible and, if so, whether it is adverse to the visitor experience, although there are differences in mandates regarding development of resources on different lands. For example, the Forest Service nonwilderness areas and Bureau of Land Management are oriented toward a multiple use objective of allowing both recreation enjoyment and development of resources simultaneously, while the National Park Service and Forest Service wilderness area objectives call for the preservation of the scenic and natural settings to leave them unimpaired for all generations to enjoy.

The research focus of the papers in this volume generally take Step 1, in Figure 1, as a given and focus upon designing and implementing methodologies for use in Steps 2 through 4. Part I first sets the stage by focusing upon the role and importance of air quality and visual resources in the recreation experience based upon surveys of current users, a historical analysis of user attitudes and behavior, and the perspective of a National Park Service interpreter.

Part II addresses human perceptions of air pollution impacts to answer questions concerning what levels and characteristics of impacts are perceptible and how they should be measured (Step 2) and to help identify the types of impacts that need to be considered in adverseness determinations (a partial input to Step 3). Part III presents a review of visual resource management systems, which attempt to categorize and rank charac-

FIGURE 1

Visual Resource Impact Analysis Process

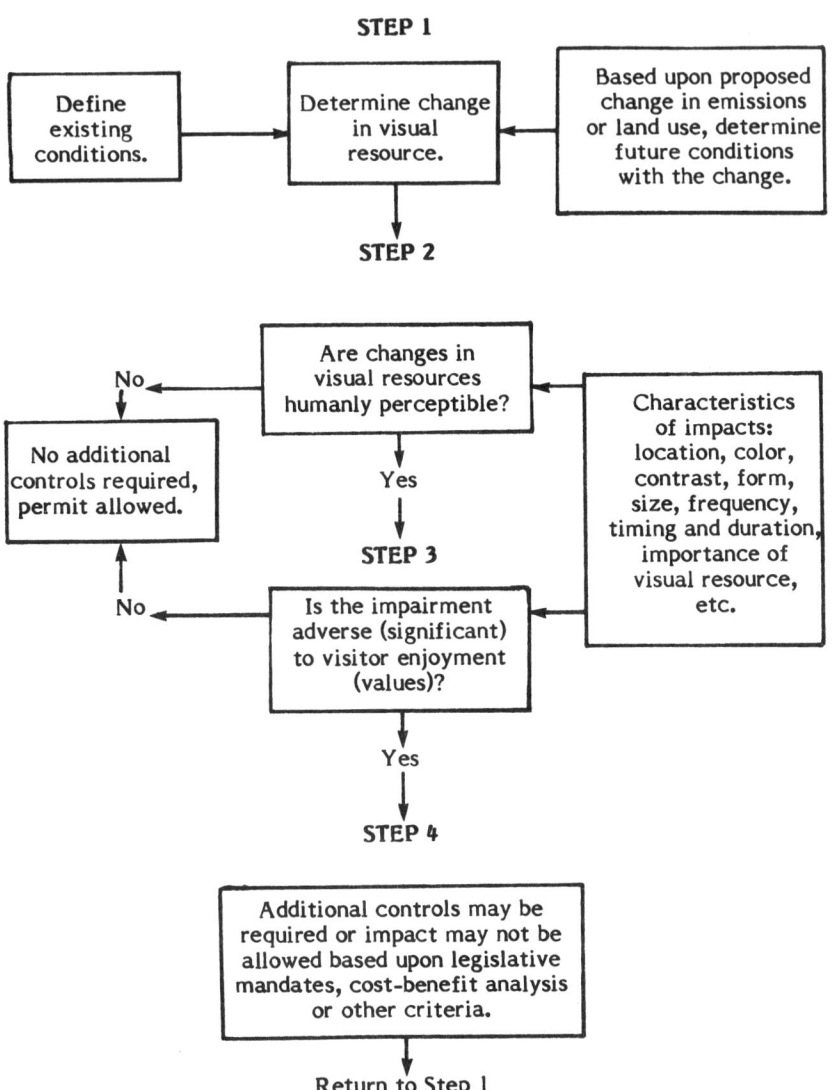

teristics of visual impacts again to assist in determining human perceptibility (Step 2) and the adverseness of the impact to visitor enjoyment (Step 3). This method can be used for visual impact assessments and, by analyzing scenic content of an affected scene, in visibility impairment decisions.

Part IV discusses issues, methodologies, and case studies concerning the use of social and psychological approaches to value assessment as a means for determining whether visual or visibility impacts are adverse to visitor enjoyment. Part V discusses economic concepts and approaches to value assessment. These economic value assessments can be used as an input to the measure of adverseness in the Step 3 decision or can be used as benefit measures in a Step 4 benefit-cost analysis of the economic feasibility and desirability of requiring additional controls. Some of the papers in this section also address the issues of nonuser benefits, such as existence values, and uncertainty about future use.

Papers in Part VI give representative perspective from individuals within industry and environmental organizations concerning the design and current implementation of the overall analysis and policy making process described in Figure 1. Two case studies at Lake Tahoe and the Flat Tops Wilderness Area are also included here as they address the difficulties in implementing the whole analysis process.

Part VII presents conclusions and directions for future research and applications based upon workshop discussions.

BIBLIOGRAPHY

Fox, D., R.J. Loomis and T.C. Green (technical coordinators). 1979. Proceedings of The Workshop in Visibility Values. U.S. Department of Agriculture, Forest Service General Technical Report WO-18. Fort Collins, CO.

U.S. Environmental Protection Agency. 1980. Visibility Protection for Federal Class I Areas. 45FR (December): 80084-80095. Washington, D.C.

PART I
Defining the Experience

Visual resources are usually enjoyed in the context of a recreation experience undertaken with family and friends. The scenic resources of the recreation areas may be the focus of the activity or may be one of several resources sought in the recreation experience. The valuation of visual resources is best undertaken with a thorough understanding of their importance to the context of the entire recreation experience and how these values have evolved. The papers in this section establish that visual resources and clean air are among the important resources to the recreation experience and set the stage for subsequent chapters that examine the value of visual resources to the recreation experience.

Perry Brown presents a careful review of the recreation literature to suggest a definition of recreation experiences and how changes in visual resources affect the realization of these experiences. He finds that visual resources (clean air and scenic beauty) are among the most important factors contributing to those recreation activities that are undertaken to gain the experiences recreationists most desire. These experiences reflect the desires to escape from pressures, to relate to nature, and to share and learn with others. He also indicates that degradations in visual resources will affect many recreation activities such that substitutions among activities will not eliminate the deleterious effect upon recreation experiences and that visitation rates will likely decrease.

John Opie indicates that visual values are a function of the human mind and gives an enlightening evolutionary account of how society came to value scenic resources using the example of the desert aesthetic. He traces the broad cultural social patterns of the desert aesthetic tradition from the desert being viewed as a vast, alien, and lifeless place, to a place perceived and used for spiritual renewal where one may escape the pressures of urban life. He describes five elements that are important to the evolution and understanding of the desert aesthetic in recreation experiences. Among these are the enlarged scales of space and time including the ability to "see forever" and the spectacular land forms and grand vistas.

While the first two papers are research oriented, Douglas Bruce McHenry provides a perspective of the importance of visual resources to national park visitors based upon a lifelong experience as a national park resident and employee. He holds a belief that visibility has been decreasing at many national parks, and that poor visibility is both noticed by and upsetting to park visitors.

1. Defining the Recreation Experience

Perry J. Brown

This volume has the purpose of examining how changes in visual aesthetic resources affect visitor experiences in parks and other national recreation areas. To me, this suggests that we need to know several things about recreation experiences, about changes in visual aesthetic resources, and about how these two might interact. The focus of this paper will be on defining recreation experience, identifying how recreation experiences are produced, showing which specific experiences are associated with different recreation activities, discussing how recreation experiences might be identified and measured, and suggesting how recreation experiences can be affected by changes in visual aesthetic resources. Discussion of specific changes in visual aesthetic resources and how these changes can be measured is left for other presentations.

RECREATION EXPERIENCE

According to Driver and Tocher (1970), recreation is a type of human experience which is based on intrinsically rewarding voluntary engagements during nonobligated time. This is a departure from the traditional definition of recreation as an activity, and it puts emphasis on the rewards or outcomes of participation in activities. This general definition of recreation fits our purposes quite well since the relevant definition of experience (our focus) is, the result of being engaged in an activity. Recreation experiences are, therefore, realizations of intrinsic outcomes from engagement in recreation activities. For example, we might say that engaging in camping leads to realizing such outcomes as enhancing understanding of nature, strengthening affiliation with one's camping group, and achieving greater development of woodsman skills. The central point is that recreation is characterized by the kinds of outcomes that are realized and that the set of salient outcomes is the recreation experience. These individual outcomes might also be termed specific experiences, which together make up the recreation experience.

Perry J. Brown is Professor and Head of Resource Recreation and Assistant Dean, School of Forestry, Oregon State University, Corvallis, OR. Presentation of this paper was supported by the Electric Power Research Institute.

Support for this way of defining recreation experiences comes from several sources including the integrative writings of Driver and Tocher (1970) as mentioned previously, other writers such as Wagar (1966), Brown, Dyer, and Whaley (1973), and Hendee (1974) who have written about quality and planning in recreation and the multiple satisfaction approach to recreation management, the expectancy-value theory of social psychology (e.g., Lawler 1973), and a model of production of recreation benefits (Brown in press, Driver and Rosenthal in press).

Expectancy-value theory has been used to suggest a relationship between recreation activities, settings, experiences, and ultimate benefits (Driver and Brown 1978). This theory clearly identifies the role of outcomes in the same way that we have used them in defining recreation experiences. When we use this theoretical construct to characterize recreation, we conceptualize experiences (sets of salient outcomes) as resulting from participation in specific recreation activities in specific settings.

In the recreation benefit production model, we view recreation experiences as intermediate products in the process of producing benefits. These experiences are produced by recreationists (not by managers) by combining their past experiences, knowledge, skills, equipment, time, and other resources with the recreation opportunities supplied by land managers. These recreation opportunities supplied by management are places where recreationists can engage in specific activities (behaviors) in specific settings made up of physical, social, and managerial attributes, and where they can have the expectation that certain experiences will be realized. This model indicated that the product of management, recreation opportunities, is a necessary but not sufficient input to production of recreation experiences, which are produced by recreationists when they participate in specific recreation activities in specific recreation settings as previously described. This model also helps expose the linkage between attributes of the recreation setting and recreation experiences, thus providing a framework for discussing the effects of changes in visual aesthetic resources on recreation experiences. The bottom line, as in recreation applications of the expectancy-value theory and in the integrative writings of many recreation researchers, is that recreation experiences are the output of engagement in recreation activities in recreation settings.

TYPES OF SPECIFIC RECREATION EXPERIENCES

What kinds of specific experiences do recreationists desire and realize? A comprehensive answer to this question is not possible, but the most highly valued specific experiences associated with ten common outdoor recreation activities are identified in Table 1.1. The data have come from several sources covering research undertaken in many sections of the United States, from New Hampshire to North Carolina to Washington to California and states in between such as Colorado and Michigan, and in national parks, national forests, Bureau of Land Management recreation areas, and selected state recreation areas (e.g., Arnold et al. 1981a, b, c, d; Brown and Haas 1980; Brown and Ross 1981; Driver and Cooksey 1977; Driver and Knopf 1976; Haas et al. 1980; Hautaluoma and Brown 1978; Manfredo 1979; More 1973; Potter et al. 1973; Roggenbuck 1980; Schreyer et al. 1976). The base data were not entirely comparable and therefore have been simplified into their most comparable format in Table 1.1, where only those specific experiences which were highly valued for at least one of the selected activities are listed. Several other specific experiences, such

TABLE 1.1
Highly Valued Recreation Experiences for Selected Recreation Activities.

Specific Experiences	Wilderness Backpacking-Western	Wilderness Backpacking-Eastern	Hiking Western	Hiking Eastern	Camping	Auto Picnicking	Sightseeing/Auto Driving	Fishing Western	Fishing Eastern	Hunting	Off-Road Vehicle Recreation	River Running	Cross-Country Skiing
Relationships With Nature	X	N	X	N	X	X	X	X	X	X	X	X	X
Escape From Physical Pressures	X	X	X	X	X	X	X	X	X	X		X	X
Escape From Social Pressures	X	X	X	X	X	X	X	X	N	X	X	X	X
Achievement/Challenge		X									X	X	
Autonomy/Independence/Freedom	X												
Reflection on Personal Values		X		X		X							
Recollection/Nostalgia						X		X	X				
Risk Taking/Action/Excitement											X	X	
Meeting/Observing Other People											X		
Use and Care of Equipment								X	N		X		
Exercise/Physical Fitness	X	X	X	X				X	X			X	X
Being With One's Recreation Grp.	X	X	X	X	X	X		X	X	X	X	X	X
Learning/Exploration	X	X	X	X	X		X	X	X		X	X	X
Family Togetherness	X		X	X	X	X	X	X			X	X	
Privacy	N	X		X			N	X					
Security				X							X		
Physical Rest					X		X	X	X				

X = Highly valued specific experiences.
N = Not measured so regional comparisons are not possible.

as harvesting big game and realizing social recognition, are not listed in the table because they were not highly valued for any of the activities listed.

Seventeen different specific experiences were identified as highly valued for one or more of the activities. For each activity, the number of highly valued specific experiences ranged from five for hunting to ten for auto camping, fishing, and off-road vehicle recreation.

Several specific experiences are common to nearly all of the selected recreation activities. <u>Relationships with nature</u> and <u>escape from social pressures</u> are common to all of the activities. <u>Being with one's recreation group</u> and <u>escape from physical pressures</u> are common to all the activities except sightseeing/auto driving and off-road vehicle recreation, respectively. <u>Learning/exploration</u> also is common to most of the activities.

While there are these commonalities in Table 1.1, some specific differences among activities are noticeable. For instance, a risk taking/ action/excitement experience was highly valued only for off-road vehicle and river running recreation, yet its value was measured for all ten actvities. Likewise, the experience of meeting and observing other people was highly valued for off-road vehicle use, yet its value also was measured for all ten activities. Also, there are a few, but not many, differences between eastern and western recreationists for the three activities for which a regional distinction is made in the table. For example, eastern backpackers and hikers value more highly an experience of reflecting on personal values than do their western counterparts.

Table 1.1 provides information which has been derived from studies done at several recreation areas and much of the individual variation among areas and recreationists within areas has been lost. One should keep in mind that the specific experiences realized from participation in specific activities will vary from area to area and among users of a particular area, and that the information in Table 1.1 presents a general case. For example, Driver and Cooksey (1977) have shown that picnickers in the mountains of northeast Pennsylvania value highly experiencing physical rest and reflecting on personal values while picnickers along Michigan's Huron River do not value these experiences so highly, but the Michigan recreationists value more highly experiencing family togetherness than do their Pennsylvania counterparts. Both groups, as one might expect, do value highly several common experiences. Brown and Haas (1980), in another study, have demonstrated that there are groups of recreationists seeking and realizing different experiences when engaging in the same activity in the same area. They identified five distinct types of backpackers in Colorado's Rawah Wilderness based upon how these backpackers valued eight specific recreation experiences.

The information in Table 1.1 also focuses only on activity-experience relationships, but as implied by the Driver and Cooksey (1977) work mentioned previously and by our understanding about how recreation experiences are produced, characteristics of the recreation setting also are related to experiences. While data on this point are not abundant, Brown and Ross (in press) have shown that certain specific experiences are linked to specific settings depicted along the recreation opportunity spectrum (e.g., Driver and Brown 1978). Manfredo (1979), in a study of users of three wilderness-primitive areas in Wyoming, also has shown that recreationists do perceive that management actions affect their realization of specific experiences. For instance, he found that zoning a wilderness is perceived to affect chances for experiencing freedom of time and movement and to affect experiencing solitude.

Relationships between activities, settings, and experiences are precisely the relationships of concern when considering the effects of changes in visual aesthetic resources on visitor experiences in parks and other natural recreation areas. Our major question is, what affect do changes in visual aesthetic resources due to air pollution have on visitor behavior (activity) and realization of desired experiences? We will return to this question later, but first digress into a consideration of how one might identify and measure recreation experiences.

IDENTIFYING AND MEASURING RECREATION EXPERIENCES

Identifying and measuring recreation experiences centers on analyzing individuals because it is individuals that actually produce and have recreation experiences. To be sure, there are social and cultural influences on recreation experiences, but individuals are the unit of production and consumption.

There are two general strategies for identifying a human phenomenon: (1) observing behavior and (2) obtaining self reports from people (Clark 1977). Which strategy is best depends upon the nature of the phenomenon being considered. For instance, if one is interested in participation in recreation activities, either strategy is applicable with observation providing description of actual behavior and self report providing description of what people say they did. If one is interested in recreation experiences, however, observing behavior is insufficient because experience is a result, and basically a cognitive result, of a behavior. Solely using observation of behavior would require inferring how a recreationist was affected. Self reports, however, would allow a recreationist to describe the effects of participation in an activity and to tell what kinds of experiences actually are realized.

The most frequently used method for identifying and measuring specific recreation experiences relies on use of a set of experience preference scales developed and tested over the past twelve years and available from the Rocky Mountain Forest and Range Experiment Station.[1] Forty-two separate scales which can be combined into twenty experience preference domains are available.

In identifying and measuring specific recreation experiences, relevant experience preference scales are included in a questionnaire and recreationists are asked to evaluate how much each scale item was realized or would be realized when participating in a specific activity in a specific setting. Sometimes respondents are also asked how much each scale item added to or detracted from their satisfaction. Usually responses are obtained using a modified Likert scale format with six to nine response points.

Once experience preference scale data are obtained from the relevant user population (or a sample of the population), they usually are subjected to a cluster analysis to discover how individual items group together and to reduce the number of specific experiences being considered. Recently this step has been occasionally omitted and the individual items have been assigned to specific experience scales which are based upon the extensive development and testing that has gone on over the past dozen years. Under either procedure, the goal has been to reduce the number of questionnaire items to a manageable, and meaningful, set of specific experiences. For each experience scale, a mean and a standard deviation are calculated. Means and standard deviations for the experience preference scales can be calculated for all subjects and for any subset which might be of interest based on participated-in activities, socio-economic characteristics, areas visited, and other classification variables. By using means, the highly valued and salient specific experiences for any group of users can be determined and then compared to any other group.

Often we desire to go beyond identification of specific recreation experiences and want to identify the recreation experience of an individual or groups of recreationists. As mentioned previously, the recreation experience is the totality of specific recreation experiences realized by the recreationist. To identify this, we can describe how each individual has evaluated (scored) each of the experience preference scales. Of more use than treating each recreationist individually, however, is identifying groups of recreationists who evaluate the experience preference scales similarly. This is done by clustering the recreationists based on their evaluations of the experience preference scales. The result is the kind of finding such as reported by Brown and Haas (1980) showing that there are five different groups of backpackers in the Rawah Wilderness based upon the kind of total experience they realized. For instance, in comparing two of the five groups of Rawah Wilderness recreationists, they showed the following quantitative evaluations for each of the specific experiences in the overall recreation experience: relationships with nature (+3.78, +2.29), escape pressures (+3.62, +2.21), autonomy (+3.38, +1.54), achievement (+3.36, +1.29), reflection on personal values (+3.24, +1.03), sharing/recollection (+2.69, +1.41), risk-taking (+1.17, -0.27), meeting/observing other people (-0.11). In this case, on the scale used, a +4.00 meant most strongly added to satisfaction and a -4.00 meant most strongly detracted from satisfaction.

Once the different recreation experiences are identified, various characteristics of them can be measured. It is possible to evaluate their monetary worth as King and Walka (1980) have done for fishing experiences at the Fort Apache Indian Reservation in Arizona, to assess their psychological value using rating and ranking techniques as Brown and Ross (1981) did for users of the Glenwood Springs Resource Area in Colorado, to assess their relationship to managerial attributes of recreational areas as notes by Manfredo (1979) in his study of wilderness users, and to do many other things.

Experiences and Changes in Visual Aesthetic Resources

Our overall concern is with how changes in visual aesthetic resources might affect realization of certain recreation experiences. Since these experiences result from people participating in specific activities in specific settings, one might expect that effects on experiences would arise by changes in settings causing changes in activities of recreationists. That is, if setting conditions change sufficiently, recreationist behavior and resulting experiences are likely to change.

Referring again to Table 1.1, we can note that some activities are more likely or appropriate in national parks than are other activities and also that some specific experiences are more likely associated with visual aesthetic resources than are some other specific experiences. From our list of ten common recreation activities we can exclude hunting and off-road vehicle recreation because they are not as common in national park areas. The specific experiences on our list most related to visual aesthetic resources are likely to be <u>relationships with nature</u>, <u>escape from physical pressures</u>, <u>recollection/nostalgia</u>, and <u>learning/exploration</u>. A new matrix showing the relationships between these activities and specific experiences is in Table 1.2.

TABLE 1.2
Common National Park Activities and Recreation Experiences Possibly Affected by Changes in Visual Resources.

Specific Experiences	Wilderness Backpacking Hiking	Camping Auto Picnicking	Sightseeing/ Auto Driving Fishing	River Running Cross-Country Skiing
Relationships With Nature	X X X	X	X X	X X
Escape From Physical Pressures	X X X	X	X X	X X
Recollection/Nostalgia		X	X	
Learning/Exploration	X X X		X X	X X

While sightseeing/auto driving is probably the only one of the activities listed in Table 1.2 which is dependent upon visual aesthetic resources, pursuit of any of the four specific experiences through any of the eight activities might be related to visual aesthetic resources. For instance, backpacking and hiking are often undertaken to experience vistas and distant scenery as part of experiencing relationships with nature and learning about and exploring different areas. Such activity is prevalent in many parks such as Yosemite, Mt. Rainier, Bryce Canyon, Rocky Mountain, Great Smokies, and many others. If visual aesthetic resources are degraded in the vicinity of these parks, opportunities to experience scenic vistas will be lessened. At some point of degradation, backpacking and hiking to experience scenic vistas would cease. Likewise, opportunities to escape from physical pressures would be lessened by degradation of visual aesthetic resources. It has been suggested, and there is some evidence to indicate, that physical pressures of the home and work environments such as crowding, noise, and air pollution are related to seeking wildland recreation experiences (Driver and Knopf 1976, Fesenmaier et al. 1981, Knopf 1976). If air quality degradation in the vicinity of national parks progresses as it has in many urban environments, these areas will no longer be able to provide opportunities to escape from air pollution related physical pressures.

Similar conclusions could be drawn about each of the activity-specific experience combinations shown in Table 1.2. But, the problem of negative changes in visual aesthetic resources is even more important because the recreation experience is made up of a combination of salient specific experiences. In reviewing Table 1.2, one can quickly note that three of the specific experiences assumed related to visual aesthetic resources are highly valued for nearly all of the activities listed. In other words, when we consider the multiple specific experiences which recreationists are attempting to realize from wildland recreation activities, we find that degradation of visual aesthetic resources can affect realization of many of the specific experiences. This observation suggests that it is not realistic to consider that if opportunity for one specific experience is lessened, recreationists will readily substitute desire for another specific experience, thus leaving park use and enjoyment at existing levels. The observation suggests that substantial parts of the recreation experience will be negatively affected, thus causing major changes in the recreation experiences realized.

CONCLUSION

This paper has briefly reviewed the concept of recreation experience, how recreation experiences are produced, which specific recreation experiences are highly valued for participants in several recreation activities, how one might identify and measure recreation experiences, and how recreation experiences arising from national park visitation might be affected by changes in visual aesthetic resources. While this paper lays out a framework for consideration of these topics, research is yet to be done which allows full exploration of them. We need to examine the relationships between additional recreation activities and specific recreation experiences. We need to continue to refine and validate techniques for identifying and measuring recreation experiences. We need to observe more carefully the relationships among specific attributes (e.g., air visibility), recreation activities and recreation experiences and we need to determine the _actual effects_ of changes in setting attributes on activities and realized experiences.

NOTES

1. Contact B. L. Driver, Research Forester, USDA Forest Service, Rocky Mountain Forest and Range Experiment Station, Fort Collins, CO 80526.

BIBLIOGRAPHY

Arnold, J. Ross, Perry J. Brown, B. L. Driver, and Steve Nachtman. 1981a. _Measuring Dispersed Use and Visitor Preferences on the Bureau of Land Management's National Resource Lands: Little Sahara Study._ Report to Bureau of Land Management. USDA Forest Service Rocky Mountain Forest and Range Experiment Station, Fort Collins, CO.

Arnold, J. Ross, Perry J. Brown, B. L. Driver, and Lynn Udick. 1981b. _Measuring Dispersed Use and Visitor Preferences on the Bureau of Land Management's National Resource Lands: Arkansas River Study._ Report to Bureau of Land Management. USDA Forest Service Rocky Mountain Forest and Range Experiment Station, Fort Collins, CO.

Arnold, J. Ross, B. L. Driver, Perry J. Brown, and Lynn Udick. 1981c. Measuring Dispersed Use and Visitor Preferences on the Bureau of Land Management's National Resource Lands: King Range Study. Report to Bureau of Land Management. USDA Forest Service Rocky Mountain Forest and Range Experiment Station, Fort Collins, CO.

Arnold, J. Ross, Steve Nachtman and Lynn Udick. 1981d. Sierra National Forest: Cross Validation User Preference Study. Report to Region 5, USDA Forest Service. USDA Forest Service Rocky Mountain Forest and Range Experiment Station, Fort Collins, CO.

Brown, Perry J. In Press. Benefits of Outdoor Recreation and Some Ideas for Valuing Recreation Opportunities. USDA Forest Service, Rocky Mountain Forest and Range Experiment Station, Fort Collins, CO.

Brown, Perry J., A. Allen Dyer, and Ross S. Whaley. 1973. "Recreation Research-So What?" Journal of Leisure Research 5(1):16-24.

Brown, Perry J., and Glenn E. Haas. 1980. "Wilderness Recreation Experiences." Journal of Leisure Research 12(3):229-241.

Brown, Perry J., and David M. Ross. 1981. Recreation Opportunity Spectrum User Preference Study for the Glenwood Springs Resource Area. Report to the Colorado State Office, Bureau of Land Management. Colorado State University, Fort Collins, CO.

Brown, Perry J., and David M. Ross. In Press. Recreation Experience Preferences as Variables in Recreation Setting Preference Decisions. USDA Forest Service North Central Forest Experiment Station, St. Paul, Minnesota.

Clark, Roger N. 1977. "Alternative Strategies for Studying River Recreationists." in Proceedings: River Recreation Management and Research Synmposium. USDA Forest Service General Technical Report NC-28, Washington, D.C.

Driver, B. L., and Perry J. Brown. 1978. "The Opportunity Spectrum Concept and Behavioral Information in Outdoor Recreation Resource Supply Inventories: A Rationale." in Integrated Inventories of Renewable Natural Resources: Proceedings of the Workshop. Gyde H. Lund et al. technical coordinators. USDA Forest Service General Technical Report RM-55, Fort Collins, CO.

Driver, B. L., and Raymond W. Cooksey. 1977. "Preferred Psychological Outcomes of Recreational Fishing." in Catch and Release Fishing as a Management Tool: A National Sport Fishing Symposium. R. A. Barnhart and T. D. Roelofs, editors. Humboldt State University, Arcata, CA.

Driver, B. L., and Richard C. Knopf. 1976. "Temporary Escape: One Product of Sport Fisheries Management." Fisheries 1(2):21-29.

Driver, B. L., and Donald H. Rosenthal, compilers. In Press. Measuring and Improving the Effectiveness of Public Outdoor Recreation Programs. George Washington University, Washington, D.C.

Driver, B. L., and S. Ross Tocher. 1970. "Toward a Behavioral Interpretation of Recreational Engagements, with Implications for Planning." in Elements of Outdoor Recreation Planning. B. L. Driver, editor. University Microfilms, Ann Arbor, MI.

Fesenmaier, Daniel R., Michael F. Goodchild, and Stanley R. Lieher. 1981. "The Importance of Urban Milieu in Predicting Recreation Participation: The Case of Day Hiking." Leisure Sciences 4(4):459-476.

Haas, Glenn E., B. L. Driver, and Perry J. Brown. 1980. "A Study of Ski Touring Experiences on the White River National Forest." in Proceedings North American Symposium on Dispersed Winter Recreation. College of Forestry, University of Minnesota, St. Paul, MN.

Hautaluoma, Jacob E., and Perry J. Brown. 1978. "Attributes of the Hunting Experience: A Cluster Analytic Study." Journal of Leisure Research 10(4):271-287.

Hendee, John C. 1974. "A Multiple-Satisfaction Approach to Game Management." Wildlife Society Bulletin 2(3):104-113.

King, David A., and Ann W. Walka. 1980. A Market Analysis of Trout Fishing on the Fort Apache Indian Reservation. Report to Rocky Mountain Forest and Range Experiment Station, USDA Forest Service. School of Renewable Natural Resources, University of Arizona, Tucson, AZ.

Knopf, Richard C. 1976. Relationships Between Desired Consequences of Recreation Engagements and Conditions in Home Neighborhood Environments. Unpublished Ph.D. dissertation, University of Michigan, Ann Arbor, MI.

Lawler, Edward E. III. 1973. Motivation in Work Organizations. Brooks/Cole Publishing Co., Monterey, CA.

Manfredo, Michael J. 1979. Wilderness Experience Opportunities and Management Preferences for Three Wyoming Wilderness Areas. Unpublished Ph.D. dissertation, Colorado State University, Fort Collins, CO.

More, Thomas A. 1973. "Attitudes of Massachusetts Hunters." in Human Dimensions in Wildlife Programs. J. C. Hendee and C. Schoenfeld, editors. Wildlife Management Institute, Washington, D.C.

Potter, Dale, John C. Hendee, and Roger N. Clark. 1973. "Hunting Satisfaction: Game, Guns, or Nature?" in Human Dimensions in Wildlife Programs. J. C. Hendee and C. Shoenfeld, editors. Wildlife Management Institute, Washington, D.C.

Roggenbuck, Joseph W. 1980. "Wilderness User Preferences: Eastern and Western Areas." in Proceedings of Wilderness Management Symposium. USDA Forest Service Eastern and Southern Regions.

Schreyer, Richard, Joseph W. Roggenbuck, Stephen F. McCool, Lawrence E. Royer, and J. Miller. 1976. The Dinosaur National Monument Whitewater River Recreation Study. Institute for Outdoor Recreation and Tourism, Utah State University, Logan, Utah.

Wagar, J. Alan. 1966. "Quality in Outdoor Recreation." Trends in Parks and Recreation 3(3):9-12.

2. Shaping the Visual Experience: Historic Origins of Wilderness and Desert Aesthetic

John Opie

The visual values of a natural area are both a state of mind and a set of physical dimensions. An outstanding example of the dialectic between subject and object is between the growing interest of visitors in the redrock arid country of southeast Utah, and the historic inhospitality of the desert to human penetration. For most of American history, few Americans saw any value to the nation's desert regions, and went out of their way to avoid them. Formally defined, a desert is any region with less than six inches of rain a year, with high heat and low humidity, where the chain of life is regulated ultimately by available water. Our attention here, however, is less of desert as an object of scientific enquiry into natural phenomena, and more upon the concrete existential human perception of desert by those who visit it.

Much is made of the human factor as a virtually unmanageable set of variables. Take five tourists, it is said, or five park service experts, and one will have as many conflicting viewpoints of the desert. But in truth the opposite is more appropriate: we habitually bind ourselves to routine visual patterns. Our visual world is more likely a world of stereotypes constantly put to work to label new places. Our visual baggage is not an incomprehensible mess. Some variables in visual values are less variable than others: patterns can be discerned. This subjective experience--"the human side of landscape"--is where humanists like historians, philosophers, experts in literature and the arts, can make significant contributions. The park visitor's state of mind is notoriously complex, diverse, and hard to quantify. But as long as visual values are taken only as highly personal mental images which remain impenetrable, little can be done in the accumulation of useful data and practical interpretation, much less planning and management. But the philosopher can report useful aesthetic principles; the historian follows human visual perceptions in a linear sequential process. We discover there are important observer constants which visitors carry when they respond to the stimuli of national parks and monuments. Important, specific, and identifiable visual values--aesthetic traditions--have been around for at least 300 years. They are surprisingly powerful and consistent despite our modern penchant for visual experimentation. Even when visual values go

---

John Opie is Professor of History, Duquesne University, Pittsburgh, PA, and Editor of *Environmental Review*. Travel funds for presentation of this paper were provided by the American Petroleum Institute.

through important changes, consistent patterns emerge. This paper explores important changes which made desert regions more visually desirable and accessible.

Not only are there immediate personal responses tied to specific visual experiences, but broader cultural and social patterns are also involved. That is, visual experiences evoke a broad range of non-visual responses. For example, are tourists visiting the western deserts to relive the frontier life? And conversely, life-styles and cultural traditions exist which are so long-term that they are virtually permanent. Historic attitudes toward desert regions fit into this category. A desert may be ecologically rich and varied, but the legendary image is an utterly hostile environment, empty of shelter and life-support. Historically it has been feared and respected as much as sailors respect the sea. The impression given is a world of nothingness, perhaps the ultimate emptiness next to the void of space. Our attention is upon desert as an invention of the human mind as much as it is a real entity.

This process psychologists call <u>cognitive mapping</u>, the creation of a geographical environment within a person's mind--"inner space"--and its collection, organization, storage, recall, and manipulation. Cognitive mapping involves "the world as people believe it to be," connected or unconnected with reality. Such an internal mental structure provides a template or map by which the individual perceives, measures, and judges external space. This familiar inner space serves as an anchor point necessary for individual self-identity and to avoid aimless wandering in a random world, Cognitive mapping thus is a directed activity, not helter-skelter, and provides important spatial information in a meaningful, useful, ordered manner. It is vital in developing a "lookahead" capacity, including a frame of reference for predictions, expectations, and guides to behavior.

While desert is a human invention, the tourist invented himself. By the 1830s, it was possible to take an American version of the English gentleman's Grand Tour. In search of educational and uplifting natural vistas, the tour worked up the Hudson to Albany, then to Lake George and the Berkshires, Mount Holyoke and the Connecticut Oxbow, and finally New Hampshire and the White Mountains. In the process one experienced all the requisite romantic scenes--rivers, waterfalls, lakes, mountain vistas, dark forests, and pastoral landscapes. The Hudson was America's Rhine journey. Lake George served as Lake Como. New Hampshire was America's Switzerland and the White Mountains did for the Alps. Guidebooks and landscape illustrations told the traveler the high points of his Grand Tour. In 1837, for example, over 3,000 people, when they visited the wonder of the Connecticut Oxbow, saw it through the expectations based on Timothy Dwight's essay and Thomas Cole's painting. At the same time, when the Erie Canal opened, Americans claimed Niagara Falls was a unique sight superior to anything that Europe had to offer.

By the end of the 1830s, tourists headed west, not far behind government expeditions and immigrant wagon trains, in search of new and more picturesque places. Railroads and steamboats opened the West, first for the well-heeled few and soon for the middle-class "vacationer." The tour, however, was not taken out of a general liking for the outdoors. Americans lived too close to the hardships of the frontier and physical labor of the farm. Outdoor life was considered primitive, unpleasant, and something to avoid for the comforts of home. A love for wilderness was not an early vacation priority. Even Thoreau's Walden Pond was free of dangerous animals and Indians, and town was a short walk away. The tourist's vacation was intended to produce a change of pace and change of scene, but not in the direction of hardship and privation. Even the rite of passage westward by train across the ocean-like Great Plains was to be rapid, comfortable, and convenient. (It is not surprising that some of the first sections of the interstate highway system crossed the same plains.) The vacation was then spent at the vast wooden hotels and lodges at Yellowstone, Manitou Springs, and El Tovar at the South Rim, lazy and relaxed, loafing on the porches, punctuated by leisurely walks to nearby vistas.

The tourist looked for a series of "master scenes," indicated by the requisite stops at "inspiration points," where the impact was required to be immediate, and could be conveniently contained in a paragraph in his guidebook, or by the steel engraving in a picture frame back home. The modern parallel is, of course, the two-week vacation, covering 6,000 miles, twelve national parks and monuments, and documented by ten rolls of color film.

The "master scenes" at Yellowstone were the geyser basin, Mammoth Hot Springs, and the Canyon of the Yellowstone. At Yosemite a single vista encompassed Half Dome and Bridal Veil Falls. But in both parks the rest of the wilderness had for tourists "a dreary and disappointing character." There was pressure in Congress to sell off the rest of Yellowstone for profit because it was so uninteresting. Both parks had originally been established to preserve "natural curiosities," not representative wilderness. If nature for the tourist was a parade of wonders for brief inspection, how were the wonders chosen and ordinary wilderness denied?

In the middle of the eighteenth century, when America was still British, Edmund Burke, the statesman and philosopher, took a look at nature, decreed it beautiful, centered his attention on wilderness, and called it "sublime" according to a set of features: obscurity (darkness of mist), power (storm, waterfall), privation (emptiness,silence), vastness (extension, remoteness), sense of infinity, eternity, difficulty of access or traversal, and magnificence (sunset, etc.). Burke also insisted that the wilderness sublime induced a personal emotional sublime. "Terrible scenes" filled the observer with "grand ideas" and raised his emotions to awe, rapture, and transcendence. Back in America, Thomas Jefferson and William Bartram, in their natural histories, applied Burke's sublime to the native wilderness. Americans paid special attention to Virginia's Natural Bridge.

As is true with most good ideas, the purity of Burke's sublime was corrupted by William Gilpin's "picturesque." Already the Englishman, Capabililty Brown, had combined park and garden into the famous "gardenpark" landscape--an idealized vista of large-scale rolling open meadows and tight clumps of trees. (See the alpine meadows of Rocky Mountain, Hayden Valley in Yellowstone, and Chesler Park in Canyonlands.) To this Gilpin added bric-a-brac which soon became requisite landscape furniture: rocky chasms, blasted trees, rustic bridges, even rustic herdsmen and lowing kine. The end appearance was a softened and accessible wilderness or a gone-to-seed rural countryside.

Another element was derived from the French landscape painter Claude Lorraine, who emphasized an idealized "art-is-superior-to-nature" approach. His emphasis upon light and atmosphere to create a poetic autumnal mood led American travelers to carry with them tinted spectacles to "improve" inspiring scenes. These "Claude glasses" were approximately rose-colored. (Was this continued into the era of photography through the use of sepia-toned prints?) Other visual traditions were also important, particularly Dutch and English rural landscape painters, America's own Hudson River School of romantic landscapes, and transplanted Europeans like Bierstadt. The result created for Americans a comprehensive mental picture of ideal landscapes. American tourists self-consciously chose travel routes to find and enjoy approximations of these ideals.

This landscape tradition openly repudiated desert regions. In 1835 Thomas Cole said every landscape without water was defective. A modern painter said it took him a long time to adjust his work to the desert: the land was vast, empty, and paralyzing, with only four or five colors. Even Georgia O'Keeffe, desert landscapist, noted while "a flower touches every heart, a red hill in the Badlands, with the grass gone, doesn't." Traditional tourist goals were hardly desert-oriented. The garden-park or bucolic rural setting was likely to be a vista framed by dark masses of green foliage, the entire setting centered upon a prominent body of water, and the atmosphere fuzzy with moisture. The scene would emphasize verdant growth, even rotting fecundity, implying layers of squandered abundance far beyond human need. Distance in the paintings was obscured by indistinct outlines and colors. The ground was entirely overlaid with grass, shrubbery, crops, and forest. Wilderness itself was represented by dense forest growth. The colors were green, brown, black; the shapes were organic and vegetative.

Reports of the American West reinforced negative attitudes about desert regions. The High Plains and the Great Basin were avoided or quickly traversed. The Great American Desert was said to begin just west of Ft. Kearney along the Platte River; at first the Rocky Mountains were included with desert regions as wasteland. As early as 1803 James Monroe reflected a habitual viewpoint when he criticized Thomas Jefferson for buying the Louisiana Purchase since it was mostly desert. This viewpoint was soon corroborated by the reports, it seemed, of Pike and others. The early tourist, conditioned by his visual expectations, took the same position. Seemingly devoid of scenery, desert travel involved more extreme physical discomfort forced by thirst and heat as well as boredom; there was nothing "artistic" to see--no pastures, plowed fields, reflected lakes or picturesque villages.

"Spiritual" or psychological risks of entering alien desert landscapes were also too high. The American traveler, dependent upon his garden-park imagery, experienced spatial disorientation triggered by the endless spaces of arid America. Successful correlation between cognitive mapping and external visual experience is important for personal well-being; disorientation brings on anxiety, shock, and even terror. In arid America the usual markers were missing, leading to a sense of total loss. Spatial uncertainty creates enormous psychological pressures; there is a massive effort to recover an orderly recognizable setting. Being lost cannot go on repeatedly for extended times without serious disorientation.

But the American traveler did gradually enter America's desert regions, and the numbers have swelled today to unmanageable proportions. Considering the obstacles already discussed, why has this taken place?

Desert climates were good for one's health. Stories would filter back East of people given six months to live in Pittsburgh or Philadelphia who found twenty years of vim and vigor in the dry sunny climate of Arizona. By the 1870s and 1880s, folks who had been condemned to be tubercular, dropsical, and scrofulous, troubled by weak hearts, disabled lungs, and worn-out nerves, colored by the palor of the consumptive, covered with disgusting skin diseases, and affected by the complications of senility, were being miraculously revitalized by the cure of arid and uninhabited regions. Not everyone was cured however. One easterner who ventured into the dry country complained that he came back an unintended twenty pounds

lighter, "skin like a chip, juices dried in me, nerves tense, and brain on fire."

The desert not only affected miraculous healings of the body, but also brought, Americans were reminded, spiritual renewal. Did not Jesus and Mohammed have their crucial interludes in desert wilderness, not to mention Moses's theophany on Mt. Sinai? Did not the admirable ascetic discipline of early monasticism get its start among desert hermits in Upper Egypt? Americans avidly read Charles Doughty's <u>Arabia Deserta</u>, and were not surprised by the desert setting for Lawrence's <u>Seven Pillars of Wisdom</u>. John Wesley Powell felt hubris (if not hutzpah) when he named unknown mesas and canyons after oriental temples as his boats roared down the Colorado through the Grand Canyon in the 1860s. This "present-at-the-creation" viewpoint was further reinforced by the intense hardships experienced in desert regions. Nowhere else did it seem that ordinary daily existence depended upon superhuman exertions. Even in visual terms, enormous energies seemed required to create something out of nothing. In effect, desert became the "environment of revelation," the archetype of "sacred space," symbolizing purity and timelessness. The western desert became America's Holy Land.

A new vogue for the southwestern desert began in the 1890s at first as a novel change from mountain and beach resorts. By the 1920s the desert acquired an aesthetic acceptability for its scenery. Since the 1930s, desert regions became major tourist attractions, based in part on air conditioning and the long distance mobility of reliable automobiles, as well, perhaps, on the invention of color photography. By mid-century, a blend of new aesthetic consciousness and the interests of wilderness preservation combined to focus interest on the canyon and badlands country upriver on the Colorado from the Grand Canyon. Bryce, Zion, Glen Canyon before it was drowned, later Arches, Canyonlands, and Capitol Reef received interest not only as a last refuge against the pressures of civilization but also for their stark harsh beauty which had once repelled earlier generations. This aesthetic worth was not for everyone: a pioneer rancher near Bryce Canyon complained that it was "a hell of a place to lose a cow!" A local horseman accompanying a geologist into Canyonlands in 1936 just said, "There's just gotta be mineral in those rocks; no piece of country could be so gosh-darned (sic) worthless."

As desert tourism came into its own in the first half of the twentieth century, there were other factors than the aesthetic. The West was no longer physically dangerous or psychologically disturbing. With the closing of the frontier, even wilderness was no longer perceived as oppressive or frightening. No longer was it a phenomenon which stretched endlessly, but now was fenced in on all sides. Wilderness no longer had to be mastered for personal and national survival.

Early American tourism was characterized by lazy and easy vacations. But Theodore Roosevelt and his generation did much to popularize the "strenuous life." A more active vacation meant camping out for a more intense wilderness experience. A contemporary of Roosevelt wrote, "The joys and profits of the camp lie in its difficulties, physical exertions, denials, and glorious distances from anybody." Roosevelt saw America's uninhabited regions as vast playgrounds; John Muir believed wilderness living, no matter if abbreviated, was essential for civilized well-being. "Roughing it" included a more elaborate, profound, sophisticated, and more integrated aesthetic. Beauty in the wilderness was more than the appreciation of natural curiosities or precious scenes. Roosevelt, Muir, William

Burroughs, and others worked toward a wilderness visual experience which demanded more than the effete snobbery of Claude glasses and Gilpin's picturesque. Roosevelt's infectious enthusiams gave wilderness beauty a wider acceptability, integrated with his emphasis on "the free, open, pleasantest, and healthiest life in America."

In the twentieth century, the new objective was "elemental contact with the reality of nature." Scenic America had become more than "scenery" to be complacently inspected. The new goal was to live for a time under the total experience and demands of wilderness. But roughing it in the backcountry was not a re-enactment of the pioneer camp, nor a recapitulation of the wilderness frontier experience. Americans were proud of western expansion and the civilization it helped to build. But there is little identity with the pressures the first explorers and settlers felt to fight and exploit nature rather than simply enjoy it. The typical modern American tourist has little in common with the historic frontiersman. Today's committed camper, backpacker, or environmentalist has even less in common.

Growing disillusionment with American industrial society increased interest in the value of wilderness. Such criticism goes back to Jefferson and Thoreau, Olmstead and Muir, and includes the "back to the country" and Garden City movements, and the general exodus to suburbia. But a peak was reached with the Depression of the 1930s, when many Americans sought reassurance in the Great American Outdoors as they found flaws in the industrial system. Despite the poor economy, tourism boomed in the 1930s. The quest for permanence, security, and foundations through the experience of wilderness seems also to have taken special hold on the American imagination during the turbulent years of the late 1960s and early 1970s. The appeal of desert regions also increased rapidly, perhaps because arid lands are the most opposite to civilized zones. Desert is "noncivilization" in the extreme.

Having progressed to the 1980s in this historical review, are there recognizable features of a contemporary desert aesthetic? What are the visual goals as the modern traveler makes his or her grand tour of the redrock, badlands, and canyon country?

Aside from the standard aesthetic and landscape principles (form, light, and color; feature, enclosing, and focal, etc.), which are important measures explored elsewhere, what distinctive elements belong to desert scenery? The result is a combination of aesthetic, cultural, historical, and environmental resources brought to bear on the subject. A brief listing must do, to open discussion and analysis of a largely-unexplored subject: (1) <u>aridity</u>: in an aesthetic sense this is an asset, not a defect, the primary reason for spectacular landforms and colorful grand vistas. (2) <u>life systems</u>: the adaptability and survival of plants and animals is far more visible and vivid in arid regions where such demands are extreme. (3) <u>geology and geography</u>: in desert regions, one has a sense of being "present-at-the-creation," surrounded by visible metaphors for the eternal and infinite. (4) <u>enlarged scales of space and time</u>: "On a clear day you can see forever;" scale and distance are monumental; time is not human time, not even seasonal time, but equally monumental. (5) <u>permanence</u>: lacking transitory ground cover as in the East, without substantial seasonal change, there is a sense of being down to irreducible basics, down to bare rock. (6) <u>human impermanence</u>: no society has historically overcome the inherent hostility of arid regions to human life. (Even classic Near Eastern civilizations have a strong nomadic aspect.) Long-term human settlement and control are unlikely; humans cannot overcome the problem of not enough water. Desert regions are vulnerable to human desecration, but not massive settlement.

Examination of a desert aesthetic can inform us about the possibilities of a new-fangled environmental aesthetic. A good deal of the environmental rhetoric of the 1970s focused on the now-classic question, "Do trees have rights?" An environmental aesthetic might focus on visual values intrinsic to a working ecosystem. Our question might be, "Does a healthy environment radiate beauty?" Desert environments are the most distant from civilized environments, and thus could supply perceptual models for an environmental aesthetic.

BIBLIOGRAPHY

Abbey, E. 1968. Desert Solitaire. Ballantine Books, New York, NY.
Abbey, E., and P. Hyde. 1971. Slickrock. Sierra Club, San Francisco, CA.
Arnheim, R. 1969. Visual Thinking. University of California Press, Berkeley, CA.
Baigell, M. 1981. Albert Bierstadt. Watson-Guptill, New York, NY.
Baigell, M. 1981. Thomas Cole. Watson-Guptill, New York, NY.
Brooks, Paul, 1964. Roadless Areas. Ballantine Books, New York, NY.
Conron, J. ed. 1974. The American Landscape. Oxford University Press, New York, NY.
Douglass, M. 1973. Natural Symbols. Vintage Books, New York, NY.
Fine, A. 1972. Frederick Law Olmstead and the American Environmental Tradition. George Braziller, New York, NY.
Glacken, C. 1967. Traces on the Rhodian Shore. University of California Press, Berkeley, CA.
Gussow, A. A Sense of Place. The Artist and the American Land. Friends of the Earth, n.d., San Francisco, CA.
Hollon, W. E. 1966. The Great American Desert, Then and Now. Oxford University Press, New York, NY.
Huth, H. 1957. Nature and the American. Three Centuries of Changing Attitudes. Oxford University Press, New York, NY.
Krutch, J. W. 1958. Grand Canyon. Today and All Its Yesterdays. William Morrow, New York, NY.
Litton, R. B. et. al. 1974. Water and Landscape. Water Information Center, New York, NY.
Lowenthal, D., and M. J. Bowden, eds. 1976. Geographies of the Mind. Oxford University Press, New York, NY.
Marx, L. 1964. The Machine in the Garden. Technology and the Pastoral Ideal in America. Oxford University Press, New York, NY.
McCloskey, M. et. al. 1964. Wilderness and the Quality of Life. Sierra Club, San Francisco, CA.
Meinig, D. ed. 1980. Ordinary Landscapes. Oxford University Press, New York, NY.
Novak, B. 1979. Nature and Culture. Oxford University Press, New York, NY.
O'Keeffe, G. 1976. Georgia O'Keeffe. Penguin Books, New York, NY.
McShine, K. ed., 1976. The Natural Paradise. Painting in America, 1800-1950. Museum of Modern Art, New York, NY.

Opie, J. 1979. "Seeing Desert as Wilderness and as Landscape." In Proceedings: National Conference on the Visual Resource, Forest Service, USDA, Berkeley, CA.

Opie, J. 1982. "Reading the Landscape of Middle America." The Great Plains Quarterly (January). Lincoln, NE.

Pomeroy, E. 1957. Search for the Golden West. Alfred A. Knopf, New York, NY.

Powell, J. W. 1957 (1875), The Exploration of the Colorado River. (condensed) University of Chicago Press, Chicago, IL.

Schmitt, P. J. 1969. Back to Nature: The Arcadian Myth in Urban America. Oxford University Press, New York, NY.

Shepard, P. 1967. Man in the Landscape. A Historic View of the Esthetics of Nature. Ballantine Books, New York, NY.

Shirakawa, Y. 1975. Eternal America. Kodansha International Ltd., Tokyo, Japan.

Stegner, W. 1954. Beyond the Hundredth Meridian. John Wesley Powell and the Second Opening of the West. Houghton Mifflin Company, Boston, MA.

Tetsuro, W. 1962. Climate: A Philosophical Study. Government Printing House, Tokyo, Japan.

Watts, M. T. 1975. Reading the Landscape of America. rev. ed. Collier Macmillan Publishers, New York, NY.

Wilmerding, J. ed. 1981. American Light. The National Gallery, Washington, DC.

Zevi, B. 1957. Architecture as Space. Horizon Press, New York, NY.

3. Interpretation and Visitor Values

Douglas Bruce McHenry

Looking is certainly the most common activity in any park, forest, or recreation area. I have lived in national parks for fifty years, and although my personal observations are not documented statistically, or even tested scientifically, I believe they are useful in examining the importance of looking at scenic resources in the park experience. My professional experience in parks like Yosemite, Rocky Mountain, Colonial, National Capital, Grand Canyon, Big Bend, Shenandoah, Everglades, and the northeastern parks, serve to substantiate my statements about looking or viewing.

DEFINING THE EXPERIENCE

Recreation activities of one sort or another frequently draw visitors to parks, forests, and recreation areas. While they are there they spend considerable time just looking around at the scenery, even if they are involved in other activities. Recreation can be divided into active and passive modes. A few examples are included in Table 3.1.

In each recreation activity considerable looking is involved. A person comes to the site to recreate in a pleasant setting. Campers will select nice looking spots near scenic features if offered the option. Fishermen constantly extoll the beauty and solitude of their "spot". Boaters of all kinds appreciate wet and wild scenery. People spend hours looking from overlooks along drives such as Skyline Drive, Trail Ridge Road, East Rim Drive. Their looking is enhanced by the clean fragrance of the area or the melodious sounds. A drive to Desert View in Grand Canyon is filled with impressions. The smell of trees and ponderosa pine trees adds to the view. The sound of wind in the trees or the call of a raven also adds to the experience.

Recreational activities are frequently individual or family activities. Many are educational in that skills are taught or learned. Often people request information which provides confidence building or approval for their activity in this setting. Some people are also very eager to join in educational programs or conducted activities designed to orient visitors. Here interpreters can enhance the visitor's experience by focusing on var-

---

Douglas Bruce McHenry is Chief of Interpretation for the North Atlantic Region of the National Park Service, Boston, MA.

ious subjects. Again, there are two general types of education experiences: active and passive. Some examples are included in Table 3.2.

TABLE 3.1
Examples of Active and Passive Recreation Activities

Active Recreation	Passive Recreation
Camping	Looking
Fishing	Listening
Boating	Smelling
Hiking	Touching
Skiing	Sitting
Snowmobiling	Driving
Hang gliding	Sun bathing

TABLE 3.2
Examples of Active and Passive Eductional Experiences in Recreation

Active Education	Passive Education
Conducted walks	Sightseeing, driving
Conducted tours	Birdwatching
Conducted programs	Using literature, flower identification
Demonstrations	Self guiding trails and tours

Interpretive programs offer a variety of educational experiences. They nearly always involve some form of looking. The conducted walk focuses attention on objects near and far which fit the basic theme. For example, a geology walk along the South Rim of Grand Canyon frequently refers to layers of stones down in the canyon, a long view. The same walk along the Franklin Cliffs in Shenandoah focuses more on rocks under foot. The same relationship is found in historic sites. The Statue of Liberty is first seen from Manhattan, a nice long view. The Edison Laboratory tour is a close-up intimate view. My experience indicates that visitors prefer a shifting from long views to close views, just as we all prefer variety in our food, clothing, and shelter and visitors are very disappointed if the long view is obstructed no matter what the cause.

WHAT VISITORS SEE

I think that we can safely say that looking is the principal activity at parks, forests, and recreation areas. It follows then that the setting is something special in the visitors' experience. Again, experience indicates that visitors don't want anything to come between them and the setting.

Visitors to Grand Canyon complain if haze obscures the view. Foreign visitors who have come great distances to see the Canyon are the most disappointed. Shenandoah has a natural haze and there are frequent complaints about obscured views. The long views of landscapes are easily altered by haze from any source.

The scene at the Grand Canyon changes hour by hour some days. This is also true day by day with the approach of haze generated by emissions we are told come from the Los Angeles area or Southern Arizona area. Visitors notice changes in visibility when it deteriorates to very poor levels. My experience indicates that historically the canyon appeared much clearer. People look at the photographs taken ten to twenty years ago and are disappointed in the real thing.

Similar responses from visitors have been heard in Shenandoah where there is a natural haze over the Blue Ridge. When this gets dense and the long view is obscured visitors complain. They don't see any obvious cause so they frequently ask if it is "pollution." Most visitors are surprised by the explanation of transpiration from trees and plants. They still voice frustration such as "when does it clear up?"

The frequency of poor visibility days is apparently increasing at these parks. In both Shenandoah and Grand Canyon the interpreters have turned the focus of their walks toward the small and close in features. The geology walks at Grand Canyon are frequently forced to substitute photographs and museum specimens for the views of distant rock formations when the visibility was poor. Some of the very interesting distant features such as mountains, canyons, rivers, lakes, and buildings are often invisible.

HOW VISITORS REACT

My experience reveals me that people do notice poor visibility in our national parks. They not only make comments, they select alternatives less affected by visibility. Regardless of its cause, poor visibility is presumed by most visitors to be caused by air pollution. Most visitors to parks like Grand Canyon and Shenandoah expect to look at canyons and mountains through pristine air. They expect park management to protect their parks, including their forests, the water, the animals, and the air.

I am saddened to see the changes which have taken place in my lifetime. As a young boy I remember looking out across the Grand Canyon and feeling that the North Rim was only a short distance away. The air was <u>so clear.</u> As a young man I remember approaching Denver from a plane and marveling at the mountains seeming so close. Salt Lake City with the Wasatch behind was an awesome sight. I have waited with my roommate while his dad, Ansel Adams, took pictures of distant mountains in places where this is no longer possible. The tragedy is that the deterioration has been so gradual that no one noticed. National parks are special! We need to develop methods to measure visual values before it is too late.

PART II
Measuring Visual Perceptions

Measuring values associated with changes in visibility conditions first requires understanding how visibility and visual impacts are perceived, how best to measure impacts and how to relate human perceptions to measures of physical characteristics of the visibility change. Certain characteristics of visibility can be measured by instruments, but we are just beginning to understand how changes in such visibility characteristics are perceived by the human observer and which changes are considered relatively more or less important. Identifying and quantifying these perceived changes is one link in the process of determining visual values and is essential in identifying the amount of change that is just noticeable by most people. The papers in this section address measuring perceptible changes in visibility characteristics and, thereby, address Step 2 and the linkage to Step 3 in Figure 1 of the Introduction.

The paper by Malm, MacFarland, Molenar, and Daniel examines how the human observer perceives and ranks changes in visual air quality when it is affected by different types of layered haze. Survey respondents ranked photographs of Navajo Mountain, as seen from Bryce Canyon National Park, with layered hazes (plumes) of various characteristics superimposed in the scene. The authors conclude that dark plumes seen against the sky are more intrusive than light plumes and that plumes that block part of the scenic landscape features have a more undesirable impact than when seen against the sky.

Latimer, Hogo, Hern, and Daniel describe the results of their work concerning visibility conditions and scenic beauty assessment. In contrast to Malm et al. visibility is not mentioned to the observers who are asked to rank the overall scenic beauty of photographs taken at selected points in several national parks and wilderness areas under different visibility and illumination conditions. The authors found that scenic beauty ratings had different sensitivities to visibility and illumination conditions at different scenic vistas due to differences in scenic features. Visibility was found to have little effect upon scenic beauty assessments unless a vista contained a distant dominant landscape feature. The authors suggest that this scenic beauty approach may be more appropriate than the visibility approach posed by Malm et al. because park visitors are not typically concerned about visibility alone. It may be, they suggest, that small changes in visibility are noticeable when the observer is asked to attend specifically to changes in air quality, but that the same small changes may not be important to the observer of the scenic resource as a whole.

Middleton, Stewart, Dennis, and Ely describe an observer-based approach used to assess visual air quality in urban areas and compare it to the above approaches used in parks and natural areas. Based upon their approach, they suggest that more attention must be paid to the operational models that explicitly link emissions to human judgments of visual air quality and to conceptual and methodological issues such as the role of perceptions and judgment in visual air quality assessments, representative design of the survey, and alternative modes of analysis.

Henry examines for the case of semitransparent layered haze and shows that there are many important refinements remaining in the theory and measurement of human perceptions of visibility impairment. The transparency of a haze is shown to affect the viewer's judgment of brightness and color of a scene consistently and significantly different from what would be expected. He suggests that the theory of color scission serves as an explanation of these differences, and that past studies using photographic reproductions may be flawed as they may not accurately reproduce transparency effects.

4. Human Perception of Visual Air Quality (Layered Haze)

William Malm
Karen Kelley MacFarland
John Molenar
Terry Daniel

INTRODUCTION

Impairment of a visual resource[1] that results from air pollution can manifest itself in two distinct ways: as layered or uniform haze. Layered haze can be thought of as any confined layer of air pollution that results in a visible spectral discontinuity between that layer and its background. Uniform haze, on the other hand, exhibits itself as an overall reduction in air clarity. The classic example of layered haze is a tight vertically constrained coherent plume. However, as the atmosphere moves from a stable to unstable condition and the plume mixes with the surrounding atmosphere, it sometimes can be difficult to differentiate between layered and uniform haze.

Quantification of visual impairment resulting from either layered or uniform haze can be separated into two categories: (1) the establishment of the amount of pollution that causes just noticeable differences in the appearance of a visual resource (Henry 1977) and (2) a determination of the functional relationship between air pollution and perceived visual air quality (Malm et al. 1981a).

The first objective is important when it is necessary to quantitatively specify if industrial emissions will be perceptible from a given observer viewpoint (impairment) while the second objective is important when trying to assess the societal value that people place on given levels of visual air quality (significance/adversity).

Previous studies on judgments of visual air quality have concentrated on uniform haze (Malm et al. 1980, Latimer et al. 1981, and Henry 1981). However, little is known about people's judgments of visual air quality as they relate to layered haze. For instance, are judgments of visual air quality affected by size and color of the haze layer? Does the position of

---

William Malm and Karen Kelley MacFarland are with the Air Quality Division of the National Park Service, Fort Collins, CO and Washington, D.C. John Molenar is with the John Muir Research Institute, Fort Collins, CO and Terry Daniel is a Professor of Psychology at the University of Arizona, Tucson, AZ. The assumptions, findings, conclusions, judgments, and views presented herein are those of the authors and should not be interpreted as necessarily representing official National Park Service policies.

the haze layer in the scene have an effect on perceived visual air quality? Some of these questions are addressed in this paper. Specifically, judgments of visual air quality are examined as a function of position of plume in the scene, plume color and plume size.

STUDY APPROACH

The research reported here follows from and extends work carried on at Grand Canyon, Canyonlands and Mesa Verde National Parks (Malm et al. 1981a and 1980). These studies established that park visitor's perceptions of visual air quality, as depicted by color slides, were consistent and reliable and revealed substantial sensitivity to measured optical parameters of air quality. These studies tended to confirm that color slides are acceptable surrogates for actual three-dimentional scenes, and that the demographic background of observers had little effect on judgments of visual air quality.

These studies also showed that cloud cover, sun-angle, landscape features, and snow conditions all played an important role in judgments of visual air quality. Thus, it was felt that if the goal of this study was to extract the effect of layered haze on perceptions of visual air quality, it would be necessary to control the background scene (including meteorological conditions) on which the plume was superimposed. Even though, through the National Park Service (NPS) monitoring program, there was a considerable number of slides showing plume impacts under a variety of illumination and meteorological conditions (Malm et al. 1981b), it was impossible to find enough plume slides that met these criteria. In fact, it was impossible to find two layered haze slides under similar atmospheric dispersion characteristics, much less plumes with the same color or in the same position relative to other landscape features. As a consequence, it was felt that the use of computer simulated slides was the best way to systematically examine the effects of plume size, color, density, and position in the scene on judgments of visual air quality. Computer simulation techniques allow for a precise control of the background scene, background air quality as well as plume size, shape, and color.

The process of computer simulation of plume impacts has been described by other researchers in the field (Williams et al. 1980). The procedure involves selecting a "base" slide on which layered haze effects will be superimposed. This image is reduced to a digital format that can be modified in accordance with modeled visual impacts of various types of layered haze and illumination conditions. Plume dispersion characteristics and plume chemistry were modeled in a similar way to those described by Latimer et al. (1978), while the plume or layered haze radiance field was calculated using a similar radiative transfer equation (Chandrasekhar 1950).[2] Physically, the radiative transfer equation is composed of three segments that can be interpreted as a gain or loss of spectral radiance in the sight path. The first segment describes the removal of radiation from the sight path that results from absorption and scattering processes. The second segment contributes to an increase in the amount of air light added to the sight path between the observer and an edge of the plume, while the final segment represents the amount of incident radiation scattered toward the observer due to the presence of a plume.

While some researchers have attempted to model background radiation fields, we chose to measure these values with a teleradiometer. Measurements of sky and target radiance were made at the same time the "base" photo was taken in the parks. These measurements were not only used as an input to the radiative transfer equation, but also were used to calibrate the density values of the base slide.

The calculated spectral radiance field of the layered haze is then "mapped" onto the base photo. This new image is hardcopied to a final slide or photograph. The density versus exposure characteristics of the copy film are adjusted to match (as close as possible) those of the original base slide.

STUDY DESIGN

The broad objective of this study was to investigate how plumes and layered hazes affect peoples' judgments of visual air quality. Specifically, it is of interest to: (1) see whether plumes in the sky are perceived differently from those that obscure scenic features; (2) determine whether the color of plumes and/or layered hazes have an effect on judgments of visual air quality; (3) establish whether contrasts levels between the center of the plume and the sky (plume centerline) or color differences are adequate to specify perceptual effects of this type of visibility impairment; and (4) compare the perceptual effects of plumes and layered hazes to those of uniform haze.

It was decided to concentrate the study on plumes that might be typically expected from a coal fired power plant. It was assumed that the nitrogen oxides (NO_x) to particle mass ratio is 5.0 and that the particulate emissions were primarily in the fine mode (Blumenthal et al. 1981). Particulate scattering to mass ratios were assumed to be 0.79×10^{-3} km^{-1}/$\mu g/m^3$, 0.65×10^{-3} km^{-1}, 0.57×10^{-3} km^{-1}/$\mu g/m^3$ at 450 nm, 550 nm, and 630 nm respectively. The corresponding nitrogen dioxide (NO_2) absorbtion to mass ratios are 1.36 km^{-1}/ppm, 0.31^{-1} km^{-1}/ppm and 0.06 km^{-1}/ppm (Latimer et al. 1978).

Plumes and layered haze slides were simulated using two different sun angles. For one case, the angle between sun and observer was assumed to be $100°$ while in a second case this angle was chosen to be $25°$. The corresponding phase functions used were 0.2 and 5.0 respectively. Scattering to mass ratios and phase function values were obtained using Mie (1980) theory. A phase function of 0.2 corresponds to a dark plume while a 5.0 will yield a white plume.

The base photo chosen for this study was of Navajo Mountain as seen from Bryce Point, Bryce Canyon National Park. Navajo Mountain is 130 kilometers (km) away from the observation point. The background air quality on the day that this photo was taken corresponded to an extinction coefficient of 0.011 km^{-1}.

Characteristics and position of plumes, layered hazes and uniform haze are summarized in Table 4.1. All plumes extend horizontally across the full extent of the scene depicted. Six dark and six light plumes corresponding to an "E" atmospheric stability condition and a wind speed of 4 meters/ second (m/sec) were placed in the sky above the mountain. Each plume was generated assuming different emission rates. These same plumes were also placed such that the plume centerline was right at the sky-mountain interface. In these "plume on mountain" photos, the top portion of the mountain was partially obscured. Another set of five dark plumes

TABLE 1
Slide Characteristics

Slide Number	Plume Position	Plume "Color"	Stability	Blue Con.	Green Con.	Red Con.
1	Sky	Dark	E	-0.34	-0.06	-0.11
2	Sky	Dark	E	-0.44	-0.24	-0.09
3	Sky	Dark	E	-0.47	-0.26	-0.11
4	Sky	Dark	E	-0.61	-0.42	-0.26
5	Sky	Dark	E	-0.62	-0.51	-0.42
6	Sky	Dark	E	-0.71	-0.64	-0.60
7	Sky	White	E	-0.12	+0.22	+0.56
8	Sky	White	E	-0.16	+0.39	+1.02
9	Sky	White	E	+0.19	+0.66	+1.84
10	Sky	White	E	+0.02	+1.01	+2.37
11	Sky	White	E	+0.13	+1.21	+2.83
12	Sky	White	E	+0.13	+0.83	+2.30
13	Top of Mountain	Dark	E	-0.14	-0.12	0.0
14	Top of Mountain	Dark	E	-0.27	-0.18	-0.11
15	Top of Mountain	Dark	E	-0.24	-0.24	-0.14
16	Top of Mountain	Dark	E	-0.24	-0.31	-0.23
17	Top of Mountain	Dark	E	-0.44	-0.54	-0.63
18	Top of Mountain	Dark	E	-0.35	-0.44	-0.39
19	Top of Mountain	White	E	-0.22	+0.05	+0.38
20	Top of Mountain	White	E	+0.03	+0.70	+1.42
21	Top of Mountain	White	E	+0.16	+0.17	+0.42
22	Top of Mountain	White	E	-0.03	+0.81	+1.58
23	Top of Mountain	White	E	+0.05	+0.39	+0.75
24	Top of Mountain	White	E	+0.04	+0.53	+1.21
25	Layered haze	Dark	stable	-0.07	-0.08	-0.08
26	Layered haze	Dark	stable	-0.20	-0.23	-0.18
27	Layered haze	Dark	stable	-0.27	-0.34	-0.23
28	Layered haze	Dark	stable	-0.26	-0.40	-0.39
29	Layered haze	Dark	stable	-0.16	-0.45	-0.49
30	Layered haze	Dark	stable	-0.16	-0.50	-0.59
31	Layered haze	White	stable	-0.06	+0.02	+0.18
32	Layered haze	White	stable	0.0	+0.06	+0.18
33	Layered haze	White	stable	-0.12	+0.25	+0.70
34	Layered haze	White	stable	-0.07	+0.59	+1.46
35	Layered haze	White	stable	-0.09	+0.56	+1.49
36	Layered haze	White	stable	-0.01	+0.69	+1.50
37	Top of Mountain	Dark	<F	-0.22	-0.20	-0.18
38	Top of Mountain	Dark	F	-0.12	-0.21	-0.19
39	Top of Mountain	Dark	F	-0.35	-0.34	-0.27
40	Top of Mountain	Dark	D	-0.31	-0.31	-0.25
41	Top of Mountain	Dark	>D	-0.25	-0.32	-0.20
42	General	Haze		0.0	0.0	0.0
43	General	Haze		-0.09	-0.11	-0.15
44	General	Haze		-0.02	-0.13	-0.18
45	General	Haze		-0.06	-0.23	-0.32
46	General	Haze		-0.11	-0.24	-0.38
47	General	Haze		-0.11	-0.33	-0.47

was generated assuming different atmospheric stability conditions, but with identical plume centerline contrasts. These plumes were also placed with their centerline at the mountaintop. An additional twelve slides were created with a layer of haze that extended from the mountain bottom to about two-thirds of the way up the mountainside. Six of the hazes were dark while the other six were light. Finally, six uniform haze slides were created. The color of uniform haze corresponded to the background sky.

These forty-seven plume, layered haze, and general haze slides were inserted into a slide tray in random fashion. After every third plume or layered haze slide, a control slide (in this case the base photo) was inserted into the slide tray. The control slide allowed for a calculation of the accuracy with which observers used the rating scale. The forty-seven evaluation slides were preceded by ten preview slides. The preview slides were used to orient the observers to the full range of visual air quality levels that they would be asked to evaluate.

Since previous studies showed that demographic backgrounds of observers were not related to visual air quality judgments, the studies were moved from a "park" setting into a laboratory environment. The study reported here was carried out on the University of Arizona campus. Students attending the University, rather than park visitors, were asked to participate in a study designed to evaluate visual air quality in the national parks. The questionnaire used in this study follows:

> It is important that everyone participating in the survey has the same information; so I will first read some standardized instructions. Congress has passed a number of laws to protect the scenic beauty and natural qualities of national parks and wilderness areas. The Clean Air Act declared a national goal of preventing future air pollution and cleaning up existing manmade air pollution in national parks and wilderness areas. The intent of this law is to preserve the scenic beauty of our environment.
>
> Visual air quality, or visibility, is the ability to see and appreciate a resource, such as an object, activity, scene, or atmospheric phenomenon. In a clean atmosphere, natural resources can be seen clearly enough to be easily identified, and appear free of discoloration so that their special natural, scenic, historic, or recreational values can be fully enjoyed. Visibility can be affected by either natural or man-caused pollution sources; resulting in either changes in color or clarity of near and distant vistas.
>
> In this study, we are trying to determine your perceptions of the visual air quality in a southwestern national park.
>
> I am going to show you some color slides of a scenic vista in the park. You will notice that the weather conditions, ground cover, and other factors such as lighting and the amount of air pollution are somewhat different in each slide.
>
> You should look at each slide separately and try to judge the VISUAL AIR QUALITY represented at the time the slide was taken. You will notice that factors such as the time of day, clouds, ground cover, and weather conditions sometimes make the judgment difficult, BUT TRY TO BASE YOUR RESPONSES ON VISUAL AIR QUALITY.
>
> Slides will be shown one at a time. Please use the rating scale shown at the top of your response sheet to indicate your judgment of each slide. The scale extends from one, indicating that you judge the

VISUAL AIR QUALITY to be poor, to ten, indicating very good visual air quality.

Before you begin rating the slides, I will quickly show you a few slides ranging from poor to good visual air quality just to give you an idea of the variety of conditions you will be evaluating.

Imagine how you would rate these slides on the one to ten scale, but please do not write down any ratings for them. (Show preview slides at about five seconds per slide.)

Now, I would like you to begin rating the scenes using the one to ten visual air quality scale. Slides will be shown rather briefly, but you will have plenty of time to judge each scene and mark your rating. You should try to use the full range of the one to ten rating scale, and BE SURE TO MARK A NUMBER FOR EACH SLIDE.

A NUMBER WILL APPEAR WHERE THE "X" IS NOW, TO HELP YOU KEEP TRACK OF WHICH SLIDE IS BEING SHOWN. (Show tray one, at about 8 seconds per slide.)

THAT IS ALL OF THE SLIDES. (Turn on the lights.)

At this point, the data sheets were collected and the purposes of the study were briefly explained to participants.

STUDY RESULTS

Past studies have shown that there is a linear relationship between perceptions of visual air quality (PVAQ) and atmospheric transmittance of the sight path that involves the most sensitive scenic landscape features (Malm et al. 1981a). PVAQs are also related to the size of sensitive landscape features: The larger the scenic feature the more sensitive the vista becomes to air pollution impact. The general equation is given by:

$$P = \sum_{i=1}^{N} f_i (P_{0i} + C) e^{-\Delta \bar{b}_{ext} R_i} - C, \qquad (1)$$

where P is perceived visual air quality, P_{0i} is the inherent PVAQ of the ith scenic element, f_i is the fraction of total area subtended by the ith scenic element, $\Delta \bar{b}_{ext}$ is the increase in atmospheric extinction, R_i is the distance to the ith scenic element and C is a calibration constant. Of course, $\exp(-\Delta \bar{b}_{ext} R_i)$ is the change in atmospheric transmittance resulting from an increase in atmospheric extinction, $\Delta \bar{b}_{ext}$, between the observer and scenic element i. Since target apparent contrast, C_r, is also proportional to $\exp(-\Delta \bar{b}_{ext} R_i)$ it is expected, and data show, that PVAQ is also proportional to C_r.

Equation 2 suggests the importance of the relationship between the amount of vista actually affected by air pollution and perceptions of visual air quality. For instance, a scene with primarily foreground features is, from a PVAQ standpoint, less sensitive to increases in air pollution than is a vista with primarily distant features. It is suggested that Equation 2 can be modified to include effects of stratified air pollution that impair portions of a given scenic element i. Let f'_i equal the fraction of the ith

scenic element obscured by a haze layer and let $\int \Delta \bar{b}_{ext,j}(r)dr$ be the optical thickness, t_{ij}, over the sight path that includes f'_j, which is defined such that $\Sigma f'_j = 1$. It is convenient to further define

$$\sum_{ij} f_i f'_j = \sum_{ij} f_{ij} = 1. \qquad (2)$$

The fraction of the vista seen by an observer through an atmosphere with an optical thickness of t_{ij} is then f_{ij}. Using this nomenclature, Equation 1 becomes:

$$P = \sum_{ij} f_{ij}(P_{oi} + C) e^{-t_{ij}} - C \qquad (3)$$

Equation 2 suggests that as more of a scenic element becomes obscured, the more judgments of visual air quality will be affected. However, Equation 3 does not account for alterations in PVAQ that result from plumes that obscure the sky.

Figure 4.1 is a plot of the average ratings of visual air quality for Navajo Mountain under uniform haze conditions, plotted as a function of apparent contrast of Navajo Mountain. The error bars around each point are the 90 percent confidence intervals while the straight line through the data points is a result of a least square calculation. Figure 4.1 shows that judgments of visual air quality when viewing Navajo Mountain are much like PVAQ ratings of other scenic vistas (Henry 1977). Even though there are many foreground features, the PVAQ ratings appear to show a linear relationship when plotted against the contrast levels of the scenic feature that is most sensitive to air pollution impact.

Figures 4.2 and 4.3 show visual air quality judgments of the same scene but with plumes of varying density and color placed at different positions in the scene. Physical characteristics of the plumes are given in Table 4.1. The reader should keep in mind that a plume with a centerline contrast of 0.02 is perceptible and one with a contrast of 0.05 can be easily seen.

Figure 4.2 is a plot of PVAQ as a function of plume contrast for "white" plumes positioned in the sky over Navajo Mountain and just above Navajo Mountain. The just above mountain plumes are placed such that the plume centerline corresponds to the mountain-sky interface. The "plume in sky" plume does not intersect or obscure any of the scenic features of the vista. Notice that these plumes all have positive contrast levels indicating they are brighter than the background sky.

The white "plume in sky" has little effect on people's judgment of visual air quality. In fact, there appears to be a slight, although not statistically significant, increase in PVAQs. However, as the plume is moved down to the mountaintop, it begins to affect the PVAQs. The "plume in sky" and "plume on mountaintop" ratings for the more dense plumes are statistically different. The "plume on mountain" slides are rated lower than "plume in sky" slides. However, the effect is not large. There is only a slight decrease in PVAQ for plumes which vary from a contrast of 0.06 to greater than 5.0. The slope of the least square line for the "plume on mountain" top is only -0.84.

Figure 4.3 shows ratings of dark plumes. These plumes were placed in exactly the same position as white plumes and they have the exact same physical dimensions. First, it should be pointed out that dark plumes are rated lower than white plumes regardless of their position in the picture.

FIGURE 4.1
Average PVAQ Ratings of Navajo Mountain as a Function of Apparent Contrast of Navajo Mountain.

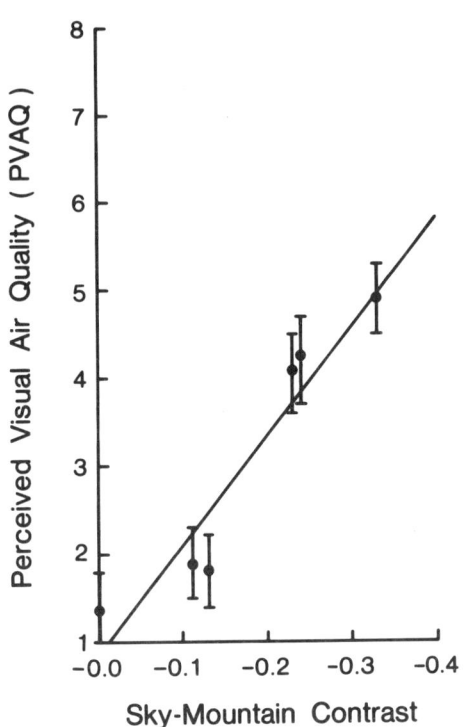

NOTE: The line through the data points is a result of a least square calculation.

Second, as the contrast and density of the dark plumes increase, the ratings of visual air quality decrease. However, there appears to be only a slight difference between plume in sky and plume on mountaintop ratings. The slope of PVAQ is slightly greater (not statistically significant) when the plume is centered on mountaintop than when placed in sky.

Figure 4.4 shows ratings of dark and light layered haze cases. In these photos, almost three-quarters of the mountain is obscured by a haze layer. It should be noted that the slope of PVAQ versus haze layer contrast line is steeper than that for any of the plume slides. A given increase in haze-sky contrast results in a significantly larger change in PVAQ. It is also worthwhile to note that for the haze layer case, it does not seem to make any difference whether the haze layer is dark or light. PVAQ ratings for dark haze are not significantly different (within 90 percent confidence intervals) from ratings of light haze layers.

FIGURE 4.2
Average PVAQ Ratings of the Navajo Mountain Vista with Varying Levels of Plume/Sky Contrast

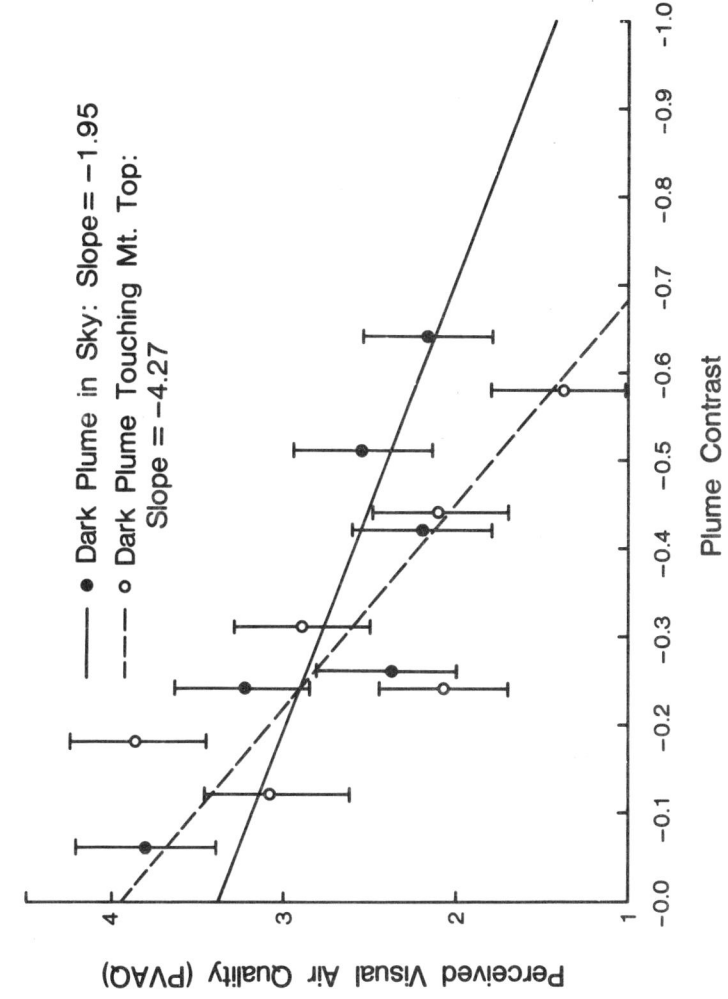

FIGURE 4.3
Average PVAQ Ratings of the Navajo Mountain Vista with Dark Plumes of Varying Levels of Plume/Sky Contrast

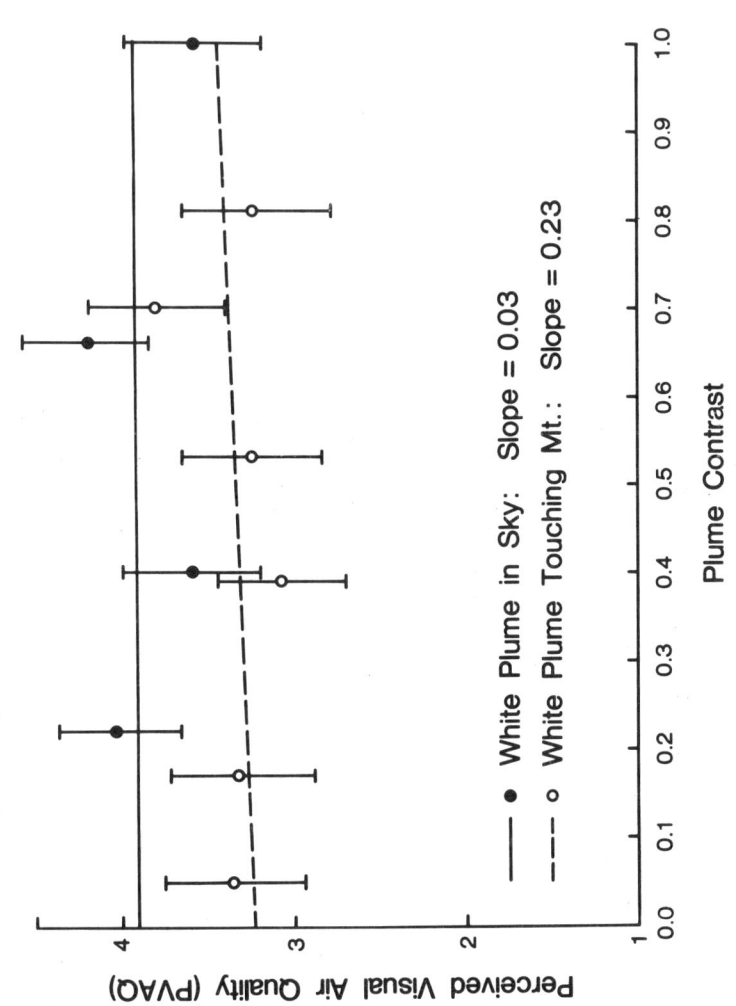

FIGURE 4.4
Average PVAQ Ratings of the Navajo Mountain Vista with Dark and Light Layers of Haze Superimposed on the Lower Two-Thirds of the Mountain

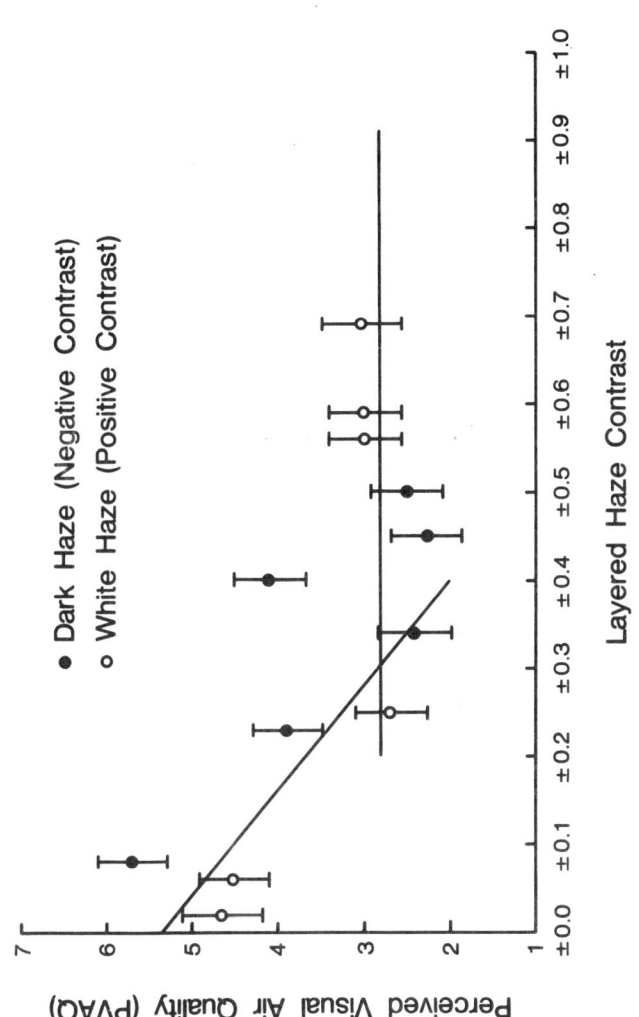

While Figure 4.4 shows what happens to PVAQ ratings for hazes that obscure the bottom of the mountain, Figure 4.5 exemplifies what happens as more and more of the mountain is obscured, starting from the top and working down. Each data point in Figure 4.5 represents PVAQ for plumes with identical centerline contrast, identical color (dark), and all positioned with their centerline on the sky mountain interface. However, σ_z, which describes how the concentrations of particulates and gases are vertically distributed, was allowed to vary. Specifically, it was assumed that the plume concentration gradient in the vertical direction obeys a gaussian distribution. σ_z is just the distance on either the top or bottom of the plume where the pollutant concentration falls to 60 percent of its centerline concentration. As σ_z increases, or as more and more of the mountain is obscured, the PVAQ decreases.

FIGURE 4.5
Average PVAQ Ratings of the Navajo Mountain Vista with Dark Plumes with Varying Vertical Dispersion

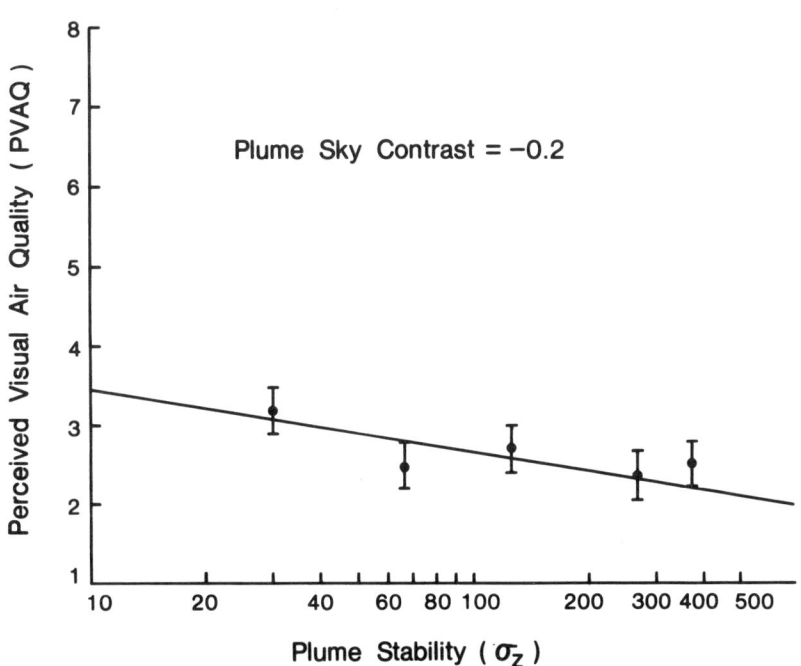

CONCLUSIONS

Examination of Figures 4.2, 4.3, 4.4, and 4.5 seems to show an important trend.

1. Haze layers or plumes positioned in the sky in such a way as to not obscure scenic features have a minimum amount of impact on perceived visual air quality.
2. When plumes are seen against a background sky, dark plumes are more intrusive than light plumes.
3. As more and more of the scenic features are obscured, the more the plume or haze layer affects judged visual air quality.
4. When the haze layer is placed totally in front of a scenic feature, there does not appear to be a difference between dark and light plumes.

A direct implication of conclusion three is that a statement of plume centerline contrast or color change, ΔE, is entirely inadequate in predicting the affect that a plume will have on perceived visual air quality. In fact, plume centerline contrast or ΔE can provide only partial indications or predictions of the impact that plumes will have on a scenic resource. More important is the plume's position in the scene and the relative amount of the vista that will be impacted. When emissions result in a plume that is seen only in the sky, it appears that its absolute color (rather than color difference) must also be considered.

NOTES

1. A visual resource is defined to be any specific vista or landscape feature that contributes to public enjoyment.
2. More details are available from the principal author.

BIBLIOGRAPHY

Blumenthal, L. W., L. W. Richards, E. S. Macias, R. W. Bergstrum, W. E. Wilson, and P. S. Bhardwaja. 1981. "Effects of a Coal-fired Power Plant and Other Sources on Southwestern Visibility." Atmospheric Environment 15:1955-1969.

Chandrasekhar, S. 1950. Radiative Transfer. Oxford University Press, reprinted by Dover Publications, 1960, New York, NY.

Henry, R. C. 1977. "The Application of the Linear System Theory of Visual Acuity to Visibility Reduction by Aerosol." Atmospheric Environment 11:697-701.

Henry, R. C., J. F. Collins, and D. Hadley. 1981. "Potential for Quantitative Analysis of Uncontrolled Routine Photographic Slides." Atmospheric Environment 15:1859-1864.

Latimer, D. A., R. W. Bergstrom, S. R. Hayes, M. T. Liu, J. H. Seinfelt, G. F. Whitten, M. A. Wojcik, and M. J. Hillyer. 1978. The Development of Mathematical Models for the Perception of Anthropogeic Visibility Impairment. EPA-450/3-78-110a, b, and c. SAI, San Rafael, CA.

Latimer, D. A., H. Hugo, and T. C. Daniel. 1981. "The Effects of Atmospheric Optical Conditions in Perceived Scenic Beauty." Atmospheric Environment 15:1865-1874.

Malm, W. C., K. K. Leiker, and J. V. Molenar. 1980. "Human Perception of Visual Air Quality." Journal of the Air Pollution Control Association. 30:122-131.

Malm, W. C., K. K. Kelley, J. V. Molenar, and T. C. Daniel. 1981a. "Human Perception of Visual Air Quality (Uniform Haze)." Atmospheric Environment 15:1875-1890.

Malm, W. C., E. G. Walther, K. O'Dell, and M. Kleine. 1981b. "Visibility in the Southwestern United States from Summer 1978 to Spring 1979." Atmospheric Environment 15:2031-2042

Mie, G. 1980. "Beitrage Zur Optik Fruber Medien Speziell Kolloidaler Metallosungem." Annals of Physics 25:377-445.

Williams, M. D., E. Treman, and M. Wecksung. 1980. "Plume Blight Visibility Modeling With a Simulated Photograph Technique." Journal of the Air Pollution Control Association 30:131-134.

5. Effects of Visual Range on the Beauty of National Parks and Wilderness Area Vistas

Douglas A. Latimer
Henry Hogo
Don H. Hern
Terry C. Daniel

INTRODUCTION

Over the past few years, the relationship between air pollution and the visual perception of scenic areas has become an important topic for research. For many years, the public has been concerned with the visual impact of air pollution in urban areas. Although people are concerned about the impact of urban air pollution on health, they are sensitive to such pollution principally because of its visual impact--the haziness it causes. More recently, the public has become concerned about the impact of pollution in relatively clean nonurban areas. Haze transported from urban areas and industrial facilities has degraded the pristine air quality of such remote locations as national parks and wilderness areas, and its effects on the beauty of landscapes and therefore the enjoyment of visitors are believed by some to be deleterious.

These latter concerns spurred Congress to enact, as part of the 1977 Clean Air Act Amendments, Section 169A, which required the U.S. Environmental Protection Agency (EPA), the states, and federal land managers (FLMs) of national parks and wilderness areas to protect and to restore visibility in these areas. In December 1980, the EPA promulgated regulations that required states and FLMs to consider, as part of their air quality management responsibility, the visual impact of well-defined plumes and layered hazes from industrial facilities on vistas in national parks and wilderness areas. The control of regional haze due to emissions from multiple sources was left for future regulations.

Because the primary impetus behind the EPA visibility regulations and section 169A of the Clean Air Act was the protection of the scenic resource of national parks and wilderness areas, numerous researchers have begun to study the relationship between human visual perception and physical measures of air pollution, such as visual range and contrast of landscape features. These researchers believe that realistic visibility protection goals cannot be implemented without developing an understanding of the effects of air qualtity and visibility on visual perception, aesthetics, and human enjoyment of scenic areas.

Douglas Latimer, Henry Hogo and Don Hern are researchers with Systems Applications Inc., San Rafael, CA. Terry Daniel is a Professor of Psychology at the University of Arizona, Tucson, AZ.

The recent research in this area performed by Systems Applications, Inc. and sponsored by the American Petroleum Institute (API) is described here. This supplements earlier reports on the first two phases of this research effort (Latimer and Daniel 1979; Latimer, Daniel, and Hogo 1980; Latimer, Hogo, and Daniel 1981).

All phases of research were aimed at investigating the relationships between air quality and visibility conditions and human perception of scenic areas. Color photographs of scenic vistas sampled randomly under a variety of air quality and visibility, illumination, and weather conditions were shown to many observers, both in groups and individually, to obtain their ratings of either visual air quality or scenic beauty.

SUMMARY OF THE FIRST AND SECOND PHASES OF THE STUDY

In the first phase of the study, measurements were taken at a number of scenic vistas in Arizona, with intensive measurements at the Grand Canyon and on Mt. Lemmon near Tucson. In the second phase measurements were made in the eastern United States, in the Great Smoky Mountains and Shenandoah National Parks. At the same time, photographs of these scenic vistas were taken under a variety of lighting, weather, and visibility conditions. Each photograph was documented with characteristic physical parameters, including measurements of visual range and atmospheric coloration using a multi-wavelength telephotometer and integrating nephelometers. Color slides were presented to groups of observers representative of both the general public and special interests; separate groups were asked, in separate tests, to rate either "visual air quality" or "scenic beauty."

From these ratings, indices of visual air quality (VAQI) and scenic beauty estimates (SBE) were calculated for each color slide. Relationships between standardized perceptual indices and physical parameters were investigated using multiple linear regression statistical analysis techniques.

The analysis led to several conclusions:

1. Physical conditions could be characterized by two principal components that were, to a large extent, mutually independent: visual range (haziness) and illumination conditions.
2. Sky color and brightness were mainly affected by the angle of illumination (scattering angle), not by visual range.
3. Observer group ratings of the color slides were quite consistent, even when groups consisted of observers with quite different interests. For example, oil company and EPA observers rated slides in a similar manner.
4. Ratings of visual air quality and scenic beauty correlated quite well. Different landscapes were rated quite differently for both visual air quality and scenic beauty; the Grand Canyon vistas rated relatively high while the Mt. Lemmon and eastern U.S. vistas rated fairly low.
5. Except for the Mt. Lemmon (Tucson area) vistas, ratings were also affected by visual range and by illumination conditions. The greater the visual range (atmospheric clarity), the higher the rating. For a fixed visual range, ratings decreased as the angle between the observer's (or camera's) line of sight and the sun (which is called the scattering angle) decreased. This reduction in ratings with decreasing scattering angle was probably caused by several factors. As scattering angle decreases, more light is scattered

into the observer's (or camera's) line of sight. This reduces the blue saturation of the sky and increases sky brightness. It also tends to make terrain objects look more hazy because of the extra light scattered toward the observer. Cumulus clouds tended to increase ratings also.
6. Scenic beauty ratings of different vistas appear to have different sensitivities to visual range. Scenic beauty ratings of the Mt. Lemmon (Tucson area) vistas were essentially insensitive to visual range; by contrast, scenic beauty ratings of the Grand Canyon and eastern U.S. vistas were sensitive to visual range. Generally, scenic beauty ratings were less sensitive to variation in visual range than were visual air quality ratings. However, scenic beauty and visual air quality ratings of the Grand Canyon and eastern U.S. vistas were equally sensitive to visual range.

From a practical standpoint, the most important conclusion of the first two phases of research is that changes in atmospheric optical properties caused by air pollution may have quite different effects on the human aesthetic experience of different scenic areas because of differences in the scenic features of the terrains of these areas.

THE THIRD PHASE OF THE STUDY

The third phase of the study was designed to collect and analyze physical and perceptual data from vistas in Bryce Canyon and Canyonlands National Parks and from some of the vistas intensively sampled in the first phase of the study (in Grand Canyon National Park and at Mt. Lemmon). A sufficient number of photographs were taken under a variety of visibility, lighting, and cloud conditions from each vista to allow separate multiple regression analyses for each vista rather than for groups of vistas, as was done in the first two phases. The emphasis in the third phase of the study was on ratings of scenic beauty rather than visual air quality. It is the authors' belief that the former rating is more indicative of the experience of the typical visitor to a scenic vista who is not concerned with air quality per se, but rather with the enjoyment of the scenic beauty of the area.

Physical data were collected for this phase of the study in a manner similar to that used in the earlier two phases. Color photographs of the selected scenic vistas were taken. Care was taken to frame each photograph of a given vista exactly the same way each time a photograph was taken. After each photograph was taken, the visual range and various horizon sky color parameters (luminance and chromaticity coordinates x and y) were measured with a telephotometer. In addition, solar zenith and scattering angles were measured and cloud cover was documented. These physical data were later included in a computerized data base that was used for statistical analysis of the relationship between perceptual indices and physical parameters.

Data were collected at Bryce Canyon and Canyonlands National Parks during the period from 27 June to 12 July 1979. Additional measurements were made at some of the sites in the Grand Canyon and at Mt. Lemmon that were measured intensively in the first phase of the study; both of these areas were sampled during the period from 2 June through 18 August 1980. The Grand Canyon vistas were further sampled during the period from 4 to 15 December 1979. These data were combined with those collected during the first phase of the study so that a large number of photographs (34-77) were available for each vista.

Mean visual ranges were as low as ninety km at the Mt. Lemmon vistas, which overlook a valley containing a copper smelter, and over 200 km at the Grand Canyon and Bryce Canyon sites. Individual measurements of visual range at each location varied considerably around the mean, with standard deviations of twenty to forty percent of the mean visual range.

The relationships among physical parameters were found to be similar to those identified in the first and second phases. The strongest relationships were found between the chromaticity coordinates x and y, indicating that sky color varies in saturation, not hue. Sky luminance (brightness) is a strong nonlinear function of scattering angle because of the scattering angle dependence of the Rayleigh and Mie scattering (phase) functions. Sky luminance and chromaticity are largely independent of visual range, but are dependent to some degree on the solar zenith and scattering angles.

A total of 1,211 color slides from the first and third phases of the study were rated by a total of 2,309 observers in seventy-one groups. These groups were asked to rate either visual air quality or scenic beauty, but not both.

The intercorrelations between the twenty baseline slides rated by all observer groups were computed to evaluate group consistency. Most of these group correlations were relatively high, averaging about 0.8; however, several groups had correlations below 0.7. To investigate the variability in group ratings of the baseline slides, histograms of VAQI and SBE ratings, an example of which is shown in Figure 5.1, were prepared for each of the baseline slides that were rated by each observer group. There is considerable variability in group ratings of a given slide. Although the distinction between high-rated and low-rated slides is fairly consistent among the groups, ratings of the same slide by different groups can differ by as much as 100 points on the VAQI and SBE rating scale. The average standard deviations of group ratings of VAQI and SBE for the baseline slides are twenty-seven and twenty-three respectively. This suggests that VAQI or SBE ratings from a single group are typically more than twenty points from the "true" value.

Many of the slides in the third phase of the study were rated by only one observer group. For these slides, the single VAQI or SBE rating given by one group could be more than fifty points (i.e., twice the standard deviation) from the true mean.[1] One must recognize that this observer group sampling error will contribute to the variance in VAQI and SBE ratings. We suspect that much of the unexplained variance in perceptual indices is related to this sampling error.

The relationships between the perceptual indices, namely, the visual air quality index and the scenic beauty estimate, and physical parameters, such as visual range, sky chromaticity and luminance, solar zenith and scattering angles, and cloud cover, were investigated using multiple linear regression. In the first and second phases of the study, more than one vista had to be lumped together to obtain sufficient points for statistical analysis. In the third phase of the study, however, sufficient data were obtained so that each vista could be treated independently in a multiple linear regression analysis.

On the basis of the regression analyses for each vista, we found that visual range explained little of the variance in SBE ratings. Other parameters, more directly related to illumination and cloud conditions, ex-

FIGURE 5.1
Example of Variability in Observer Group Ratings of Scenic Beauty (SBE) of a Baseline Slide.

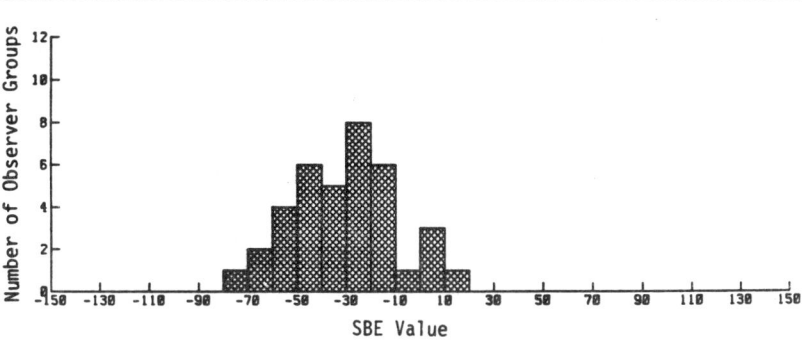

plained most of the variance. Only in the case of vista 5, which is a view of distant San Francisco Peaks from Lipan Point at the south rim of the Grand Canyon, did visibility explain more than 10 percent of the SBE variance.

Since the emphasis in the third phase of the study was on scenic beauty rather than on visual air quality ratings, sufficient VAQI data were not obtained to perform a regression of VAQI separately for each vista. However, both VAQI and SBE regressions were performed for various groups of vistas as was done in the first two phases of research. These regression analyses indicate that visual range accounted for little or none of the variance in SBE. Visual range explains a significant portion of the VAQI variance for the Canyonlands and Mt. Lemmon vistas. None of the variance in VAQI for the Bryce Canyon vistas was explained by visual range; for the Grand Canyon vistas only five percent of the VAQI variance was explained by visual range. These results for the Grand Canyon vistas differ from those obtained in the first phase of the study, which, with a smaller sample of photographs, found considerable sensitivity of both VAQI and SBE to visual range.

THE FOURTH PHASE OF THE STUDY

In the fourth phase of this effort, all slides from the intensively sampled vistas, as well as more than 400 additional slides taken in the Flat Tops Wilderness Area, were presented to observers for ratings of scenic beauty.

Photographs of various vistas from Big Marvine Peak in the Flat Tops Wilderness Area located in northwestern Colorado were taken during July 1981. Over 500 photographs and corresponding physical data were taken during this period. Six directions of view were sampled. For three of these directions, photographs were taken with a 150 mm telephoto lens as well as the normal 50 mm lens used in all the previous studies. The three sets of telephoto shots were treated as separate vistas.

During the month of measurements the visual range varied from about 100 to 230 km, and the mean was about 165 km. Other physical parameters, such as horizon sky chromaticity and luminance, cloud cover, and sun angle, were also measured, as was the case in the other phases of research.

A procedure was developed for the fourth phase that enabled the presentation of slides to each observer in a different random order. The procedure used a projection system based on a random-access slide projector controlled by a microcomputer. This apparatus was used in conjunction with a rear projection screen and an enclosure that housed the system and the observer. For each observer, the computer recorded the date, time of day, tray number, observer number, slide presentation sequence, and rating for each slide.

The slide projector used an eighty-slide tray, which had four instruction slides and seventy-six rating slides (of which twenty were baseline slides). Instruction slides explaining the program and introducing ten preview slides (previewing the types of scenes to be presented) were shown first. These preview slides were not rated. After the instruction and preview slides, the seventy-six slides for rating were presented randomly by the computer-controlled slide projector. The projected image provided a forty-two degree field of view, matching that which the observer would experience in the field. Approximately twenty observers rated each tray of slides.

The computer commanded the presentation order of all the slides according to a random-number table that it had generated before the sequence began. This table was constructed with certain constraints, so that the search time between slides would be limited to less than 1.5 seconds. Each slide was shown for eight seconds. The entire presentation took fifteen minutes.

Scenic beauty ratings on a ten-point scale were obtained from presentations to 440 observers in San Francisco, Los Angeles, and Denver. From these ratings, the SBE index was calculated for each of 1,327 slides for the four Grand Canyon vistas, five Mt. Lemmon vistas, five Bryce Canyon vistas, four Canyonlands vistas, and nine Flat Tops vistas (including six normal lens and three telephoto lens views). In total, twenty-seven vistas were sampled, with an average of forty-nine photographs each.

On the average, the Mt. Lemmon vistas were rated the lowest and Bryce Canyon the highest. The telephoto views at Flat Tops were rated, on the average, thirty-nine points lower on the SBE scale than their normal lens counterparts. The standard deviation of SBE around the mean for each vista averaged twenty-four. Thus, assuming the range in SBE values to be $\pm$ 2 standard deviations, the SBE for a given vista will vary over a 100-point range depending on air quality and visibility, illumination, and weather conditions.

It is important to point out that some of this variability in SBE may be due to sampling error. The uncertainty in the estimate of SBE of a given slide by one group is almost as large as the variability in the ratings of all slides of a given vista. Even with the randomized slide presentation, we found considerable variation in the SBE values (with an average standard deviation of seventeen points on the SBE scale) around the mean for each baseline slide. Since the SBE value for each slide (except for the baseline slides) is based on the ratings of only about twenty observers, the SBE value for a given slide could be as much as fifty points above or below the "true" SBE value of the slide, simply because of sampling error associated with the limited number of observer ratings of a given slide. This sampling error could be decreased simply by increasing the number of observers of a given slide.[2]

As in other tests of the correlation of group ratings of baseline slides, the correlations are generally high (> 0.8). This means that observer groups consistently distinguish between high-rated and low-rated baseline slides, even though, as noted above, there is considerable variability in group ratings of a given baseline slide (on the order of $\pm$ fifty points on the SBE scale).

Multiple linear regression analysis was used to evaluate the relationships between the SBE values for individual slides of a given vista and the physical parameters that characterize that individual slides. The sets of perceptual and physical data for individual vistas were sufficiently large (twenty-two to sixty-nine conditions or slides for a single vista) to permit separate analyses for each vista. Physical parameters used as independent variables in the regression analyses included visual range, horizon sky chromaticities, horizon sky luminance, cumulus and cirrus cloud cover, scattering angle, and solar zenith angle. In addition, subjective ratings on a five-point scale of certain aesthetic properties of the slides were used as independent variables in the regression analyses. These properties included such factors as foreground and background illumination intensity, and subjective ratings of the beauty of clouds, terrain shadows, and cloud shadows. These subjective ratings are related to the aesthetic effects of individual illumination and cloud conditions.

On average, 41 percent of the variance in SBE for individual slides of a given vista is explained—27 percent by physical parameters and 14 percent by the subjective ratings of illumination conditions. Table 5.1 summarizes the fraction of SBE variance explained for typical vistas sampled in the four phases of the study. Visual range explains little (< 6 percent) or none of the variance in scenic beauty for most vistas studied in the western U.S., including those of the Grand Canyon.[3] Exceptions to this general statement are vistas that have dominant landscape features at distant ranges, such as the view south from the Grand Canyon with its view of the San Francisco Peaks (for which visual range explains 15 percent of SBE variance); vistas in Flat Tops with long views of the Maroon Bells-Snowmass range and Battlement Mesa (for which visual range explains 16 to 23 percent of the variance), and the telephoto views of distant landscape features (for which visual range accounts for 25 to 29 percent of the variance). As in the previous research phases, illumination and cloud conditions explain most of the variance in SBE. Considering the substantial SBE sampling error discussed previously, it is not surprising that, on average, only 41 percent of SBE variance is explained by visual range and the physical and subjective factors related to illumination and cloud conditions.

Visual range accounted for a significant fraction of variance in SBE for each of the vistas in the eastern United States sampled in the second phase of the study. For such vistas in the Shenandoah and Great Smoky Mountains, visual range explained 17 to 39 percent of the variance in SBE.

It is important to point out that though the scenic beauty ratings of several vistas were found to be largely insensitive to visual range for the conditions sampled in this study, it should not be concluded that the SBE of these vistas is totally insensitive to air quality and visibility. It is quite possible that air quality and visibility conditions worse than those sampled in this study could significantly affect ratings of scenic beauty.

TABLE 5.1
Percentage of Scenic Beauty (SBE) Explained by Visual Range and Illumination-Cloud Conditions

Vista	% of SBE Explained by			% Unexplained and Sampling Error
	Illumination /Clouds (Subjective Parameters)	Illumination /Clouds (Physical Parameters)	Visual Range	
Typical Grand Canyon Vista (#4)	15%	38%	0%	47%
Grand Canyon Vista with Long View to San Francisco Peaks (#5)	12%	17%	15%	56%
Mt. Lemmon Vista across City of Tucson (#13)	5%	57%	5%	33%
Typical Bryce Canyon Vista (#43)	16%	22%	2%	60%
Typical Canyonlands Vista (#46)	16%	25%	0%	59%
Typical Flat Tops Vista (#71)	4%	40%	0%	56%
Flat Tops Vista with Long View to Maroon Bells/Snowmass (#72)	20%	0%	16%	64%
Telephoto of Flat Tops Vista with Long View to Maroon Bells/Snowmass (#79)	0%	4%	25%	71%
Views of Shenandoah and Great Smoky National Parks	0%	33%	39%	28%

VISITOR EXPERIENCE AND STUDY RELEVANCE

A note concerning the typical experience of a visitor to a national park or wilderness area is in order because it is important when considering the design and interpretation of perception experiments. Our study and others (Malm, Leiker, and Molenar 1980; Malm et al. 1981) indicate that human observers can be quite sensitive to visual range in perception experiments, particularly if they are asked to attend to air quality factors and are shown repeated examples of the same vista under different air quality and visibility conditions. The question is, however, whether a given perception experiment is relevant to visitor experience. A typical visitor to a scenic area probably comes to see the scenery, to enjoy the beauty of a natural environment. The visit is usually brief, lasting only a few hours to a few days. Although the visitor enjoys a variety of things, including solitude, direct experience of a natural environment, and the change of pace from the routine daily environment, the visual enjoyment of beautiful landscapes is probably one of the most important experiences.

Because a typical visitor does not go to a scenic area specifically intending to evaluate visual air quality or to directly attend to air quality, we believe that it is most appropriate to use the rating of scenic beauty as a measure of visitor appreciation of a vista. If a person is directed to rate the "visual air quality" of a scene, one would expect that he or she would become more sensitive to air quality and visibility conditions than would be the case for the typical visitor to a scenic vista. Similarly, an observer who is asked to detect the presence of air pollution will have a quite different predisposition from that of a person who is prepared simply to "enjoy the view."

Visitor experience is probably conditioned by expectations and by the variety and change of scenery from that normally experienced in the urban environment. If photographs of a given vista under various air quality and visibility and illumination conditions are presented one after the other in a perception experiment, an observer can be expected to be more alert to differences in the photographs because he or she has an opportunity to compare conditions that are not available to the typical visitor. A typical visitor experiences a given vista only for a brief period of time during a vacation that may include visits to a variety of vistas of different landscape types.

The interpretation of perception study results is critical. Would air pollution that is visually detectable to an observer who is specifically looking for its presence also be detectable to an observer who is not? If air pollution is detectable, does it affect visitor perception and enjoyment? If a given vista is rated poor in a perception experiment when compared with other conditions of the same vista, would it be rated poor by the typical visitor on a trip through a variety of scenic areas? If visitors' experiences are adversely affected by air pollution, what options do we have to control air pollution, and what will it cost? Are the costs worth the benefits? These are just some of the questions that need to be addressed.

CONCLUSIONS

The most signficant result of this work was the finding that the sensitivity of scenic beauty to visual range differed considerably among vistas. Ratings of scenic beauty of a given vista were independent of visual range (visibility) unless there was a dominant, distant landscape feature in the landscape scenery. Scenic beauty ratings of most vistas appear to be more strongly affected by conditions related to illumination, cloud cover, and the order of the slide presentation than by visual range. This result suggests that the scenic beauty of the relatively short vistas within most Class I areas is insensitive to visual range changes of the order sampled in this study. Those vistas with long views of landscape features possess scenic beauty that appears to be sensitive to visual range only when the distant landscape feature is dominant and central to the inherent scenic beauty of the scene.

Another important conclusion of this work is that the context in which a person observes a given scene can significantly affect that person's perception. For example, if a person is asked to rate visual air quality, ratings generally become more sensitive to visual range than if the person is asked to rate scenic beauty. In such a context, the person focuses on visual range more because his or her attention is directed to air quality.

NOTES

1. These results for the Grand Canyon, as well as results from the third phase of research, differ from those of the first phase of the study. Except for the long view of the San Francisco Peaks, the SBE of the Grand Canyon vistas was insensitive to changes in visual range of the order sampled. SBE was largely affected by illumination conditions.

2. The sampling error is proportional to $\sqrt{(1/N)}$ where N is the number of observers.

3. For slides rated by many groups, such as the baseline slides, however, the mean of the VAQI of SBE ratings by individual groups is closer to the true mean. Indeed, the standard error of the mean is simply the standard deviation of ratings from all groups divided by the square root of the number of groups.

BIBLIOGRAPHY

Latimer, D. A., and T. C. Daniel. 1979. "Preliminary Results of a Study of Human Judgements of Visual Air Quality." Air Pollution Control Association Specialty Conference on Visibility, 27-29 November, Denver, Colorado.

Latimer, D. A., T. C. Daniel, and H. Hogo. 1980. "Relationships Between Air Quality and Human Perception of Scenic Areas." Publication No. 4323, American Petroleum Institute, Washington, D.C.

Latimer, D. A., H. Hogo, and T. C. Daniel. 1981. "The Effects of Atmospheric Optical Conditions on Perceived Scenic Beauty." Atmospheric Environment 15:1865-1874.

Malm, W. C., K. K. Leiker, and J. V. Molenar. 1980. "Human Perception of Visual Air Quality." Journal of Air Pollution Control Association 30(2):122-131.

Malm, W. C., K. Kelley, J. Molenar, and T. C. Daniel. 1981. "Human Perception of Visual Air Quality (Uniform Haze)." Atmospheric Environment 15(10/11):1875-1890.

6. Implications of NCAR's Urban Visual Air Quality Assessment Method for Pristine Areas

Paulette Middleton
Thomas R. Stewart
Robin L. Dennis
Daniel Ely

INTRODUCTION

Although visual air quality (VAQ) is an important issue in regional air quality management, the ability to forecast effects of potential air quality management plans on VAQ has not yet been realized. Neither a widely accepted metric for measuring visual air quality nor appropriate techniques for linking potential policy actions with such an index has been developed. The need for such an index has been most pronounced for pristine area air quality management where protection of Class I areas from significant deterioration is a prominent regulatory issue. VAQ in urban areas, although not yet in the regulatory arena, is an important quality of life issue to be considered in any assessment of air quality management plans in urban areas, particularly in the west.

Over the past several years, extensive pristine area VAQ assessment research has been conducted primarily by the Environmental Protection Agency, the National Park Service, and their collaborators (Malm et al. 1981) any by Systems Applications, Inc. and the University of Arizona (Latimer et al. 1981). Hereafter, these studies are referred to as EPA/NPS and SAI, respectively. The authors' research conducted at the National Center for Atmospheric Research (NCAR) has focused on urban area VAQ assessment (Mumpower et al. 1981, Middleton et al. 1981). The purpose of this paper is to outline the three VAQ assessment methods and compare the methods with respect to objectives, conceptual framework, study design and data analysis. Implications of the NCAR method for pristine area assessments conclude the discussion.

The authors are researchers with the Environmental and Societal Impacts Group of the National Center for Atmospheric Research (NCAR), Boulder, CO. NCAR is sponsored by the National Science Foundation. Travel funds to present this research were provided, in part, by the American Petroleum Institute.

OUTLINE OF VISUAL AIR QUALITY METHODS

NCAR Method

The major goal of the NCAR research effort is to develop a VAQ index that (1) is based upon human judgments of visual air quality, (2) can be related to easily monitored properties of the physical environment, and (3) can be used in forecasting levels of VAQ expected from alternative air quality management plans.

Development of this index is based on the premises (1) that a forecast model should be based as much as possible on established theory, not data, and (2) that development of the model is facilitated by separating the system into components that are related by theory. The first premise is important because the index is intended to be used for forecasting the future effects of policies. An index based entirely on data can only be used under conditions that are similar to those represented in the data. Forecasting often requires generalizing beyond the restrictions of presently available data, and only sound theory can provide the justification for such generalization. The second premise is important because of the complexity of the link between emissions and judgments of VAQ. Urban VAQ can be described by five components: (1) emissions, (2) concentrations of visibility-reducing pollutants and their precursors, (3) physical characteristics of the visual environment, (4) perceptual cue judgments used in making judgments of VAQ, and (5) individual's judgments of overall VAQ. Figure 6.1 illustrates the components of the NCAR study.

The relationship between emissions and pollutant concentrations are described with a three dimensional dispersion model. The relationship between concentrations and properties of the visual environment, such as contrast, and the relationship between these properties and perceptual cues are derived using radiative transfer theory. The relationship between the human judgments of VAQ and perceptual cues, whether calculated or modeled, has yet to be described by any established theoretical model. Therefore, this relationship is derived from observational data.

The first step in the development of an index of VAQ then is to verify that the human judgments of VAQ can be explained by the hypothesized set of perceptual cue judgments and to verify that these cue judgments are significantly related to measured environmental properties. Once these relationships are verified, the year/season VAQ index development proceeds by:

1. calculating perceptual cues using a three-dimensional visibility model which relates emissions to perceptual cues for a set of characteristic days in the urban area being studied;
2. validating the cue prediction part of the model by comparing the calculated cues with observer judgments for a variety of different atmospheric conditions;
3. deriving a simple visual air quality index using parameterization of the three-dimensional visibility model results for the characteristic days; and
4. producing a weighted VAQ index for the year/season based on estimates of the average frequency of occurrence of each characteristic day over a year or a season.

FIGURE 6.1
VAQ Components and Models

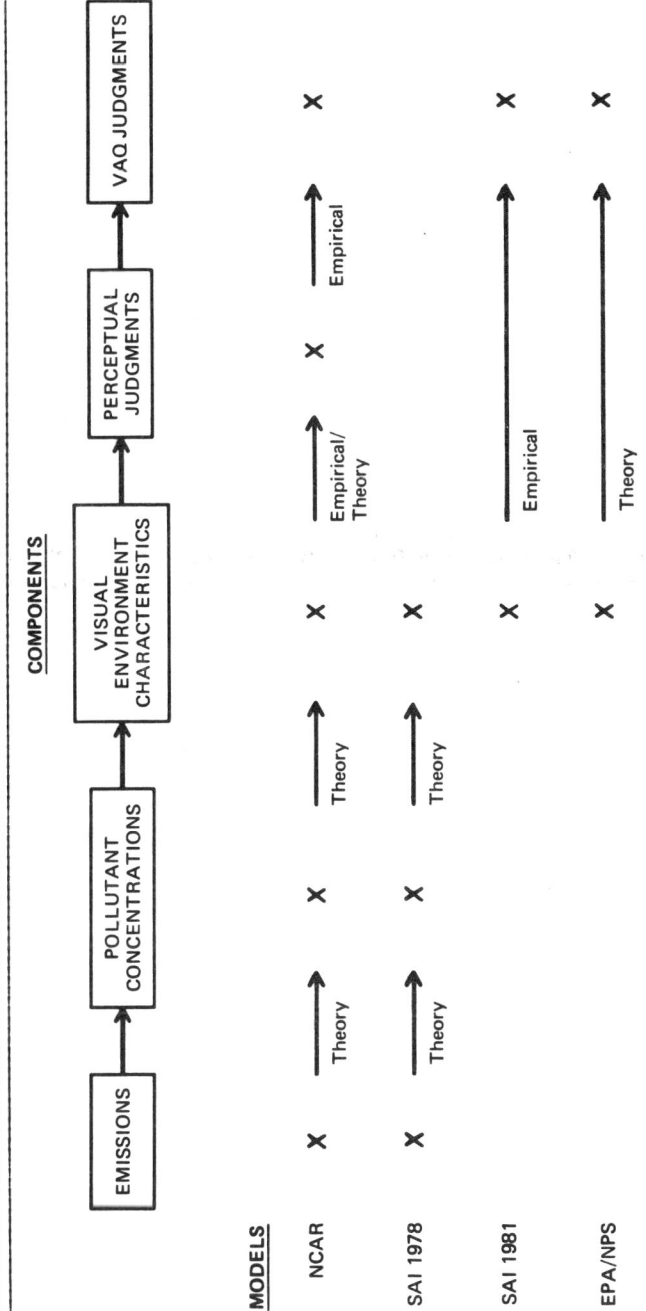

In order to develop an index using this approach several data bases and models are required. First, a data base of human judgments and simultaneous environmental measurements is needed to establish the existence and extent that the assumed relationships can be modeled. Second, in order to obtain a weighted average index, a minimum of one year of environmental measurements must exist as a basis for selecting characteristic days and their frequency of occurrence. Finally, an emissions inventory is required.

Two types of models are used in the index development. A dispersion-radiative transfer model is needed to relate emissions to concentrations to perceptual cues. Judgment models are empirically derived to describe the relationships among the different judgments and among the judgments and environmental measures (Mumpower et al. 1981, Middleton et al. 1981).

Observational Procedure

The observational procedure used to obtain the data base of VAQ judgments can be summarized as follows. Observers were hired for a period of approximately one month. A total of five separate observation studies have been conducted, in the summers of 1979 and 1980, and the winters of 1980-1982. The first study used two observers; the other four employed five to seven observers. Each observer was sent to a number of sites in a random sequence at specified times during the study period. At the site, the observer was instructed to observe views in several different directions and to give his or her impression of the visual air quality in each direction. Then the observer rated the <u>clarity</u> with which designated targets in each direction could be seen. Generally, two target objects were associated with each view. These targets, which were usually buildings or mountains, were between .8 and 117 km from the observation sites.

Other judgments made by the observers for each of the directional views included the <u>color</u> of the air and sky (as compared to Munsel color chips), the sharpness of a <u>border</u> between discolored air and clean air and the imputed <u>source</u> of visibility degradation--natural or human. The clarity of target objects, color, and border judgments describe aspects of the views which have been found to be important cues to VAQ (Mumpower et al. 1981).

VAQ was rated using a seven-point rating scale. Observers are not instructed to make their judgments in a particular way, but rather are encouraged to use the rating scale to express their subjective impression of the VAQ at a given time. Thus, individual differences in judgments of visual air quality are allowed to emerge. During the first four NCAR studies, the representativess of the paid observer judgments was verified by comparing their judgments with judgments of random passersby taken during the same observation period.

SAI Approach

The main objective of this research is to investigate the relationship between the visual environment as described by physical parameters either directly measured or derived from atmospheric optical measurements (visual range, horizon sky chromaticity and luminance, solar zenith and scattering angles, and cloud conditions), and human judgments of visual air quality and scenic beauty (Figure 6.1).

The approach to investigating the relationship between environmental measures and judgments consists of (1) photographic documentation and atmospheric optical measurements of the characteristics of the visual environment, (2) ratings of VAQ and scenic beauty from photographs by groups of observers, and (3) statistical analysis of relationship between physical measurements and human judgments. The first step required taking color photographs and atmospheric optical measurements, using telephotometers and nepholometers, in the western USA (Grand Canyon National Park and Mt. Lemmon near Tucson, Arizona) and in the eastern USA (Great Smoky Mountains and Shenandoah National Parks). Next, over 1,300 individuals rated the color slides for either visual air quality or scenic beauty using a ten-point rating scale. Ratings were then transformed to indices based on ratings of a set of baseline studies.

EPA/NPS Approach

This program focuses on the examination of the relationship between properties of the visual environment and human judgments of VAQ. A model based on measured optical properties is derived to predict perceptions of visual air quality (Figure 6.1).

The approach is summarized as follows. Visitors rated vistas depicted by color slides using a one to ten scale of VAQ. The slides used were selected from about 1,000 slides taken by the National Park Service. Telephotometer measurements were taken along with each slide to provide measures of the visual environment used in the model. Variables such as sun angle, snow cover, cloud conditions, and landscape elements were also monitored and used in the model derivation. A theoretical model linking atmospheric transmittance to perceived contrast of elements of the scene and to the VAQ was developed. The data gathered were used to establish certain functions used in the model and to evaluate the model results.

COMPARISON

As is noted in the previous section, the NCAR method attempts explicitly to link emission changes to measurable properties of the visual environment and then to link these variables to human judgments of VAQ. The EPA/NPS approach addresses the second half of the overall visual air quality assessment method as defined by the NCAR study, i.e., the link between the visual environment and judgments of VAQ. SAI addresses aspects of the first part in earlier work (Latimer et al. 1978) and the second half but does not attempt to establish a complete relationship. Thus, the following comparison of the three methods focuses only on their common aspect--the relation between the visual environment and judgment.

The three studies of VAQ, while similar in some important respects, differ with regard to (1) the conceptual framework for studying VAQ, (2) research design and data collection procedures, and (3) data analysis. Each of these areas will be discussed in turn.

Conceptual Framework

The studies are different in their treatment of the role of perception and judgment in visual air quality judgments. The SAI and EPA/NPS studies use the terms "perception" and "judgment" interchangeably. In both reports, one can find instances of the use of both "perceptions of visual air quality" and "judgments of visual air quality." In the NCAR studies, VAQ has always been considered a judgment: "We conceptualize visual air quality as a judgment based upon a number of perceptual cues..." (Mumpower 1981, p. 6).

The distinction between perception and judgment is an important one in psychology. The two processes have been studied by different investigators, using different methods, and focusing on different problems. Perception may be considered a type of judgment that is based primarily on sensory input and involves little of the mental processing we call "thinking." For example, judgments of the color and brilliance of a diamond are primarily perceptual, while the judgment of its value involves integration of information to form a judgment, employing cognitive processes that are not perceptual. The importance of this distinction has been emphasized in the writings of Brunswik (1954).

The conceptual framework of the NCAR studies is based upon an adaptation of Brunswick's (1956) lens model. In this adaptation, the visual scene itself is linked to an observer's judgment of VAQ by (1) a set of objective attributes of the scene (X_1, X_2, X_3) that are determined by measurements or models and (2) a set of subjective attributes of the scene (Y_1, Y_2, Y_3) that depend on the attention and visual perception of the person viewing the scene. The subjective attributes are related to VAQ by a judgmental (not perceptual) process. The judgment of VAQ may well be influenced by nonperceptual elements such as the observer's expectations with regard to visual air quality, or the observer's judgment about the source of visibility impairment as well as aesthetic values obtained through culture and experience.

The framework clarifies the role of perception and judgment in VAQ and suggests that answers to several questions are required for understanding and prediction of VAQ:

1. What are the objectively measurable attributes of scenes that are associated with judgments of VAQ?
2. What are the perceived subjective attributes that are associated with judgments of VAQ?
3. How are the objective attributes related to the subjective attributes?
4. What are the relations among objective attributes and among subjective attributes?
5. How are the subjective attributes related to VAQ?
6. How do people differ with regard to perceptions of the subjective attributes?
7. How do people differ with regard to judgments of VAQ?

The answers to these questions are influenced by the specific conceptual framework used in a particular study, as well as by the research design and data analysis procedures. Figure 6.2 summarizes the structural models underlying the three studies. The distinguishing feature of the SAI study is that subjective attributes are not explicitly measured. Measured objective attributes are related directly to VAQ (or the judgment of scenic beauty) by multiple regression analysis without the intermediate step of relating objective attributes to subjective attributes. The SAI results, then are purely descriptive and reveal little about the process relating objective

FIGURE 6.2
Structural Models for the Three Studies

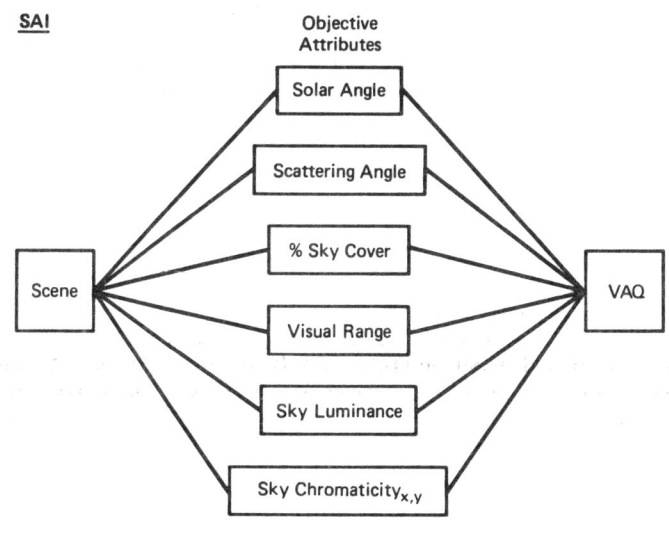

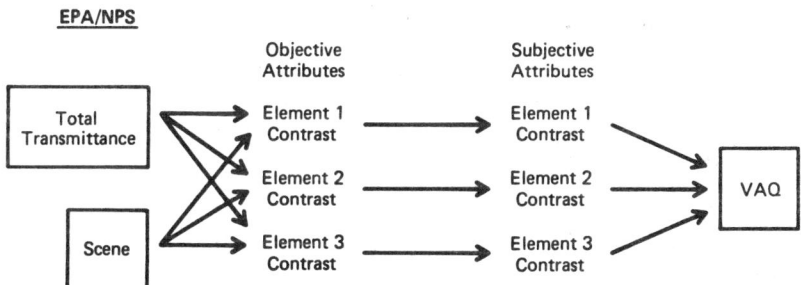

FIGURE 6.2 (continued)

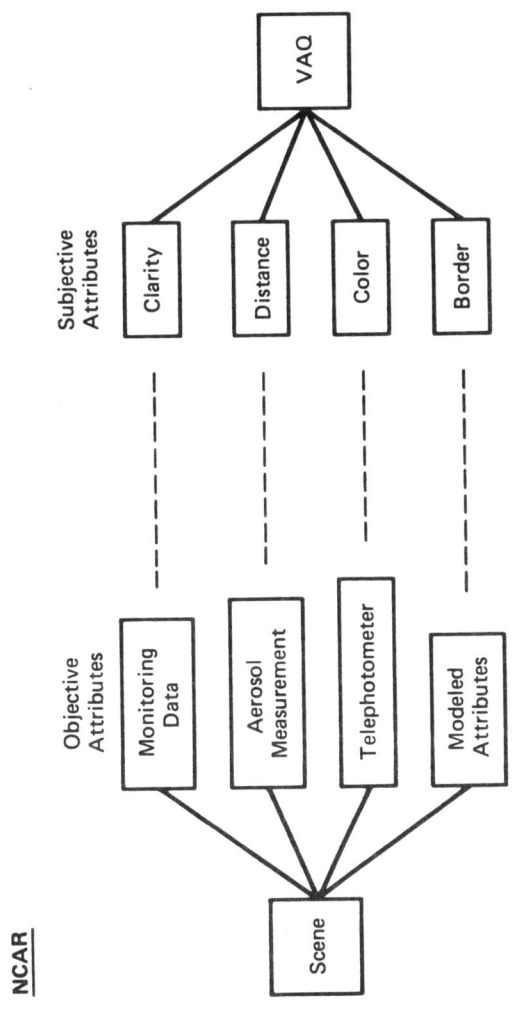

attributes to judgments of VAQ. Questions 2, 3, 4, 5 and 6 above are not addressed.

In the EPA/NPS approach, there is a direct correspondence between subjective and objective attributes (a correspondence which was tested empirically). The relevant attributes are the contrasts of visual elements of the scene (distant mountains, canyon walls, trees, etc.). The visual elements are derived from the effect on the scene of total transmittance of the atmosphere through which the scene is viewed. The relation between these elements and the judgment of VAQ is specified by a theoretical model involving the size, distance, and beauty of the element.

For the NCAR study, a variety of objective attributes have been examined, including routine and special measurements as well as modeled attributes based on emissions. The subjective attributes are not scenic elements but general characteristics of views which are relevant to the judgment of VAQ. The link between subjective and objective attributes is not one to one, as in the EPA/NPS study, but is a complex relation.

Research Design and Data Collection Procedures

The studies differ with regard to (1) method of presenting scenes for judgment, (2) selection of scenes for judgment (including control of certain attributes of scenes), (3) method for obtaining objective attributes, (4) method for obtaining subjective attributes, and (5) method for selecting observers.

1. Method of presentation. The scenes were presented as photographs (all studies), field observations (NCAR) and framed field observations (EPA/NPS). The frame for the EPA/NPS field observations was provided by a window in a mobile trailer which restricted the observer's view to correspond closely with the photographs.

2. Selection of scenes. The scenes in the NCAR and the SAI studies were selected to be representative of a range of scenes within the domain of interest. The SAI study included a number of scenes photographed at various times of day and under a variety of atmospheric conditions. The NCAR study included a variety of urban scenes viewed at different times of day and in different seasons. Scenes in the EPA/NPS study were selected with the aim of controlling factors that were known to be related to VAQ in order to isolate the effect of atmospheric pollution factors.

3. Obtaining objective attributes. The data on objective attributes were obtained through the use of optical measuring devices (all studies), measurement of pollutant concentrations (NCAR), meteorological observations (all studies), and theoretical model calculations based on emissions data (NCAR,SAI).

4. Obtaining subjective attributes. Subjective attributes were omitted (SAI), calculated from a theoretical model (EPA/NPS) or judged by the observers at the time of the observation (NCAR).

5. Selecting observers. Volunteer observers were used in the SAI and EPA/NPS studies. Since they were available for only a short time, they judged a series of slides and, in the case of the EPA/NPS study, made one observation from the trailer window. In the NCAR study, observers were hired for a period of about one month to make a series of field observations. Each observer made seventy to ninety field observations and numerous slide observations (250 in latest study). The hired observers also interviewed passersby who served as volunteer observers.

FIGURE 6.3
Data Matrix

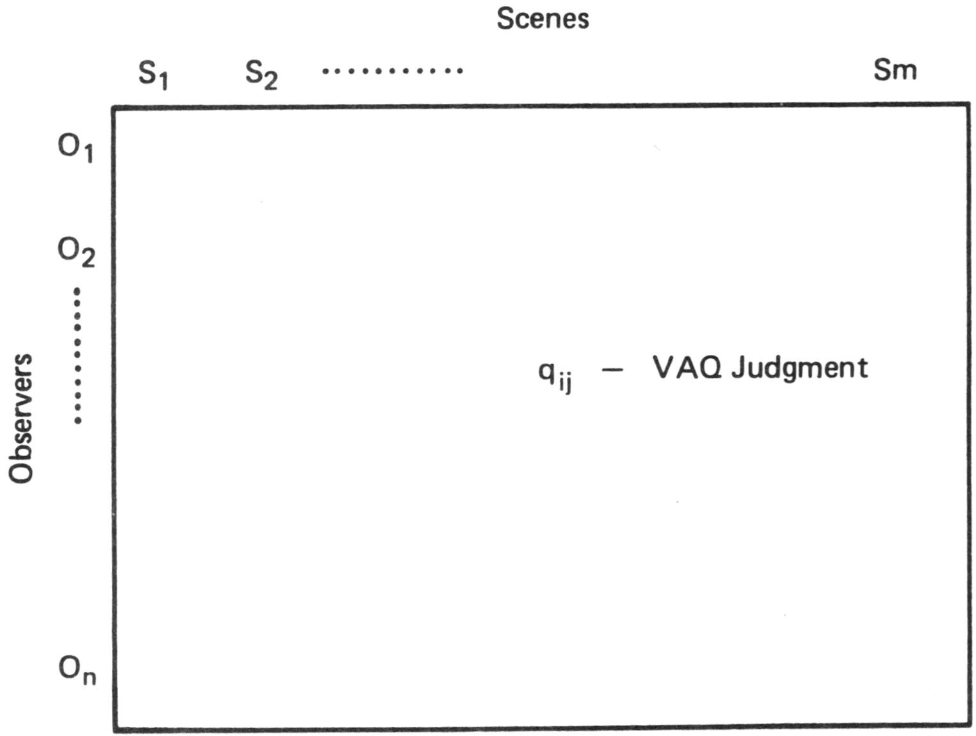

Figure shows a data matrix with Scenes ($S_1, S_2, \ldots, S_m$) along the top and Observers ($O_1, O_2, \ldots, O_n$) along the side, with entries q_{ij} — VAQ Judgment.

Data Analysis

The data obtained from each study could be organized into a matrix like the one in Figure 6.3 which has a row for each observer and a column for each scene. If observer i judged scene j, then the judged visual air quality (q_{ij}) is entered in the appropriate cell. Note that in field studies, such a matrix will contain many empty cells because each observer judges only a subset of the total number of scenes. In photographic studies, the data matrix will generally be full because all observers judge all photographs.

A critical choice in judgment research is whether to aggregate data first over rows of this matrix (e.g., by averaging over observers) or to aggregate over columns (e.g., by analyzing the judgments of scenes separately for each observer) and then aggregate over rows. The former approach is call nomothetic and the latter is called idiographic. As Hammond et al. (1980) indicate, there is no correct form of analysis, and each has its own strengths and weaknesses. Furthermore, most studies contain characteristics of both forms of analysis.

Two of the studies reviewed here have adopted an essentially nomothetic approach (SAI, EPA/NPS). VAQ judgments in both studies were first averaged within interest groups or demographic categories and then, since no important group differences were found, the judgments were aggregated across groups and the analysis proceeded based on VAQ judgments averaged across people in the sample. The NCAR study used a more idiographic approach. The VAQ judgments are analyzed separately for each observer. In effect, a separate model is derived for each observer. The models for the individual observer are compared and, if two observers' models are similar, they are aggregated.

The advantages of the nomothetic approach are greater ease of analysis and, because a larger sample size can be used, greater generality of results to the general population. The idiographic approach is more likely, however, to provide an understanding of individual differences in judgments. The SAI and EPA/NPS studies examined individual differences along one dimension--membership in various interest or demographic groups. Individual differences not related to these groups would be washed out in the analysis. The NCAR study examines individual differences by doing a separate analysis for each individual and comparing them.

For a more complete discussion of idiographic and nomothetic approaches, see Hammond et al. (1980). Two points they make are relevant here. First, the nomothetic approach has the danger that the results that apply to a group average may apply to no individual member of the group. Second, nomothetic and idiographic analysis of the same data are likely to yield different results.

A second difference among the studies with regard to data analysis is in the reliance on theoretical models in the treatment of the data. In this respect, the studies could be roughly ordered along a continuum. At one extreme, no explicit theory is invoked--a predictive relationship is derived directly from observer data (SAI). At the other extreme, an elegant model is derived based on a few simplified theoretical relationships and data are used to test predictions based on the model (EPS/NPS). In the middle, empirical and theoretical models are used and the results compared to take advantage of the strengths of both approaches (NCAR).

IMPLICATIONS

The comparison of existing pristine area studies with the NCAR approach suggests several considerations for pristine area assessments. An index which will predict the impact of emissions changes on VAQ in a pristine area requires the development of an operational model or set of models that link emissions to human judgments of VAQ. Because VAQ is a nonlinear system (i.e., emission changes are not necessarily linearly related to either properties of the visual environment or judged VAQ) models connecting only properties of the visual environment to VAQ judgments are not adequate for predictive VAQ assessment. The link to emissions must be explicitly established. Without this explicit connection, a VAQ model cannot be used in a regulatory function such as granting power plant siting permits. Researchers in pristine area VAQ assesment should be encouraged to pursue this difficult, but necessary, regulatory model development.

The comparison of these studies which focused on the visual-environment-to-human-evaluation part of the problem has exposed a number of important and complex issues in the method and design of research. We cannot give these issues the discussion they deserve, but all have been treated extensively in the research methods literature. At this time, we would like to review what seems to us to be the three most critical issues in the design of research on VAQ.

1. Choice of a conceptual framework. The differences illustrated in Figure 6.2 will influence nearly every decision made in the research and will certainly affect the results. In particular, it is unlikely that a variable omitted during the conceptual stage will be discovered to be important later on.

It is particularly important to recognize the distinction between perception and judgment and to sytematically investigate the role of subjective attributes which intervene between objective attributes and judgments of VAQ. Procedures for measuring these subjective attributes, developed in the NCAR research, should improve our understanding and prediction of VAQ judgments.

2. Representative design vs. control. Representative design can be frustrating for the scientist because the relationship of interest is embedded in an entangled web of other relations. Brunswik (1956) has argued, however, that this entangled web is exactly what we should be interested in. As shown by Hammond and Steward (1974), artificial control of variables creates a new situation that may affect the judgment of the observer. The generality of results obtained under controlled circumstances must be carefully examined in more realistic settings.

3. Idiographic vs. nomothetic. If VAQ is a judgment that can be influenced by aesthetic values, then the possibility of important individual differences exists (Craik and Zube 1976), and understanding of VAQ requires careful examination of such judgments. This is particularly important if the research is to be used in support of regulation. The nomothetic approach produces an impression of consensus with regard to VAQ which may not be the case. The reality of this consensus should be explored more fully using idiographic techniques.

There is strength in the diversity of approaches to the study of VAQ which should be enhanced by future comparative work and cooperation among researchers in the field. Understanding and prediction of VAQ can only be improved by an understanding of the theoretical and methodological choices available to researchers in the field. Only comparative research using a variety of methods can assure that the results of studies are not created by the use of a particular method.

BIBLIOGRAPHY

Brunswik, E. 1954. "Reasoning as a Universal Behavior Model and a Functional Differentiation between "Perception" and "Thinking." Paper presented at the International Congress of Psychology, Montreal. Reprinted in Hammond, K. R. (Ed.), The Psychology of Egon Brunswik. Holt, New York.

Brunswik, E. 1956. Perception and Representative Design of Experiments. University of California Press, Berkeley, California.

Craik, K. H., and E. H. Zube. 1976. "The Development of Perceived Environmental Quality Indices." In K. H. Craik and E. H. Zube (Ed.), Perceiving Environmental Quality: Research and Applications. Plenum, New York.

Hammond, K. R., G. H. McClelland and J. Mumpower. 1980. Human Judgment and Decision Making. Praeger, New York.

Hammond, K. R., and T. R. Stewart. 1974. The Interaction between Design and Discovery in the Study of Human Judgment. University of Colorado Center for Research on Judgment and Policy Report 152. Boulder, CO.

Latimer, D. A., R. W. Berstron, S. R. Hays, Liw Mei-kao, J. H. Seinfeld, G. Z. Whitten, M. A. Wojcik, M. J. Hillyer. 1978. The Development of Mathematical Models for the Prediction of Anthropogenic Visibility Impairment. EPA-450/8/78-110a. U.S. Environmental Protection Agency, Washington, D.C.

Latimer, D. A., H. Hogo and T. C. Daniel. 1981. "The Effects of Atmospheric Optical Conditions of Perceived Scenic Beauty." Atmospheric Environment 15:1865-1874.

Malm, W., K. Kelley, J. Molenar and T. C. Daniel. 1981. "Human Perception of Visual Air Quality (Uniform Haze)." Atmospheric Environment 15:1875-1890.

Middleton, P., R. L. Dennis and T. R. Stewart. 1981. "Urban Visual Air Quality: Modelled and Perceived." In Proceedings of the 12th International Technical Meeting (NATO/CCMS) on Air Pollution Modelling and its Applications. Palo Alto, California.

Mumpower, J., P. Middleton, R. L. Dennis, T. R. Steward and V. Viers. 1981. "Visual Air Quality Assessment: Denver Case Study." Atmospheric Science 12:2433-2441.

7. Psychophysics, Visibility, and Perceived Atmospheric Transparency

Ronald C. Henry

INTRODUCTION

As used in this paper, the term visual psychophysics refers exclusively to the relationship between physical stimuli and sensory impressions. Psychological reactions to the sensory data are not considered. Visual psychophysics occupies a key position in understanding the chain of processes that define questions of visibility degradation. It is the link between the external and internal worlds of the observer. It relates perceived levels of lightness, color, form, depth, and texture to the physical antecedents of light intensity distribution over space and wavelength as they enter the eye. Only visual psychophysics can answer questions of perceptible thresholds in contrast or color, or can predict just-noticeable differences in scenic vistas caused by changes in optical properties of the atmosphere.

Recent advances in visual psychophysics have made possible quantitative models of parts of the eye-brain system that can be usefully applied to atmospheric visibility questions. Two previous papers discussed some possible applications (Henry 1979a and 1979b); however, to date, no application of these or other modern psychophysical theories to atmospheric visibility has been made, presumably because of emphasis on the physical side of the problem on one hand and the psychological side on the other. A basic motivation of this paper is the belief that modern psychophysical theories provide the only hope of finding a scientifically sound, quantitative relationship between observed or calculated optical effects of pollution and perception.

This paper examines one area with important application to visibility issues, the perception of transparency. The most important objects of visibility studies, plumes and haze layers, are often seen subjectively as layers with varying transparency. This has a profound effect on their perceived brightness and color and that of objects seen through them.

Ronald C. Henry is a senior scientist with Environmental Research & Technology, Inc., Westlake Village, CA. This work was encouraged and supported by Dr. George Hidy, who was one of the observers in this experiment. Assistance with travel funds was provided by the American Petroleum Institute.

PERCEPTION OF TRANSPARENCY

Regional or layered haze is usually seen as one or more semi-transparent layers of a different brightness and/or color than their background. However, current visibility modeling technology relies entirely on the C.I.E. methodology which is based on judgment of colors and brightness of opaque surfaces and ignores perceptual transparency. In this work, the term perceptual transparency describes the situation in which the observer has the subjective impression that he is looking at a background feature through a distinct, semi-transparent surface. Depth perception is important to perceptual transparency because the transparent layer must seem to the observer to be in front of the observed background. It is stressed that perceptual transparency, as defined here, is a subjective phenomenon. The observer in a natural environment is always looking through a more or less transparent atmosphere (in the physical sense). However, atmospheric haze is not always seen as a distinct, perceived layer through which objects are observed. The psychophysical implications of perceptual transparency to achromatic brightness perception are discussed below.

The work of Metelli (1974) has revived interest in the perception of transparency. Using the principle of color scission, Metelli derives simple brightness conditions that are necessary for the appearance of transparency in achromatic situations. This theory of transparency and color scission is explained by a typical example, as shown in Figure 7.1. According to this theory, the brightness of the region corresponding to the distant mountains is subjectively split between the brightness of the white haze layer and the brightness of the mountains. (Note that here brightness refers to the subjective judgment of white to black, not luminance as measured by a radiometer.) This splitting of brightness in transparency perception is necessary because the eye needs to see two surfaces, the semi-transparent surface over the opaque lower surface, so depth perception is also involved. In symbolic form, brightness splitting between two surfaces is expressed as:

$$B_R = (1 - \alpha) B_H + \alpha B_m \tag{1}$$

where:

B_R = brightness of the mountain and haze if seen as one surface, i.e., without perceived transparency,
B_H = brightness of the haze,
B_m = brightness of the mountains as perceived through the semi-transparent layer, and
α = index of transparency, a number between 0 and 1

The value of α in Equation 1 is related to the degree of brightness splitting between the mountain and the haze. If $\alpha = 1$, then all the brightness is attributed to the mountain and $B_R = B_m$, i.e., there is no apparent haze. This corresponds to perfect transparency. If $\alpha = 0$, then all the brightness is seen as an opaque haze and the mountain disappears behind the haze. So the empirically identified parameter is an index of transparency whose asymptotic limits of zero and unity yield cases that are consistent with physically measurable conditions using optical instruments.

One consequence of Equation 1 is that, assuming the haze is brighter than the mountain, then the mountain seen through a perceptually transparent haze will look darker than if the haze and mountain were seen with-

FIGURE 7.1
Schematic of Typical Appearance of Transparent Haze

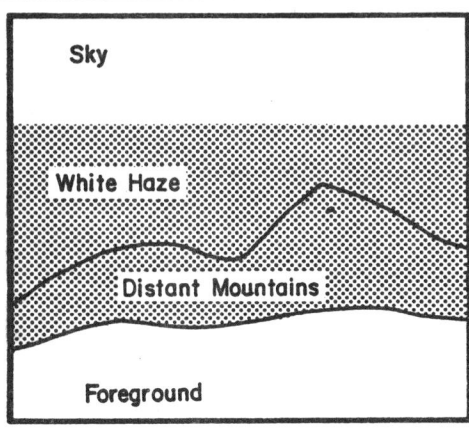

out the subjective appearance of transparency. This is demonstrated by dividing Equation 1 by B_M to obtain

$$\frac{B_R}{B_m} = \alpha + (1 - \alpha) \frac{B_H}{B_m} \qquad (2)$$

Since $0 < \alpha < 1$ and the haze is lighter than the mountain, i.e., $(B_H/B_m) > 1$; then the right hand side of Equation 2 is always greater than 1. B_R/B_m equals B_H/B_m for $\alpha = 0$ and declines to 1 for $\alpha = 1$. Therefore, the predicted effect of perceptual transparency is to darken the mountains seen through a semi-transparent haze.

EXPERIMENTAL PROCEDURE

A simple experiment was carried out that supports the above analysis, using simultaneous observations of relative physical brightness and perceived brightness. The apparent brightness of a mountain and nearby sky was estimated by comparison with a thirty-two step Munsell, matte finish gray scale. The brightness of the mountain and sky was recorded to the nearest 0.1 in the Munsell value notion where 10 is a pure white and 0 is pure black. A telespectroradiometer was used to measure the radiance of the mountain and the sky at 550 nm. Note was made of the weather, sky condition, time, and direction of the observation.

Observations were carried out by two observers (RH and GH) at two locations in southern California with good mountain views, and by observer GH at different locations, with a range of lighting conditions in the Sierra Nevada Mountains of California and in the Colorado Rockies. Therefore, a wide range of realistic scenic vistas was sampled. Approximately fifty observations were recorded. No obvious differences between the observers were noticed, but no systematic comparison was made as this was a feasibility study only.

In retrospect, observations of the perceptual transparency of the atmosphere should have been made. When the measurements were taken, the importance of transparency was not realized.

DATA ANALYSIS AND RESULTS

The original purpose of this experiment was to determine if the relationship between subjective brightness and physical luminance in the natural environment is the same as that derived from laboratory studies. A large body of psychophysical laboratory data supports a power law dependence of subjective brightness on luminance (Wyszecki and Stiles 1967). This is expressed as:

$$B = aL^b \tag{3}$$

where B is a number proportional to perceived brightness and L is the luminance. The exponent, b, of the power law is variously taken to be between 0.3 and 0.5. The best fit for the Munsell system is a value of 0.426. The value of a is somewhat arbitrary, in the present case it is taken to be:

$$a = B_s L_s^{-0.426} \tag{4}$$

where B_s is the observed Munsell value of sky, and L_s is the observed radiance of the sky at 550 nm. Then the calculated Munsell value (brightness) of a mountain is:

$$B_R = B_s (L_m/L_s)^{0.426} \tag{5}$$

where B_R is the calculated Munsell value of the mountain, and L_m is the observed radiance of the mountain at 550 nm.

Thus, the original goal of the experiment was equivalent to comparing the brightness calculated from Equation (5) and teleradiometer measurements, and the brightness as directly estimated by an observer using a Munsell gray scale. The results were very surprising. The calculated brightness was very often much lighter than the observed brightness as shown in Figure 7.2. Only one value lies above the line of equal values, showing that the observed brightness of distant mountains is usually less than predicted by teleradiometer measurements and the accepted psychophysical relationship between luminance and subjective brightness. This indicated that the well established power law relating brightness and luminance does not hold in the natural environment.

There are compelling physiological and psychophysical reasons for accepting the power law relationship, so it could not be dismissed lightly. Obviously, something must be affecting the perceived brightness in the natural environment that is absent from the laboratory studies. As shown above, perceived transparency can explain the observed apparent darkening of the mountains. Consequently, analysis of the data from the experiment has focused on perceptual transparency and quantitative predictions of the theory of color scission, as outlined in the previous section.

FIGURE 7.2
Brightness of Mountains Calculated from Teleradiometer Readings Versus Observed Brightness using a Munsell Gray Scale.

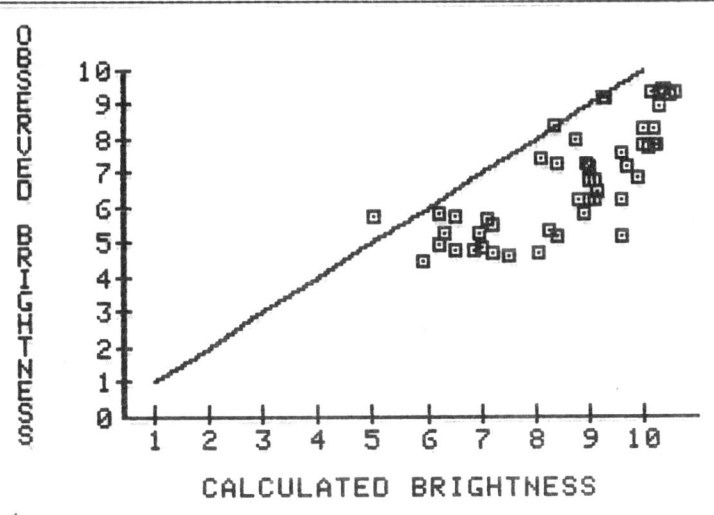

CALCULATED BRIGHTNESS

The color scission theory of transparency predicts that the ratio of the actual, observed brightness to the calculated brightness of the mountain is:

$$\frac{B_R}{B_m} = \frac{B_s}{B_m} (L_m/L_s)^{0.426} \qquad (6)$$

and that this ratio should always be greater than or equal to 1. (See Equations 2, 4 and 5 for definition of symbols.) Also, theory allows one to define a coefficient of transparency, α, from Equation 2 as

$$\alpha = \left(\frac{B_R}{B_m} - \frac{B_s}{B_m}\right) \Big/ \left(1 - \frac{B_s}{B_m}\right) \qquad (7)$$

Equation 7 is obtained from Equation 2 by setting $B_H = B_s$ and solving for α. The values of these quantities calculated from the experimental data are discussed next.

Figure 7.3 shows the histogram for the ratio as calculated by Equation 6 for fifty observations. Seventy-eight percent of the observations show the result predicted, i.e., a ratio greater than one, indicating an apparent darkening of the mountain seen through a semi-transparent layer.

FIGURE 7.3
Frequency of Occurrence of the Ratio of Predicted to Observed Brightness for Distant Mountains.

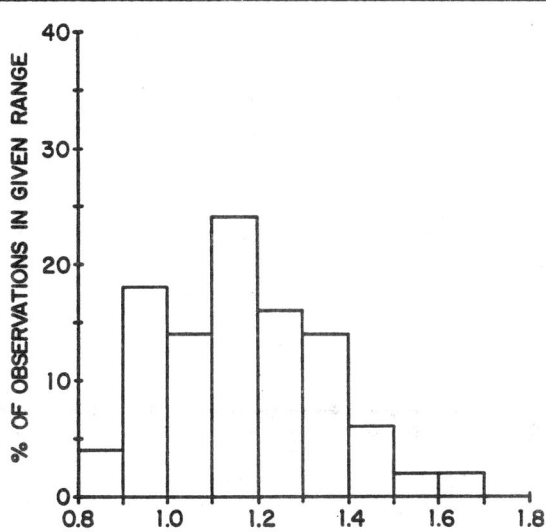

twenty-two percent of the ratios lie below one, most only marginally. This number of unrealistic ratios are to be expected since subjective brightness judgments by an observer in the field have a large potential for uncertainty. An error analysis of Equation 6, assuming a standard error of 0.5 for brightness judgments and ignoring errors in radiance, gives an expected random error of $\pm$ 20 percent, ninety percent of the time. Thus, values as low as 0.8 are within the range of random error if the true value is near 1.0. On the other hand, 40 percent of the observations are greater than 1.2. This many large values cannot conceivably be due to random error. Thus, the observations support the hypothesis that mountains seen through a partially transparent medium appear to be darker than they should be given no perceptual transparency effect. In fact, 54 percent of the observations showed that the mountains were 10 percent to 40 percent darker than they should appear. This shows that the apparent darkening of mountains through haze is not a small effect on observers' judgments of brightness.

Figure 7.4 displays a histogram of the transparency index, α, as calculated using Equations 6 and 7. Values of α greater than 1 are not physically meaningful, but some will occur because of the effects of large random errors in the brightness judgments used to calculate α. Transparencies range from nearly clear (α close to one) to rather opaque (α near 0.2). During the limited period of these observations, α was fairly uniformly distributed.

It is emphasized that the frequency of occurrence as shown in Figures 7.2 and 7.3 are intended only to show the range of values that were obtained during this informal, limited experiment. A larger, more systematic

FIGURE 7.4
Frequency of Occurrence of Subjective Transparency Index, α.

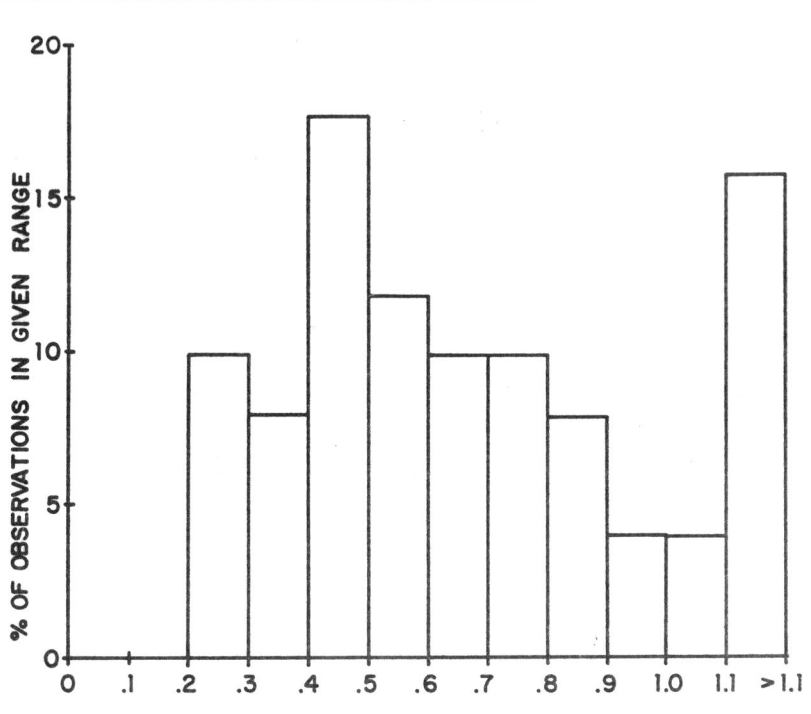

study is needed at specific sites to determine the frequency of occurrence of these quantitative, psychophysical measures of transparency and visibility.

DISCUSSION

The above results show that perceptual transparency has a large effect on the perceived brightness of a mountain seen through haze. The theory of color scission leads to the definitions of a new quantitative, psychophysical visibility index, α, the perceptual transparency index. The value of α is set by the eye-brain system. It will depend on all visual clues related to transparency and depth perception.

It is interesting to examine the relationship between the calculated perceptual transparency index and physical contrast. Intuitively, one would expect that large, negative contrasts would be associated with high perceptual transparency values and low perceptual transparency with small, negative contrast. Figure 7.5 shows that this expectation is fulfilled, with some exceptions. (Note that in Figure 7.5 transparency values plotted as 1 may

FIGURE 7.5
Scatter Diagram of Perceptual Transparency Index Versus Physical Contrast.

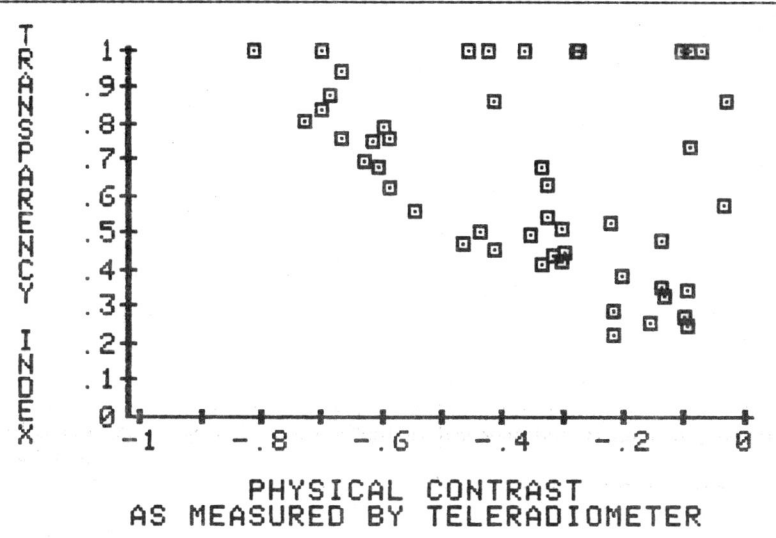

be greater than one. This was done to avoid distorting the vertical scale of the plot to include a few large values.) There are no cases of large, negative contrast with low transparency, but there are times when low or medium contrast is associated with high transparency. This is probably caused by a breakdown of the conditions necessary for perceptual transparency, most likely a loss of depth perception for some reason or reasons not yet understood. For small, negative contrasts, a loss of depth perception can be explained by the appearance of the mountain barely visible in an almost opaque haze. With no visible texture or other visual clues, the mountain, under these conditions, can be expected to appear to be at the same distance as the haze. Therefore, perceptual transparency and color scission theory would not apply to the low contrast case.

Figure 7.6 shows the results of editing out observations for which the perceptual transparency index was greater than 1 and/or the physical contrast was greater than -0.05. These values may be indicative of conditions not covered by the theory used to calculate the transparency index. It is admitted that this editing is not strictly justified; it is done only to illustrate an exploratory hypothesis. Figure 7.6 shows a strong linear relationship between physical contrast and the perceptual transparency index. The correlation coefficient is about -0.8. This relationship, if it holds, could explain the close linear relationship between subjective visual air quality and physical contrast reported by Malm et al. (1981). It is reasonable to speculate that subjective visual air quality estimates are strongly correlated with perceptual transparency of the atmosphere.

For purposes of modeling human perception in the natural environment, it is extremely important to quantify the relationship between physi-

FIGURE 7.6
Scatter Diagram of Transparency and Physical Contrast with Some Points Edited Out.

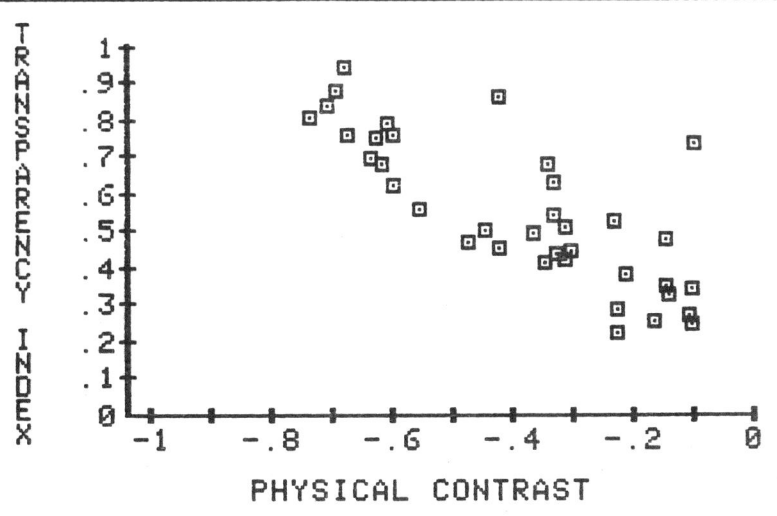

cally measurable parameters and the perceptual transparency index, α. One visual clue for perceptual transparency, other than contrast, is the amount of visible texture on the mountain. This can be quantified by either the modulation depth or Fourier analysis as recommended in Henry et al. (1981). It is suggested that current ongoing visibility studies be expanded to include some measurements of subjective brightness as described in this paper and scanning teleradiometer readings. This would provide the necessary data for a study of transparency effects and their relationship to texture as determined by Fourier analysis of the scanning teleradiometer data. Fourier analysis can also be used in conjunction with modern theories of the human visual system to estimate subjective appearance of the scene and the effect of haze, see Henry (1979a,b) for a brief description.

An important unanswered question concerns the ability of photographs to reproduce the transparency effects noted in this paper. Several studies of visibility perception have been based on evaluations of photograhic slides. If slides do not faithfully reproduce perceptual transparency effects, then their use in visibility studies will need to be reevaluated. Preliminary studies by the author indicate that transparency effects are not well reproduced by ordinary photographic slides, but further work is needed before any definite conclusions can be reached.

IMPLICATIONS TO VISIBILITY STUDIES AND SUGGESTIONS FOR FURTHER RESEARCH

Perceptual transparency effects as described in this paper have a significant effect on the appearance of regional and layered haze. The existence of these important psychophysical effects has not been previously recognized. A basic goal of visibility research is to find reliable relationships between physical measurements of visibility and the corresponding appearance of the science as described by the human observer. Failure to account for perceptual transparency effects appears to defeat this basic goal. The ability of the eye to perceive a separate haze layer cannot be inferred from contrast or color difference calculations or measurements alone. This paper suggests that a psychophysical parameter, a subjective index of transparency, is necessary to describe human perception of haze.

The perceptual transparency index is, in effect, a subjective variable that controls the appearance of the mountain and haze. Changes in the optical thickness of the haze may not be perceived if the visual system adjusts the transparency factor, α, to compensate for the change. It is possible that the evolutionary use of transparency effects is to provide yet another mechanism by which the visual system stabilizes our view of the natural world. From a practical standpoint, this implies that calculated decreases in contrast of a mountain may not be perceived as such. Consequently, actual visibility may differ significantly from that projected from optical measurements or model calculations.

Perceptual transparency intuitively must have a strong relationship to perceived visual air quality studies. It is not known what aesthetic effects perceptual transparency may have. Considering the popularity of nature photos featuring fog and rain, perceptual transparency effects may not be all adverse.

There is also the question of just what conditions are necessary for perceptual transparency. Metelli has given the answer for laboratory conditions. A set of answers for the natural environment needs to be found.

Much further work is needed to define how the index of perceptual transparency is related to physical parameters so it can be included in air quality model calculations expressing visual effects. Ongoing visibility monitoring programs could easily, and at small additional expense, include in their measurements the subjective brightness measurements needed for perceptual transparency calculations. More intensive special studies using Fourier analysis techniques are necessary to define the interaction between texture and transparency perception. Finally, the important case of discolored transparent layers needs study to develop appropriate theoretical and practical models to predict the appearance of discolored layers due to nitrogen oxides from modern power plants.

BIBLIOGRAPHY

Henry, R.C. 1979a. "Psychophysics and Visibility Values." In Proceedings of the Workshop in Visibility Values. ed. by D. Fox et al., U.S. Dept. of Agriculture General Technical Report WO-18. Fort Collins, CO.

Henry, R.C. 1979b. "The Human Observer and Visibility - Modern Psychophysics Applied to Visibility Degradation." In View on Visibility - Regulatory and Scientific, Air Pollution Control Association Specialty Conference, 26-27 November, Denver, CO.

Henry, R.C., J.F. Collins, and D. Hadley. 1981. "Potential for Quantitative Analysis of Uncontrolled Routine Photographic Slides". Atmospheric Environment 15:1859-1867.

Malm, W., K. Kelley M. Molenar, J. and T. Daniel. 1981. "Human Perception of Visual Air Quality (Uniform Haze)." Atmospheric Environment 15:1975-1890.

Metelli, F. 1974. "Achromatic Color Conditions in the Perception of Transparency." In *Perception Essays in Honor of James J. Gibson*, ed. by R.B. Macleod and H.L. Pick, Jr., Cornell University Press, Ithaca.

Wyszecki, G., and W.S. Stiles. 1967. *Color Science*, John Wiley & Sons, New York.

PART III
Visual Resource Management Systems

Visual resource management systems (VRMs) are generally used to provide land managers with information concerning the visual content and quality of a landscape, as well as the sensitivity of the scene to potential changes. For changes in land use these methods use principles of landscape architecture to determine what characteristics of the change will be humanly perceptible and rank potential changes in terms of their potential impact to visitor experiences or changes in scenic content.

These tools could also be usefully integrated into visibility analyses. The papers in the preceding section revealed initial findings that visible air pollution that blocks the view of specific landscape features will be less desirable than if the same pollution were not blocking the view of those features. Visibility researchers could use the principles of VRM techniques to more accurately describe and quantify the landscape characteristics associated with specific impacts to improve analyses on Steps 2 and 3 of Introduction Figure 1.

Specht reviews the Bureau of Land Management's (BLM) VRM system, which is used to describe and manage visual resources on an equal basis with other resource values for BLM lands. This approach considers visual quality, viewer sensitivity to changes, use volume and distance between the viewer and the scene to classify landscapes into categories called management classes. The potential impacts for different management activities in each management class are then examined and given a contrast rating that is used to judge the acceptability of potential impacts.

The paper by Smarden, Feimer, Craik, and Sheppard evaluates the reliability, validity and generalizability of selected observer based visual impact assessment (VIA) methods with application to a broad array of landscape and land use contexts. Two components of observer based VIA methods are specifically examined: the descriptive and evaluative dimensions, which serve as the basis for landscape ratings; and the rating procedure itself. This paper also focuses upon recent refinements in the BLM method.

Newkirk presents an alternative VRM that he has applied in Canada. This approach applies concepts of landscape diversity and measures of landscape associations, vegetation cover, cultural features, terrain characteristics, viewability, and visibility using available mapped inventory data. It next employs moving surface fits and neighborhood scanning techniques to produce detailed computer maps of visual values.

8. The Bureau of Land Management's Visual Resource Management System

Stanley V. Specht

INTRODUCTION

The quality of the visual environment has become increasingly important to the American public. The Bureau of Land Management (BLM) is managing visual resources on an equal basis with all other resources as it continues to put public land to productive use.

Visual Resource Management (VRM) has two program purposes: (1) it manages the quality of the visual environment, and (2) it reduces the visual impact of development activites. At the same time program effectiveness is maintained in all bureau resource programs. VRM also identifies scenic areas that warrant protection through special management attention. It is a specific process that can be mapped and incorporated into design planning for projects ranging from siting transmission lines to minerals management.

Managing the visual aspects of changes to the natural landscape is particularly important for the Bureau of Land Management because most activities taking place on bureau lands involve some degree of alteration.

The bureau's responsibilities for visual management are spelled out in key passages of recent federal legislation which follow.

The Federal Land Policy and Management Act of 1976 (FLPMA), often referred to as the "organic" act for the bureau, requires that:

> public lands be managed in a manner that will protect the quality of scientific, scenic, historical, ecological, environmental, air and atmospheric, water resource, and archaeological values; that, where appropriate, will preserve and protect certain public lands in their natural condition; . . .; and that will provide for outdoor recreation and human occupancy and use, . . .

Stanley V. Specht is Landscape Architect, U.S. Department of the Interior, Bureau of Land Management, Denver Service Center, Denver, CO. This paper is based upon U.S. Department of Interior, Bureau of Land Management, Visual Resource Management Program, U.S. Government Printing Office, Washington, D.C., 1980.

The Act also states that the Secretary of the Interior shall:

> prepare and maintain on a continuing basis an inventory of all public lands and their resources and other values (including but not limited to outdoor recreation and scenic values).

The Act, for the first time, places scenic resources on an equal basis with other resources. It also makes inventorying and managing scenic and other environmental values an explicit criterion that must be applied throughout the land management activities of the bureau.

The National Environmental Policy Act of 1969 (NEPA), an earlier and very important piece of environmental legislation, states that it is the federal government's responsibility to:

> assure for all Americans safe, healthy, productive, and aesthetically and culturally pleasing surroundings.

The Act further says that:

> all agencies of the Federal Government shall... identify and develop methods and procedures... which will insure that presently unquantified environmental amenities and values may be given appropriate consideration in decision-making along with economic and technical consideration.

Significant aspects of these federal laws are their increased emphasis on environmental and scenic values and their requirement that the long-term and short-term consequences of all resource commitments receive equal consideration. It is therefore bureau policy that visual resource considerations be included in environmental assessments, in land use planning decisions, and in the implementation of resource projects.

Since it was put into effect in 1975, the VRM program has helped set standards for bureau policy that visual resource considerations be included in environmental assessments, in land use planning decisions, and in the implementation of resource projects.

Since it was put into effect in 1975, the VRM program has helped set standards for transmission line location, timber harvesting, recreation development, range management, mining activities, and highway placement.

Because the scenic value and management objectives of public lands vary, it is not practical to provide a uniform level of visual management for all areas administered by the bureau. The agency has therefore developed a system for evaluating the visual resources of a given area and for determining what degree of protection, rehabilitation, or enhancement is desirable and possible. This bureau-wide system provides an interdisciplinary approach to managing visual resources. The integration of VRM into the bureau's procedures for planning and environmental analysis ensures maximum coordination between a proposed land use and the existing visual conditions.

SYSTEM CONCEPTS

The VRM system is an analytical process that identifies, sets, and meets objectives for maintaining scenic values and visual quality.

The system is based on research that has produced ways of assessing aesthetic qualities of the landscape in objective terms. What had been considered extremely subjective (aesthetic judgment, particularly concerning the landscape) was found to have identifiable, consistent qualities that can be described and measured. Whatever the terrain (and whoever the observer), perception of visual quality in a landscape seems to be based on several common principles:

1. Landscape character is primarily determined by the four basic visual elements of form, line, color, and texture. Although all four elements are present in every landscape, they exert varying degrees of influence.
2. The stronger the influence exerted by these elements, the more interesting the landscape.
3. The more visual variety in a landscape, the more aesthetically pleasing the landscape. Variety without harmony, however, is unattractive, particularly in terms of alterations (cultural modifications) that are made without care.

The bureau incorporates these and other principles in its broad program for managing visual resources.

The VRM system functions in two ways. First, for management purposes, the bureau conducts an inventory that evaluates visual resources on all lands under its jurisdiction. Once inventoried and analyzed, lands are given relative visual ratings, or Management Classes. The development of Management Classes is not project-specific, but rather a general process to identify broad visual objectives for all public lands.

Second, when development is proposed by the bureau itself (through its planning process), or by other agencies or the private sector, the degree of contrast between the proposed activity and the existing landscape is measured.

These combined steps constitute the VRM process, which has a number of applications. The process can help make the visual imapct of proposed activities more acceptable while these activities are still in the design stage. Graphic simulations of proposed activities help illustrate the extent of potential visual impact. Modifications may be suggested. During project construction, monitoring assesses actual visual impact. In both instances, VRM plays a support role.

THE VRM PROCESS

The inventory/evaluation process in VRM consists of three steps: (1) assessment of the visual or scenic quality of the landscape, (2) the sensitivity of the people to change(s) in the landscape, and (3) the viewing distance. Although the details of the evaluation are intricate, the process itself is straightforward.

Scenic Quality

Scenic quality is perhaps best described as the overall impression retained after driving through, walking through, or flying over an area of land. In the VRM process, rating scenic quality requires a brief description of the existing scenic values in a landscape. This step identifies (1) areas that must be protected, (2) opportunities for enhancement and rehabilitation, and (3) opportunities for improvement by reducing the contrast of cultural modifications.

When inventoried, an area is first divided into subunits that appear homogeneous, generally in terms of landform and vegetation. Each area is then rated by seven key factors: landform vegetation, water, color, influence of adjacent scenery, scarcity, and cultural modification. A standardized point system assigns great, some, or little importance to each factor. The values for each category are calculated and, according to total points, three scenic quality classes are determined and mapped:

1. Class A: Areas that combine the most outstanding characteristics of each rating factor.
2. Class B: Areas in which there is a combination of some outstanding features and some that are fairly common to the physiographic region.
3. Class C: Areas in which the features are fairly common to the physiographic region.

Sensitivity Levels

Although landscapes do have common elements that can be measured, there is obviously still a subjective dimension to measuring these elements. Each viewer brings perceptions formed by individual influences: culture, visual training, familiarity with local geography, personal values.

To measure regional and individual attitudes in the evaluation of a landscape, visual sensitivitiy is determined in two ways:

1. Use Volume: Frequency of travel through an area (by road, trail, river) and use of that area (for recreation, camping, events) are tabulated. The area is then assigned a high, medium, or low rating according to predetermined classifications.
2. User or Public Reaction: Public groups are familiarized with the area (if necessary) and asked to respond to activities that will modify that particular landscape. The concern they express about proposed changes in scenic quality is also rated high, medium, or low.

The various combinations of Use Volume and User Reaction for each area are rated by a matrix to an overall sensitivity rating of high, medium, or low. A map is then developed that illustrates final sensitivity levels.

Distance Zones

The visual quality of a landscape (and user reaction) may be magnified or diminished by the visibility of the landscape from major viewing routes and key observation points. In the VRM system, thus, distance plays a key part in visual quality management.

A landscape scene can be divided into three basic Distance Zones: foreground/middleground, background, and seldom-seem. Because areas that are closer have a greater effect on the observer, such areas require more attention than do areas that are farther away. Distance Zones allow this consideration of the proximity of the observer to the landscape.

Selection of the key viewing points and accurate assessment of Distance Zones require some judgment. Where several routes exist, what is foreground from one route may be background from another. (The more restrictive designation is used.) Atmospheric conditions may also modify the perception of distance.

For small projects, in-field photographic assessment of Distance Zones is usually sufficient. For large projects, however, or projects that require evaluation from many key viewpoints, an alternative method for generating data is to use a computer graphic modeling technique, such as the VIEWIT system.

The process culminates in the preparation of a final Distance Zone map.

Management Classes

Management Classes describe the different degrees of modification allowed to the basic elements of the landscape. Class designations are derived from an overlay technique that combines the maps of Scenic Quality, Sensitivitiy Levels, and Distance Zones. The overlays are used to identify areas with similar combinations of factors. These areas are assigned to one of five Management Classes, or VRM Classes, according to predetermined criteria. The resulting map of contiguous areas sharing the same Management Class is an important document for all bureau land use planning decisions, and it is also used to assess the visual impact of proposed development.

FIGURE 8.1
Management Class Matrix

Visual Sensitivity		H			M			L
Special Areas		1	1	1	1	1	1	1
Scenic Quality	A	2	2	2	2	2	2	2
	B	2	3	3	3	4	4	4
	C	3	4	4	4	4	4	4
Distance Zones		FG MG	BG	SS	FG MG	BG	SS	SS

Note: Class 5 areas are those that have been identified in the VRM planning system which require rehabilitation or enhancement and, therefore, are not included in this chart.

Matrix for Determining Management Classes. The Management Classes are defined as follow:

1. Class 1: Natural ecological changes and very limited management activity are allowed. Any contrast created within the characteristic landscape must not attract attention. This classification is applied to wilderness areas, wild and scenic rivers, and other similar situations.

2. Class 2: Changes in any of the basic elements (form, line, color, texture) caused by a management activity should not be evident in the characteristic landscape. Contrasts are seen but must not attract attention.
3. Class 3: Contrasts to the basic elements caused by a management activity are evident but should remain subordinate to the existing landscape.
4. Class 4: Any contrast attracts attention and is a dominant feature of the landscape in terms of scale, but it should repeat the form, line, color, and texture of the characteristic landscape.
5. Class 5: The classification is applied to areas where the natural character of the landscape has been disturbed to a point where rehabilitation is needed to bring it up to one of the four other classifications. The classification also applies to areas where there is potential to increase the landscape's visual quality. It would, for example, be applied to areas where unacceptable cultural modification has lower scenic quality; it is often used as an interim classification until objectives of another class can be reached.

THE CONTRAST RATING SYSTEM

To evaluate specific proposed projects, the Contrast Rating System is used to measure the degree of contrast between the proposed activity and the existing landscape. This score is compared with allowable levels of contrast for the appropriate Management Class. The comparison will determine if mitigation is required to reduce visual impacts.

The process first segregates a landcape into its major features (land/water surface, vegetation, structures) and each feature, in turn, into its basic elements (form, line, color, texture). The Contrast Rating quickly reveals the existing features and their respective elements that will be subject to the greatest visual impact. A total contrast score for each feature may then be used to define the overall contrast according to the following general categories:

1. Contrast can be seen but does not attract attention.
2. Attracts attention and begins to dominate.
3. Demands attention and will not be overlooked by the average observer.

This score is then compared to the appropriate Management Class to determine if contrast totals are acceptable. If the proposed project exceeds the allowable contrast, then a bureau decision is made to (1) redesign, (2) abandon or reject, or (3) proceed, but with mitigation measures stipulated to reduce critical impacts.

Since each activity proposed for bureau-administered land must pass through this evaluation, it has proven useful to identify and mitigate extreme contrasts to scenic quality in the planning/design stage of a proposed activity prior to submittal for approval. This pre-evaluation can save time and money because it forestalls a potentially lengthy revision process. The Contrast Rating compares the proposed activity with existing conditions element by element, feature by feature, according to the degree of contrast.

SUMMARY

The Bureau of Land Management has developed a system to assure consideration of the quality of the visual environment when conducting its land management responsibilities on the nation's public lands. The Visual Resource Management system is based upon the scenic quality of the landscape being managed, coupled with the sensitivities of the people viewing these landscapes. Together, these inputs form a basis against which proposed changes which could potentially be created by a management activity could be judged as being acceptable or unacceptable. The Contrast Rating System is used for this determination. The management activity, if judged to be unacceptable, could then be redesigned before implementation, or mitigation measures could be suggested to lessen undesirable visual contrast which would be created by the project.

9. Assessing the Reliability, Validity and Generalizability of Observer-Based Visual Impact Assessment Methods for the Western United States

Richard C. Smardon
Nickolaus R. Feimer
Kenneth H. Craik
Stephen R. J. Sheppard

INTRODUCTION

Incorporation of VIA in Decision-Making

The development and use of visual impact assessment (VIA) methods has proceeded rapidly in the last decade. Arising largely from a confluence of legal mandates, governmental administrative policies (Smardon 1979), and the progressive accumulation of a significant body of research on landscape perception (Craik and Feimer 1979; Elsner and Smardon 1979; Zube 1976), these methods are generally intended to provide land use managers with objective information concerning the impact of land use activities upon the aesthetic quality of the landscape. That information can then be incorporated into the decision-making process, with aesthetic factors taking their place alongside the other environmental, economic, and social factors which are inevitably of importance where land use options are concerned.

An important assumption underlying the inclusion of aesthetic factors through VIA systems in the decision-making process is that they will foster more effective, judicious decisions. That goal can only be attained if the information provided by VIA methods is accurate and systematic. This issue is critically significant, where land management is concerned, since decisions involving land use often have long-term consequences. Thus, the underestimation of the visual impact of a land use might result in unneces-

---

Richard Smardon is a Senior Research Associate in the School of Landscape Architecture, S.U.N.Y., Syracuse, New York; Nickolaus Feimer is Assistant Professor of Psychology, Virginia Polytechnic Institute and State University; Kenneth Craik is Professor of Psychology, University of California, Berkeley; and Stephen Sheppard is a Research Landscape Architect, University of California, Berkeley. Parts of this research were funded under Cooperative Research Agreements PSW-36, PSW-46, PSW-62, PSW-80-0005 and PSW-80-0006 with the Pacific Southwest Forest and Range Experiment Station, U.S. Forest Service and U.S.D.I., Bureau of Land Management, Washington, D.C. Portions of this chapter were previously published in Landscape Research 6(1):12-16. Support for presentation of this work at the Visual Values Workshop was provided by the Electric Power Research Institute.

sary degradation of the visual quality of the landscape, whilst an overestimation of the visual impact of the same activity might result in modification, curtailment or disallowance of the activity which, in turn, could cause considerable social and economic disruption. To avoid these pitfalls, VIA methods of sufficient technical quality should be employed (Craik and Feimer 1979). Minimally, the technical performance of VIA systems must be evaluated so that decision makers know of the margin of error inherent in the information upon which their decisions are based.

Closely related are the legal issues of (1) the adequacy of visual analysis given the context of existing laws and policy and (2) the soundness and defensible rationale of the basic methodology. Many existing federal statutes and some state statutes call for explicit consideration and treatment of aesthetic or visual resources for certain federal/state actions or within certain land areas administered by federal/state agencies (Smardon 1978). Visual resource methodologies are being more closely scrutinized in courtrooms and administrative hearings as to their basic adequacy and soundness. The ability of any VIA methodology to stand up to such legal tests is strongly related to the methodological properties of reliability, validity, and generalizability.

Issues of Reliability, Validity, and Generalizability

The quality and utility of a measurement method is largely a function of three properties: reliability, validity, and generalizability. Reliability refers to the consistency and precision of measurement; it reflects the degree to which the obtained measures are replicable in the same or highly similar circumstances, as well as the attainable level of discrimination among the objects of interest. In the context of VIA, reliability represents the degree to which a measure accurately reflects variations among landscape and land use conditions. Validity refers to the degree to which a measure represents the construct or variable of interest. In VIA, validity provides an estimate of the degree to which a method is able to capture meaningful variations in the aesthetic quality of the landscape and to predict the impact of land use activities upon it. It should also be noted that the reliability of a measure limits its attainable validity. Finally, generalizability refers to the range of the conditions for which the attained levels of a reliability and validity are representative. In VIA, factors which could constrain generalizability might include variation in the physiographic landscape and land use conditions, background characteristics of observers used in the VIA procedure, media of presentation of landscape and land use conditions, and the extent of pertinent landscape and land use information available to VIA users confronted with specific problems.

The research reported here is directed at an evaluation of the reliability, validity and some aspects of the generalizability of selected observer-based VIA methods. The emphasis has been on VIA methods with a potentially wide range of application to a broad array of landscape and land use contexts. Related findings on the reliability of VIA methods were reported by Feimer et al. (1979).

Two components of observer-based VIA methods were under examination: first, the descriptive and evaluative dimensions which serve as the basis for landscape ratings; and second, the rating procedure. Selection of landscape dimensions and rating procedures was based primarily upon their prominence in the research literature and their potential utility for application. The landscape dimensions selected for study included ambiguity, color, compatability, complexity, congruity, form, importance (of an element), intactness, line, novelty, scenic beauty, severity (of visual impact), texture, unity, and vividness. The rating procedures selected for study are direct and contrast ratings. Direct ratings entail a simple rating of landscape dimensions for a landscape scene. Contrast ratings require a comparison of landscape scenes both before and after the imposition of new land-use activities to obtain a rating of the degree of change in the dimensions of interest. The aforementioned variables as well as direct and contrast VIA ratings were utilized in our experimental design to find a VIA method that had acceptable levels of reliability, validity and generalizability for field application. The following sections outline our study and progress phase by phase to obtain this goal.

PHASE 1: LANDSCAPE CLASSIFICATION AND VISUAL SAMPLE

The following sections describe (1) the development of a landscape classification and (2) the selection of scenes and preparation of visual simulations for use in the psychometric analysis.

Landscape Classification of the Western United States

The objective of this part of the research project is to identify and map characteristic regional landscapes of the Western states. A classification of this type, based on visual characteristics of the land, is necessary in order to obtain photographs of scenes representative of the range of landscapes managed by BLM for use in analyses and research participants responses to typical scenes and activities.

Approach. The landscape classification is based on the system developed by Litton et al. (1978) for the Northern Great Plains. The system identifies a hierarchy of landscape scales:

1. continuity: an extensive area (thousands of square miles over which a broadly similar or repetitive type of landscape prevails).
2. province: a tract of land (hundreds to a few thousands of square miles) occurring within a landscape continuity but distinguished by a combination of features which contrast with its surroundings.
3. unit: a unified spatial enclosure (tens to hundreds of square miles) which forms one of many distinct local subdivisions within the continuity or province.
4. setting: the immediate surroundings (tens of square miles) of a site or scenic feature.

The approach taken for the Western U.S.A. is to identify the continuities which dictate regional patterns of landscape, delineate their boundaries, and map the distinct provinces within them. Because of the vast area involved (eleven states and parts of five others) and because of the broad nature of the classification, smaller scale landscapes (units and settings) are not mapped, although these scales may be useful for analysis of specific sites and views used in the sample of visual simulations.

Data Sources. The basis for the landscape inventory is the set of fifteen 1:1,000,000 scale topographic maps used to integrate data at a regional level from sources of smaller and greater scale. Inconsistencies between 1:1,000,000 maps in presentation of contour information made it essential to locate landscape boundaries first on 1:250,000 topographic maps, then transfer to 1:1,000,000 scale. The more detailed maps are also useful in displaying forested, non-forested, and agricultural areas.

Three land classification systems for the U.S.A. were used to identify preliminary landscape continuities. They are Fenneman's (1931) classification of the Western U.S.A. into physiographic provinces and sections; Hammond's (1964) classification of land-surface form; and Kuchler's (1969) classification of potential natural vegetation. Also at the national scale, supplementary data were obtained from classifications of regional geomorphology (Thornbury 1965) and natural regions (Hunt 1974), and from the National Atlas of the U.S.A. (Hammond 1970). Where further analysis and clarification were needed, smaller scale classifications were used. Sources of more detail on physiography included Hinds (1937) for California, and Franklin and Dryness (1973) for the Northwest. Additional information on vegetation was found to be particularly important and came principally from the Department of Landscape Architecture, U.C. Berkeley (1976), Franklin and Dryness (1973), Tidestrom (1925), Benson and Darron (1945), and the Arizona chapter of the Soil Conservation Society of America (1973). Landscape descriptions in various visual analysis reports by BLM and others were drawn upon where available; the mapping already carried out by Litton et al. (1978) in the Northern Great Plains was incorporated directly, although refined for consistency with the rest of the study area.

Since no special field inventory was carried out, heavy reliance has been placed upon in-house (PSW, BLM, and U.C. Berkeley) color slides and upon illustrations in the sources listed above. It would not have been feasible to check map boundaries in the field or by aerial photograph stereoscopy within the time limits of the study. The landscape classification represents a general division into regions; emphasis is laid on reliable identification of distinctly different regions which internally are broadly consistent, rather than on precise location of region boundaries.

Procedure. From the sources noted above, landscape continuities were identified in terms of their prevailing topographic and vegetative character. Together, these elements determine much of the visual character typical of large tracts of landscape.

For each continuity considered to be a discrete landscape entity, topographic, vegetative, and visual characteristics were noted on a standard form. Prevailing aspects of spatial character, water forms, vegetative mosaics, scenic features, human modifications, and temporal effects were recorded. Major and extensive variations in either vegetation or topography were treated as separate continuities, while isolated or limited areas of contrasting character were treated as provinces and described on a separate form.

Boundary location was delineated on 1:250,000 maps, with reference to 1:1,000,000 maps. Wherever possible, distinct physical features, with significance to viewers on the ground, have been used as boundaries. They include the slope-toe where plain and mountain continuities meet; the tree line between forested and unforested continuities; the plateau rim of a subdued tableland surface above dissected terrain; the line of a water course separating different terrain types; the ridge-crest dividing one landscape pattern from another; and the imaginary but comprehensive line across valley mouths where open plains extend as narrow basins between ridges.

In many locations, though, adjacent continuities merge gradually over a distance of miles, or a definite edge becomes highly digitate or complex. Where no single conspicuous feature can be correlated with the boundary, an arbitrary line (dashed to indicate an indistinct or transitional boundary) has been drawn through the midpoint or along the approximate edge of the area in question.

Boundaries of provinces were also delineated and marked by a separate symbol. (See Table 9.1 and Figure 9.1).

Visual Sample: Criteria for Its Selection

In order to analyze research participants' responses to visual impact, a sample of color photographs is needed, representing scenes before and after imposition of a development or management activity. The criteria for selection of the sample included:

1. representation of landscape continuities and types.
2. representation of typical land use activities within BLM holdings.
3. suitable photographic availability and quality.

The sample size is limited to approximately twenty sets of photographs, about the maximum number on which judges may be tested at one time. Hence, it is not possible to represent every landscape continuity. To ensure that both major activities and major landscapes under BLM jurisdiction are represented, a broad classification of both is required.

The landscape types are groupings of landscape continuities not by region, but by basic visual similarities in topography and vegetation. On the basis of visual analysis and Hammond's (1964) land-surface form classification, topographic character has been crudely subdivided into four types:

1. RUGGED: hill and mountains
2. SUBDUED: plains and gently sloping low hill
3. PLAIN AND MOUNTAIN: repeated and extensive rugged landforms interspersed with expanses of subdued terrain
4. TABLELANDS: expanses of subdued topography separated by very steep slopes and canyons

Employing land use information, visual analysis, and Kuchler's (1969) vegetation maps, vegetative character of continuities has been crudely subdivided into three visually significant types:

1. FOREST: largely continuous woodland prevail over most of the continuity
2. OPEN: shrub and/or grass vegetation and/or agriculture dominates the land surface
3. MOSAIC: a conspicuous mixture of open and forested vegetation prevails or is repeated over large areas of the continuity

Most of the thirty-six landscape continuities fall neatly into one or another of the twelve possible combinations of topography and vegetation. Two of these combinations (rugged/open and subdued/mosaic) do not occur over a whole continuity and may be omitted for simplicity. It is not suggested that visual management solutions for one continuity automatically apply to another of the same landscape type, since regional differences in climate, plant species, soils, etc. are dramatic. It is argued, however, that

FIGURE 9.1
Map of Landscape Continuity and Provinces

TABLE 9.1
Landscape Classification of Western United States

	Landscape Continuity	Landscape Provinces
1	Olympic Mountain (OM)	
2	Oregon Coast Ranges	
3	Redwood/Evergreen Forest	
4	California Coast Ranges	
5	Los Angeles Ranges	Los Angeles Basin (lab)
		Riverside Basin (rb)
6	Puget Trough	
7	Williamette Valley (WV)	
8	Great Valley	Marysville Buttes (mb)
9	Salton Trough (ST)	
10	West Cascades	
11	High Cascades	
12	Sierra Nevada	
13	Yakima Ranges (YR)	
14	Columbia Canyonlands	
15	Palouse	
16	Blue Mountains	
17	Payette Plains	Great Sandy Desert (gsd)
		Owyhee Mountains (om)
18	Snake River Plain	
19	Great Basin	Great Salt Lake Desert (gsl)
20	Sonoran Desert	
21	Mexican Highland	
22	Uinta Basin	
23	High Plateaus	
24	Canyonlands	
25	Navajo Plateau	Grand Canyon (gc)
		Chuska Mountain (cm)
		Mt. Taylor Plateau (mt)
26	Mogollon Plateau	
27	Mogollon Mountains	
28	Northern Rockies	
29	Middle Rockies	Yellowstone Plateau (yp)
30	Wyoming Basin	
31	Southern Rockies	North Park (np)
		Middle Park (mp)
		South Park (sp)
		Gunnison Valley (gv)
32	San Luis Valley (SLV)	
33	Northern Great Plains	Sweetgrass Hills (sh)
		Bearspaw (bp)
		Little Rocky (lr)
		Highwood (h)
		Snowy Mtns. (s)
		Musselshell Rise (mr)
		Tongue River Uplands (tru)
		Bighorn Mountains (bm)
		Black Hills (bh)

TABLE 9.1 (continued)
Landscape Classification of Western United States

	Landscape Continuity	Landscape Provinces
34	Southern Great Plains............	Nebraska Sand Hills (ns)
35	Raton Plateau	
36	Pecos Trough	

in a limited visual sample, basic visual similarities and differences must be considered in addition to criteria of physiographic or administrative regions.

The photographs used in the sample were placed within the matrix of major selection criteria. Eleven of the landscape continuities are represented, and six of the seven important landscape types are covered. Timber harvesting and recreational impacts are the only major activities not represented. The most important activities (e.g., surface mining) and landscape types (e.g. plain and mountain country) under BLM jurisdiction are represented by a range of photographs. In addition, three off-shore energy developments are included to represent BLM's jurisdiction over the continental shelf.

A particularly limiting constraint was the availability of suitable photographs. High quality original photographs were not available for some landscape continuities and activities. Most surprising of all, virtually no sets of before and after photographs were obtainable from BLM district offices. A system of landscape control points was proposed by Litton (1973) for proposed development sites which could have provided photographs for routine monitoring of visual impacts. Instead, because of the lack of photographs, "before" and "after" sets were created by simulation.

Simulation Procedures. For most of the sample sets, a photograph of a site after a facility or activity had been developed ("after" photo) was selected and a "before" view simulated by retouching the photograph to "remove" all traces of the activity. In a few cases, a proposed project was added to a "before" photograph to create the "after" image. In general, the process was found to give good quality, convincing images when one of two 7" by 10" high quality color prints was made from the original "after" slide and is retouched to remove the activity. Both the altered and unaltered prints are rephotographed to produce slide sets, thus ensuring that the only difference between them is due to the presence or absence of the activity, not the artifacts of film processing (See Feimer et al. 1979 for details).

PHASE 2: INITIAL PSYCHOMETRIC ANALYSIS OF VIA RELIABILITY AND VALIDITY

The following sections summarize the results of psychometric analysis, using the visual simulations described above, reprinted more fully in Feimer et al. (1981).

Landscape - Land Use Stimuli

Nineteen pairs of landscape scenes were employed to assure the effectiveness of the rating procedures. One member of each pair depicted the landscape before the imposition of a given land use activity and the other after the imposition of that activity. Either the before or after version of each pair had been simulated.

Research Participants

Research participants were drawn from three populations: (1) graduate and undergraduate students (n = 54) from the Berkeley and Davis campuses of the University of California; (2) U.S. federal agency administrative personnel not trained in visual landscape analysis (n = 87) and (3) landscape architects (n = 41) from the U.S. Department of Agriculture's Forest Service.

Procedures

Ratings were obtained through three quasi-experimental treatment conditions. In one (PREPOST condition), thirty-nine members of the student subsample were first presented with the before version of each scene, and completed direct ratings for all of the landscape dimensions previously enumerated except importance and severity (which implicitly apply to impacts) immediately after viewing each scene. Next, they were presented with the after version of the scene and completed contrast ratings as well as the importance and severity ratings.[1] In a second treatment (POST condition), the remaining fifteen participant students were presented with only the after version of each scene, and subsequently completed direct ratings on all landscape dimensions except importance and severity.

A two-hour training period preceded both the PREPOST and POST conditions to familiarize judges with the rating procedures, and with the contrast rating method in particular. In addition, a subsample of participants was given feedback on their reliability levels periodically during the data collection period. However, no differential effects were found in conjunction with feedback and, hence, the subsamples were collapsed into one group for subsequent data analysis.

In the third treatment (GLOBAL condition), the entire U.S. federal agency and BLM/Forest Service samples were simultaneously presented with both the before and after version of each scene, with the order of presentation counterbalanced for subgroups within the condition. Immediately after viewing each version of the scene, scenic beauty ratings were completed; and after viewing both versions of each scene, severity (of visual impact) ratings were completed. After all ratings were completed, participants in this condition were asked to reflect on and then rank order the criteria they employed for judgments of both scenic beauty and severity of visual impact. Due to time constraints, they completed only fourteen of the nineteen pairs of scenes.

The PREPOST and POST conditions were employed to provide visual impact ratings and independent before and after direct ratings. The GLOBAL conditions served primarily to provide an independent set of criterion data on evaluations of aesthetic quality. This allowed assessment of how generalizable the direct and contrast ratings were to observer groups who were either untrained in VIA (U.S. federal agency sample) or trained but with differential training and experience (BLM/Forest Service sample).

Results

Reliability. Intraclass correlation (Ebel 1951) was employed to assess the reliability of ratings. The intraclass correlation is the average reliability of a single rater. It is derived from a one-way analysis of variance where scenes (n = 19) are a random variable which constitute the main effect and the residential variance is the error term. Due to missing observations for some research participants on various scenes and rating dimensions, it was also necessary to use an average value for the number of raters when calculating the reliability estimates. The appropriate value (n) was obtained by an application of Snedecor's (1946) formula. The results of these analyses are given in Table 9.2. It is apparent that the reliability coefficients vary substantially within each rating condition. The average reliabilities for before direct and after direct are 0.26 and 0.21 respectively. Nonetheless, even for direct ratings, the obtained coefficients are clearly below acceptable standards (generally coefficients of 0.70 and higher are desirable). However, it must be stressed that these coefficients represent reliabilities for a single rater, and while single raters are often used in applied settings, higher reliability is generally obtained when composite ratings from panels of independent judges are employed (Craik and Feimer 1979; Feimer et al. 1981; Zube 1976). In the current context, for example, applying the Spearman-Brown prophecy formula (Guilford 1954) to the average reliabilities of the respective rating procedures reveals that a panel of ten independent judges would increase the average reliability to above 0.70 for both sets of direct ratings.

Validity. Change in scenic beauty was employed as a criterion measure to represent change in aesthetic quality resulting from the imposition of land use activities. It was obtained by subtracting the average after direct rating of scenic beauty from the corresponding average before direct ratings. This criterion measure for each subsample was then intercorrelated with change scores for each of the direct ratings of other landscape dimensions (again subtracting the average after from the average before ratings). Since the average score for each rating dimension was used, the reliabilities of the dimensions employed in the analysis were at an acceptable level (an average reliability above 0.70 for all rating procedures). The intercorrelation of change in scenic beauty with direct rating change scores is given in Table 9.3. Four direct rating dimensions (compatibility, congruity, intactness and form) are significantly correlated with change in scenic beauty for two of the three samples. These variables indicate that changes in the character and coherence of the landscape seem to be associated with perceived changes in aesthetic quality. Changes in land mass features (form) appear to be an important component of the resulting incongruity.

Criterion Rankings. In order to gain more insight about which variables may be important for explaining change in visual quality or severity of visual impact, a separate qualitative criterion analysis was done. After subjects had finished their quantitative ratings, they were asked to list and rank criteria that they had used in rating before scenes for scenic quality. Second, they were asked to list criteria in the same fashion for assessing severity of visual impact as seen in both the before and after scenes. It was assumed that after the subjects had judged some nineteen sets of before and after scenes, they would have had some criteria in mind when judging the slides.

TABLE 9.2
Average Single Rater Reliabilities for Direct Ratings

Dimension	Rating Procedure	
	Before (n = 29)	After (n = 17)
Ambiguity	.19	.07
Color	.13	.25
Compatibility	.07	.28
Complexity	.49	.13
Congruity	.17	.25
Form	.45	.14
Importance	-	.27
Intactness	.34	.31
Line	.19	.05
Novelty	.31	.22
Scenic Beauty	.18	.20
Severity	-	.21
Texture	.41	.24
Unity	.21	.25
Vividness	.26	.24
Mean	.26	.21

NOTE: n is the average number of raters used in computation of reliabilities and follows Snedecor (1946).

Some 143 sets of rank ordered criteria were obtained from sixty-six federal agency personnel (not trained in VIA), thirty-eight students (primarily in landscape architecture), and thirty-nine architects (U.S. Forest Service and BLM). These criteria were then sorted into categories of physical, aesthetic, and global criteria for assessing scenic quality; and into categories of visual impact. Within these categories, criteria were listed with their mean rank order and number of times mentioned. Criteria were only grouped together if, by content analysis, they were very similar. A number of subcategories were then collapsed into the major categories. Only the major criteria, with their number of times mentioned and mean rank order, were judged to be significant criteria.

The major finding from this criteria analysis is that there are major variables which are not presently included in BLM's visual contrast rating system. Some of these variables are those that can be related to the observed physical properties of landscapes and some are not. Global non-physically-related variables do not have utility for visual impact assessment purposes because the effect cannot be identified on the physical site and, therefore, cannot be mitigated. Most often mentioned as aesthetic factors related to severity of visual impact were the naturalness, fittingness, compatibility, and appropriateness of the intrusion. The most

TABLE 9.3
Scenic Beauty Change Scores Correlated with Direct Rating Change Score

	Change in Scenic Beauty		
Direct Rating Dimensions (Student Sample)	Student Sample (n=19)	U.S. Federal Agency Sample (n=14)	BLM/Forest Service Sample (n=14)
Ambiguity	.38	.27	.08
Color	.04	.04	-.13
Compatibility	.67**	.38	.72*
Complexity	-.06	.19	.15
Congruity	.56*	.53	.67**
Form	.59**	.47	.78**
Intactness	.31	.62*	.71**
Line	.47*	-.07	.23
Novelty	.25	.30	.34
Texture	.06	.26	.20
Unity	.66**	.09	.52
Vividness	.06	.08	.23

NOTE: Correlations are based on average ratings of respective samples completing ratings. n is the number of scenes.
* $p\ 0.05$
** $p\ 0.01$

prominent physical criteria cited were changes in color and form qualities and magnitude of the intrusion.

Thus, as in the correlation analysis, continuity in the general form of the landscape and the resultant compatibility of the land use activity seem to be the most salient factors in the psychological appraisal of visual impacts. It must be stressed again, however, that this analysis of rankings is only tentative. The reliability of the categories employed in this latter analysis and the consequent tallies has not yet been fully appraised.

Prototypical Manual Development

By way of responding to the quantitative testing results, the qualitative criterion results previously discussed, legal considerations, the concerns of BLM landscape architects in the field and VRM administrative program coordinators, the visual contrast rating procedure was changed and a new manual was developed to explain the changed system (Sheppard and Newman 1979). Figure 9.2 illustrates the old rating sheet and Figure 9.3 illustrates the new. The approach taken in the manual was to present the concepts and procedure in as much detail as possible using graphics to aid understanding.

FIGURE 9.2
Old BLM Rating Sheet

VISUAL CONTRAST RATING WORKSHEET

1. PROJECT INFORMATION

District ______________ Planning Unit ______________

PROJECT NAME ______________________ Date ______________

Activity ____________ Location: T___R___sec__________

Describe Operation: ______________________________

Critical Viewpoint: #___x___y___z___VRM Class____

Name of Evaluator: ______________________________

2. CHARACTERISTIC LANDSCAPE DESCRIPTION

LAND/WATER
FORM
LINE
COLOR
TEXTURE

VEGETATION
FORM
LINE
COLOR
TEXTURE

STRUCTURES
FORM
LINE
COLOR
TEXTURE

3. PROPOSED ACTIVITY DESCRIPTION

DESCRIBE IN TERMS OF FORM, LINE, COLOR, AND TEXTURE INTRODUCED OR MODIFIED. REFER TO BLM MANUAL 1791 AND 6320 FOR PROPOSED DESCRIPTIONS AND REQUIREMENTS.

LAND/WATER
FORM
LINE
COLOR
TEXTURE

VEGETATION
FORM
LINE
COLOR
TEXTURE

STRUCTURES
FORM
LINE
COLOR
TEXTURE

4. CONTRAST RATING ( ) SHORT TERM ( ) LONG TERM

INSTRUCTIONS: (1) RATE CONTRAST OF INTRODUCED OR MODIFIED LANDSCAPE ELEMENTS AND FEATURES AGAINST CHARACTERISTIC LANDSCAPE ELEMENTS AND FEATURES. (2) CIRCLE ONE SCORE ON EACH ELEMENT LINE FOR EACH FEATURE (LAND/WATER BODY, VEGETATION, STRUCTURES) TO INDICATE THE DEGREE OF CONTRAST FOR THAT FEATURE/ELEMENT COMBINATION. (3) ADD TOTAL SCORES IN EACH FEATURE AND ENTER IN TOTAL.

FEATURES:	LAND/WATER BODY				VEGETATION				STRUCTURES			
DEGREE OF CONTRAST:	STRONG (3x)	MODERATE (2x)	WEAK (1x)	NONE (0x)	STRONG (3x)	MODERATE (2x)	WEAK (1x)	NONE (0x)	STRONG (3x)	MODERATE (2x)	WEAK (1x)	NONE (0x)
ELEMENTS: FORM (4x)	12	8	4	0	12	8	4	0	12	8	4	0
LINE (3x)	9	6	3	0	9	6	3	0	9	6	3	0
COLOR (2x)	6	4	2	0	6	4	2	0	6	4	2	0
TEXTURE (1x)	3	2	1	0	3	2	1	0	3	2	1	0
	TOTAL___				TOTAL___				TOTAL___			

CIRCLE ELEMENTS OF GREATEST CONTRAST (FORM, LINE, COLOR, TEXTURE) EACH FEATURE

HIGHEST DEGREE OF CONTRAST [F|L|C|T] [F|L|C|T] [F|L|C|T]

5. SUMMARY AND RECOMMENDATION

INSTRUCTIONS: INSERT BELOW THE MAXIMUM ALLOWABLE CONTRASTS FROM BLM MANUAL 6320.11 FOR THE VRM CLASS.

INSERT BELOW THE MAXIMUM FEATURE AND ELE- SCORES FROM SECTION 4 INDICATE FEATURE/ ELEMENT CAUSING HIGHEST SCORES.

DOES PROJECT DESIGN MEET VRM REQUIREMENTS? ( )yes ( )no

IF CONTRACT RATING IS OVER MAXIMUM ALLOWABLE FOR ANY FEATURE OR ELEMENT, REDESIGN PROJECT CONCENTRATING ON FEATURE/ELEMENT OF GREATEST CONTRAST. IF CONTRAST RATING IS ACCEPTABLE, THIS DOES PRECLUDE ADDITIONAL MITIGATING MEASURES, PROPOSE AS STIPULATION S

FIGURE 9.3
VIA Detailed Procedure

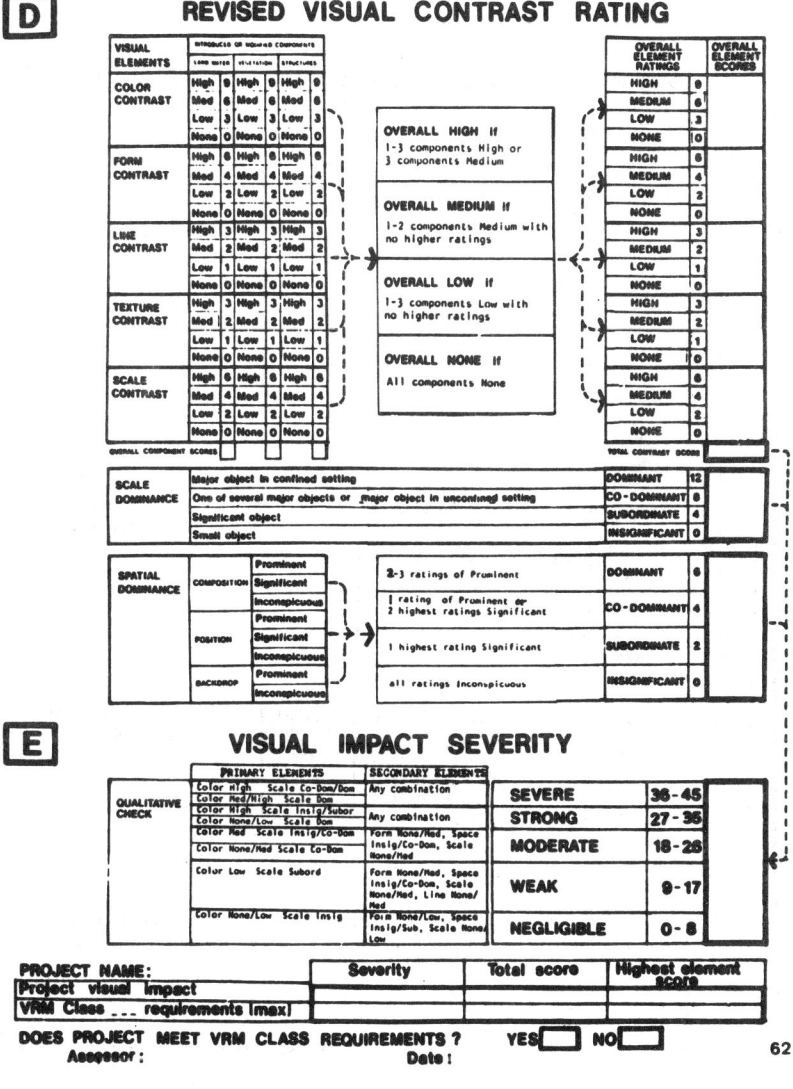

PHASE 3: FINAL ANALYSIS OF VIA RELIABILITY AND VALIDITY

The objectives of this last phase of research were to:

1. Determine whether validity and reliability levels could be significantly increased using a modified VIA training method and materials (Sheppard and Newman 1979);
2. To develop a generic checklist of visual impacts for different types of visually impacting land uses; and
3. Based on the results, attempt to improve the VIA method itself and training for use of the method.

Scenes, Research Participants, and Procedures

Twenty-five pairs of landscape scenes were employed to assess students' ability to use the modified VIA method. Thirty-five senior undergraduate and graduate students were trained to use the modified VIA method and used the manual developed by Sheppard and Newman (1979).

Similar to the testing in previous phases, the participants were shown the before photoslide, asked to describe the existing landscape, then shown the after scene together with the before scene and asked to describe and rate the visual impact, using the modified contrast rating forms (see Figure 9.3). Again, the visual stimuli were simulated. Simulation entailed either removing or imposing the land use activity by means of retouching and painting techniques (BLM 1980b). The added landscape scenes and land use activities were introduced to create a more representative crosssection of visual stimuli than before. To this end, the new scenes were taken primarily of Great Basin, Canyonland, Great Northern Plains and Interior California landscapes with surface mining, coal fired power plants, and geothermal energy development land use activities.

Results on Reliability. Use of detailed visual contrast rating variables still falls below acceptable levels (<.70) of reliability between individual raters. The consistency of rater behavior using these detailed contrast rating variables did improve significantly, if one compares results from previous testing. The additional guidance as provided in the prototype manual is useful, but multiple raters are needed if significant levels of reliability are to be obtained.

Results on Validity. Ratings taken from the same S.U.N.Y. Syracuse sample were correlated with change in scenic beauty ratings for the same visual stimuli. Those variables that react in the same way as scenic beauty change include texture contrast for structures, scale contrast for both land/water bodies and structures, and overall spatial dominance. Near significant correlations with change in scenic beauty include: color contrast for structures, form contrast for structures, scale contrast for vegetation, scale contrast overall, and spatial dominance. Scale and spatial dominance variables are highly intercorrelated with each other.

The results from the correlations and intercorrelations partially reinforce what has been found in other recent studies and our own previous testing. First, it is much easier for people to judge the visual impact of structures than land form/water bodies or vegetation. Second, the variables that most consistently behave similarly to changes in scenic beauty are scale contrast, spatial dominance, for all situations; and texture, form, line and color contrast for structures only.

FIGURE 9.4
Sample Rating Form

VIA BASIC PROCEDURE WORKSHEET

VISUAL ELEMENTS	VISUAL SUBELEMENTS	INDICATORS/CLUES	ELEMENT RATINGS		ELEMENT SCORES
LANDSCAPE COMPATIBILITY	COLOR	Significantly different color, hue, value, chroma	Severe	3	
			Moderate	2	
			Minimal	1	
			None	0	
	FORM	Incompatible 2/3 dimensional shape with landscape surroundings	Severe	3	
			Moderate	2	
			Minimal	1	
			None	0	
	LINE	Incompatible edges, bands, or silhouette lines introduced	Severe	3	
			Moderate	2	
			Minimal	1	
			None	0	
	TEXTURE	Incompatible textural grain, density, regularity or pattern	Severe	3	
			Moderate	2	
			Minimal	1	
			None	0	
			SUBTOTAL →		
SCALE CONTRAST		Major scale introduction/intrusion	Severe	12	
		One of several major scales or major objects in confined setting	Moderate	8	
		Significant object or scale	Minimal	4	
		Small object or scale of activity	None	0	
			SCORE		
SPATIAL DOMINANCE	LANDSCAPE COMPOSITION SITUATION BACKDROP	Object/activity dominates or is prominent in whole landscape composition; or is prominently situated within the landscape; or dominates landform, water, or sky backdrop	Dominant	12	
			Co-Dominate	8	
			Sub-ordinate	4	
			Insignificant	0	
			SCORE		

TOTAL VISUAL IMPACT SEVERITY ——→

Severe	27-36
Strong	26-18
Moderate	17-9
Weak or Negligible	8-0

CONCLUSIONS AND MANAGEMENT RECOMMENDATIONS

1. The sophistication of VIA should be comparable to the complexity, importance or controversy of the project in question. For most projects, a simple one-page rating form should suffice, especially if the project is typical and is structural in nature. Ideally, for all activities or structures, multiple independent (four to five) raters should be involved. If the acitivity involves extensive modification of land form, water bodies, or vegetation, then experienced VRM practitioner(s) should form these panels.

2. For all typical projects/activities, the variables of <u>landscape compatibility</u> (Benson and Darrow 1945), <u>scale contrast</u>,[2] and <u>spatial dominance</u>,[3] should be used as shown by the sample rating form in Figure 9.4. This one-page form should be supplemented by project description, location, and viewpoint delineation. Total weightings for all three variables should be equal in the absence of firm evidence to support any weighting system. A recommended revised form needs to be tested in actual VRM field work.

3. Diagnosis of more complicated projects/activities by qualified VRM practitioners could proceed in one of two ways: (a) use of the VIA checklist to identify specific aspects of the project which account for the unwanted severity of visual impact and which can be redesigned; or (b) use of a more detailed procedure as shown in the new VIA manual (Smardon 1982) for a "reanalysis" of the project or activity in question. Then a multiple independent panel could make VIA judgments and detailed mitigation solutions could be evolved.

4. All new or experienced VRM practitioners should use some type of visual documentation for each VIA rating and visual simulation method as outlined in the BLM Manual (1980b) whenever and wherever possible. Simulation should be used for any visually complex or controversial project or activity.

5. Photographs used in visual documenting before and after views and simulations can then be used as "marker" scenes for each region (Figure 9.1) and its own attendant families of activities. Use of marker scenes will facilitate training, create a similar base of judgment, and provide examples of visually compatible and incompatible activities for that landscape region.

6. All VRM practitioners engaged in VIA should strive to keep themselves abreast of the professional and academic literature (Smardon et al. 1982), in order to benefit from research results and techniques that are germaine to their respective landscape regions and types of projects which they have to assess.

NOTES

1. Subsequently, the U.S. Bureau of Land Management's Visual Contrast Rating Method (BLM 1980a) was also completed. Due to space limitations, it has not been included in this discussion. See Stanley Specht's discussion in Chapter 8.

2. Note that the scale contrast is a bi-polar variable. Scale contrast can increase both with extremely small or large activity introductions to the given landscape. This (we think) accounts for the negative correlation between scale structures and scale land/water bodies; spatial dominance and scale land/water bodies. This variable must be carefully handled by VRM practitioners.

3. General background of all these concepts and terminology are provided by the Prototype VIA Manual (Smardon 1982).

BIBLIOGRAPHY

Benson, L., and R. A. Darrow. 1945. A Manual of Southwestern Desert Trees and Shrubs. Biological Sciences Bull. No. 6, University of Arizona, Tucson, AZ.

Craik, K. H., and N. R. Feimer. 1979. "Setting Technical Standards for Visual Impact Assessment Procedures." In Proceedings of Our National Landscape: A Conference on Applied Techniques for Analysis and Management of the Visual Resource. USDA General Technical Report PSW-35, USDA Pacific Southwest Forest and Range Experiment Station, Berkeley, CA.

Department of Landscape Architecture. 1976. "A Transect of Central California Vegetation Types." Prepared by students of the Department of Landscape Architecture, Instructors, R. A. Betty and O. Zebroski, Berkeley, Ca.

Ebel, R. L. 1951. "Estimation of the Reliability of Ratings." Psychometrika 16:407-424.

Elsner, G. H., and R. C. Smardon (technical coordinators). 1979. Proceedings of Our National Landscape: A Conference on Applied Techniques for Analysis and Management of the Visual Resource. USDA General Technical Report PSW-35, USDA Pacific Southwest Forest and Range Experiment Station, Berkeley, CA.

Feimer, N. R., K. H. Craik, R. C. Smardon, and S. R. J. Sheppard. 1979. "Appraising the Reliability of Visual Impact Assessment Methods." In Proceedings of Our National Landscape: A Conference on Applied Techniques for Analysis and Management of the Visual Resource. USDA General Technical Report PSW-35, USDA Pacific Southwest Forest and Range Experimental Station, Berkeley, CA.

Feimer, N. R., R. C. Smardon, and K. H. Craik. 1981. "Evaluating the Effectiveness of Observer Based Visual Resource and Impact Assessment Methods." Landscape Research 6(1):12-16.

Fenneman, N. M. 1931. Physiography of the Western U.S. McGraw-Hill, New York, NY.

Franklin, J. F., and C. T. Dryness. 1973. Natural Vegetation of Oregon and Washington. Forest Service General Technical Report. PNW-8. USDA Pacific Northwest Forest and Range Experimental Station, Portland, OR.

Guilford, J. P. 1954. Psychometric Methods. McGraw-Hill, New York, NY.

Hammond, E. H. 1964. "Analysis of Properties in Land Form Geography: An Application to Broad Scale Land Form Mapping." Annals of the Association of American Geographers 54:11-23.

Hammond, E. H. 1970. Classes of Land-Surface Form. In U.S. Department of the Interior, Geological Survey, The National Atlas, U.S. Government Printing Office, Washington, D.C.

Hines, M. E. A. 1937. Evolution of the California Landscape. California Division of Geology and Mines, Sacramento, CA.

Hunt, C. B. 1974. National Regions of the U.S. and Canada. W. H. Freeman and Co., San Francisco, CA.

Litton, R. B., Jr. 1973. "Landscape Control Points: A Procedure for Predicting and Monitoring Visual Impacts." USDA Forest Service Research Paper PSW-91, USDA Pacific Southwest Forest and Range Experimental Station, Berkeley, CA.

Litton, R. B., R. J. Tetlow, J. Imai, and L. Diamond. 1978. "A Conceptual Framework for Landscape Inventories: Scenic Elements of the Northern Great Plains." USDA Forest Service Research Paper PSW-135. USDA Pacific Southwest Forest and Range Experimental Station, Berkeley, CA.

Kuchler, A. W. 1969. Manual to Accompany the Map, Potential Natural Vegetation of the Coterminus United States. Special Publication No. 36, American Geographical Society, New York, NY.

National Vegetation Committee, Arizona Chapter, Soil Conservation Society of America. 1973. Landscaping with Native Arizona Plants. The University of Arizona Press, Tucson, AZ.

Sheppard, S. R. J., and S. Newman. 1979. Prototype Visual Impact Assessment Manual. Department of Landscape Architecture, University California and School of Landscape Architecture, College of Environmental Science and Forestry, S.U.N.Y., Syracuse, New York, NY.

Smardon, R. C. 1978. "Law and Aesthetics or When Is the Pig in the Parlor? A Legal/Policy Overview of Legal Factors' Influence on Visual Landscape Policy." Department of Landscape Architecture, University of California, Berkeley, CA.

_____. 1979. "The Interface of Legal and Aesthetic Considerations." In Our National Landscape: A Conference on Applied Techniques for Analysis and Management of the Visual Resource. General Technical Report PSW-35, USDA Pacific Southwest Forest and Range Experimental Station, Berkeley, CA.

_____. 1982. Visual Impact Assessment Manual. School of Landscape Architecture, S.U.N.Y., Syracuse, NY.

Smardon, R.C., M. Hunter, J. Resue, and M. Zoelling. 1982. Our National Landscape: Annotated Bibliography and Expertise Index. Special publication 3279, Agricultural Sciences Publications Division of Agricultural Sciences, University of California, Berkeley, CA.

Snedecor, G. W. 1946. Statistical Methods. 4th edtion. Iowa State College Press, Ames, IA.

Thornbury, W. D. 1965. Regional Geomorphology of the U.S. John Wiley and Sons, Inc. New York.

Tidestrom, I. 1925. Flora of Utah and Nevada, Contributions from the United States National Herbarium. Vol. 25. Smithsonian Institute, National Museum, Government Printing Office, Washington, D.C.

U. S. Department of the Interior, Bureau of Land Management. 1980a. Visual Resource Management Program. Government Printing Office, Washington, D.C.

_____. 1980b. Visual Simulation Techniques. Government Printing Office, Washington, D.C.

Zube, E. H. 1976. "Perception of Landscape and Land Use." In I. Altman and J. F. Wohlwill, eds., Human Behavior and the Environment: Advances in Theory and Research. Vol. 2. Plenum Publishing Co., New York, NY.

10. Objective Evaluation of Visual Values

Ross T. Newkirk

INTRODUCTION

It has become increasingly important to identify the intrinsic value of different landscapes as part of major planning and development exercises. This is particularly true in the context of utility route selection and planning and development associated with national parks and resource areas. Very large tracts of land must be analyzed with good local accuracy. Until recently the explicit consideration of aesthetic factors has not been required of most impact studies. The methodologies reported here may prove useful in determining intrinsic visual values of all landscape-- whether of special, natural, or more subtle acculturated character.

THE PLANNING/ANALYSIS CONTEXT

Project Directed Emphasis

With the expansion of project initiated environmental impact assessments, requirements to avoid impact areas other than visual have tended to force developments into wild or natural areas and parks. For example, the Saskatchewan Research Council (1979) notes: "routings must avoid cultivated lands, residential areas, and recreational areas." In its major utility routing report for the Resource and Land Investigation Program, MITRE (1975) states: "Routing with the least incompatability of land use must be based upon the relative value of land for agricultural purposes in contrast with cultural and recreational purposes," and further "the economics of visual mitigation is very costly." Agency mandates are influential in shaping assessment orientations. For example, the Corps of Engineers environmental assessment system (Fittipaldi and Novak 1980, Riggins and Novak 1976) focused toward specific project orientations. The Corps' approach to visual assessment appears the least developed relative to other areas of concern (such as economics) consisting primarily of a series of do's and don'ts. An example of this narrow perspective is associated with the development of a highway through an Alberta Provincial Park, where a government fish and wildlife expert wrote "it is possible to use both culverts

---

Ross T. Newkirk is the Director of the Methods and Design Centre for Environmental Studies at the University of Waterloo, Ontario, Canada.

and bridges for crossing waterbodies without creating any particular problems except perhaps aesthetics" (Alberta 1974). This is the only explicit consideration of aesthetics in the whole report. In general, since there is no agency directly involved with visual concerns, visual aspects have been mainly ignored in impact studies and regional plans.

Rational Prescreening Approach

Most visual analyses of project development have led to studies that tend to emphasize the value of the landscape "with the project in it" and associated remedial measures. While this may be useful, it is desirable that methods be perfected that allow the preproject assessment of landscape aesthetics. The author feels that visual impact assessments should be completed in three distinct phases.

PHASE I: Analysis and assessment of the visual values given the current nature of the area.
PHASE II: Analysis and assessment of the visual value changes associated with project development.
PHASE III: Assessment of impact of change, recommendation on remedial measures (if any) and project suitability.

The method outlined in this paper is structured to provide mapped and tabulated evidence to support all three phases. Phase I should be used to develop a visual master plan, to be associated with an overall regional or master developmental plan. Such plans have major impacts. For example, initial planning by the Highways Department in a Provincial Park (Alberta 1974) immediately accommodated the existing plan. Early input of visual values in the over-all planning process is important.

METHOD CONSIDERATIONS

Many researchers and government officials have been quick to identify consideration of visual values as the impossible search for the understanding of beauty and have, for example, observed "few people are endowed with . . . a high degree of aesthetic sensibility" (Lewis et al., 1973). Similar to many areas in Social Impact Assessment (Waiten 1981), absolute rating values are impossible due to different perceptions by race, age, geographic location, social, and economic conditions, etc. In general, aesthetics assessment is very contextual; individual perspective exacerbates the problem. For example, how does one develop proper visual value assessments for a remote glacier park area where some would like to be able to view from the comfort of their auto or camper and others would like to view undisturbed via hiking trail access only?

In this context, it is important that methods are developed that at least enumerate the basic characteristics which may be the basis for visual values in areas. The method discussed in this paper provides quantifiable screening of key features suitable for mapping selected area characteristic items and associations. Elsewhere, I review ten major scoring or quantitative approaches to, or requirements of, aesthetic value determination (Newkirk 1982). In general, quantifiable measures of visual values may be conveniently divided into:

1. Measures of properties inherent within a specific area
2. Measures of comparison between an area and others nearby.

The former can be quantified quite readily based on available local inventory data while much of the latter can be quantified using the approach of computer searching. The latter concept is important in its ability to assist in approaching the concepts of visual harmony (Fittipaldi and Novak 1980, Battelle and Jones 1974, Lewis et al. 1973, MITRE 1975, MacLaren 1979, Vaughan 1974, and Canada Environmental Protection Service 1978), contrasts in form and vegetation (Vaughan 1974, Battelle and Jones 1974, Saskatchewan Research Council 1979, Saskatchewan Environment 1980) and determination of edge conditions and landmarks.

OBJECTIVE EVALUATION APPROACH

Initial Work: Diversity Mapping of Land Units

An examination of approaches by Battelle and Jones (1974), Vaughn (1974), and others reveals a number of common features which are thought to contribute to visual values. In an attempt to avoid absolute value ranking of individual components, the author with others (Newkirk et al. 1974) introduced a basic diversity index to approach an initial measure of the "natural" features of an area. This avoided the absolute value calculation by Battelle and Jones with the model: $VQ=1/3 (I+V+U)$ where: VQ is visual quality, I is intact fullness of scene, V is vividness of scene, U is intercompatibility. However, Battelle and others' key components were applied. For example, components such as natural condition, human encroachments, skyline boundary identification, extreme topographic relief, vegetation, and presence of water were identified and used in a general diversity type model. This diversity approach was also merged with an estimation of relative viewability to develop an evaluation index of land units.

A key function of the approach was to conduct the analysis on a large number of small fixed size areas (for example, approximately 50,000 45-acre parcels in a 2,700 square mile area). The results were mapped at this level of resolution to permit the spatial identification of areas of high visual value. Table 10.1 shows the basic general diversity components applied in the initial work. Further details and a field comparison study to a "traditional" landscape assessment are found in Newkirk et al. (1974). The results indicated that for larger regional studies, a computer based general diversity approach is at least as effective as the traditional field survey approach of landscape architects.

Applying a standard classification scheme (Newkirk 1979), the results of the local area diversity assessments can be mapped by a computer controlled plotter as shown in Figure 10.1.

The general diversity approach was further developed by adding extra information contributed by neighboring areas. The approach of local area searching and inventory checking has been mainly supplemented by some additional diversity components (Newkirk 1982). An important feature of both the initial and expanded approach is the reliance upon standard resource inventories and classifications. This permits quick data assembly and analysis with consistency within and across studies.

TABLE 10.1
Landscape Diversity Components

Component	Description
Average Terrain Slope	Greater than five degrees
Terrain Roughness	More than six contour crossings
Standing Water	Ponds, lakes, etc.
Landform Type	Any of: till moraine, kame, moraine, drumlin, esker, spillway, sandplain, dunes
Soil Character	Any of: muck, peat, organic, alluvial, or bottom land
Drainage	Either excessive or poor
Limitations on Use for Agricultural Purposes	Topography, excess water or flooding, stoniness, exposed bedrock, etc.
Recreational Potential Re: Water	Angling, viewing, shoreland w. swimming, boating, rapids, waterfalls, pools, and riffles viewing wetland wildlife
Recreation Potential Re: Uplands	Vegetation with rec. potential gathering/collecting, interesting landforms, viewing wildlife, hiking, nature study, vantage points
Streams and Rivers	
Woodland	Woodlots larger than six acres
Rough (slash) pasture	Areas larger than six acres

Evaluation of Diversity Mapping Approach

The diversity mapping approach has been subjected to two comparison studies (Newkirk et al. 1974 and Moss and Nickling 1980). In the first, after a diversity mapping study was applied for a client over a large area, the client retained a firm of landscape architects to perform a "traditional" evaluation of a selected sub area of the study. The comparison showed good general agreement between the diversity study and the landscape architects study (Newkirk et al.1974). A ground survey comparing the two revealed that the computer based diversity study detected a number of transition zones and isolated high value areas not identified by the landscape architect.

Moss and Nickling (1980) conducted two controlled studies comparing the Newkirk et al. (1979), Linton (1968), and Leopold (1969) riverscape analysis methods. Although they did not include Newkirk's relative elevation component which affected their results in hilly areas, they observed that the Newkirk method was more sensitive to local value change than Linton's. On a spatial comparison, approximately twenty percent of the unit areas studied were assigned assessments of different value, with the Newkirk method yielding higher values especially in lower lying terrain. The Leopold system, which stresses uniqueness, was applied and is fully discussed in Moss and Nickling (1980). In their summary, they observed that while results of all three methods were generally comparable, there was a marked difference in efficiency. Field work requirements, in terms

FIGURE 10.1
General Landscape Diversity

of mapping days, were calculated to be fourteen days for the Leopold approach, three days for the Linton approach, and only two days for the Newkirk approach.

They correctly note that the Newkirk and Linton methods are slightly more subjective than Leopold's due to the nature of the criteria items used. The major difference between the Newkirk approach and the others is that the degree of scenic and cultural diversity of a given area is considered as a positive and important attribute of the landscape. They comment: "The most marked difference between the three schemes is shown in their ranking of two lagoon areas of the two creeks. Both the Linton and Leopold evaluation schemes rank the lagoon areas considerably lower than does the Newkirk method." Their observation is that the Leopold approach was really oriented to riverscapes of scenic grandeur with wide low valley sites being considered less aesthetic. Their summary conclusion was that both Newkirk and Linton methods showed promise for adaptability to special needs and these methods should be chosen over Leopold's due to efficiency.

VIEWABILITY AND SEARCHES

While general diversity itself is a major contributor to the intrinsic aesthetic value of an area, it is clear that viewability (the ability to be seen or see from) is also important. An additional consideration is whether the terrain and vegetation can screen man-made alterations (for example, power plants, strip mines, gas/oil or electric transmission corridors, etc.) Battelle and Jones(1974) shows the importance of considering local relief. The use of a fine grid network of area information permits the calculation of relative elevation by a series of moving surface fits (Newkirk et al. 1974). This is done by having the computer scan the records for the immediately neighboring areas to determine, for each local area, how far above or below it lies from the trend of its surrounding neighbors.

To complete the calculation of relative viewability, one needs the computer's ability to make quantitative comparisons between a number of adjacent areas. This transcends the capability of the essentially cartographic point comparison methods of, or derived from, Ian McHarg's (1971) work. In computer searching, a suitable neighborhood radius is pre-selected (often 2-4 km in radius) and the computer successively scans around each local area, does a surface fit and calculates the residual. Figure 10.2a shows a sequence of fits across one west-east row of a study area. After a series of such west-east scans, a relative visibility value can be calculated for every local area as shown in Figure 10.2b. In usual practice, the actual residual values would be recorded in a computer data file.

In the context of developing a planning system for minimum impact routings for utilities (such as major electrical transmissions, pipelines, comunications, and transportation), Newkirk (1979) extended the viewability measure based on Battelle, MITRE, and Vaughan's observations. A second neighborhood scan is completed, similar to the west-east scanning for local surface fits, this time to analyze the vegetation and roughness screening potential by neighboring areas. This makes it possible to adjust downward areas of potential high viewability if they are screened by neighboring areas. The final adjusted viewability assessment can be mapped by the computer as shown in Figure 10.3.

The general diversity (Figure 10.1) and viewability (Figure 10.3) maps can then be combined to provide a synthesis assessment which highlights

FIGURE 10.2a
Neighborhood SCANS For Relative Viewability

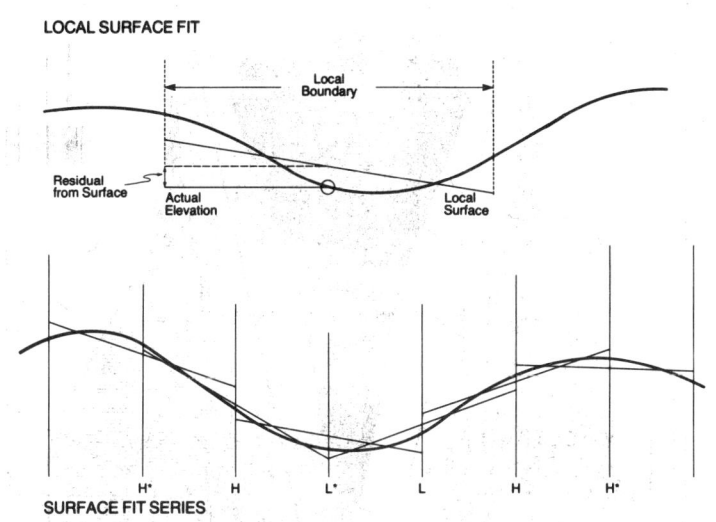

FIGURE 10.2b
Results of West-East SCANS

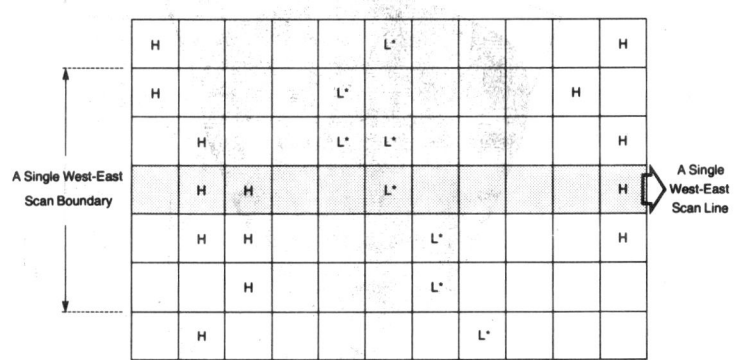

110

FIGURE 10.3
Relative Viewability

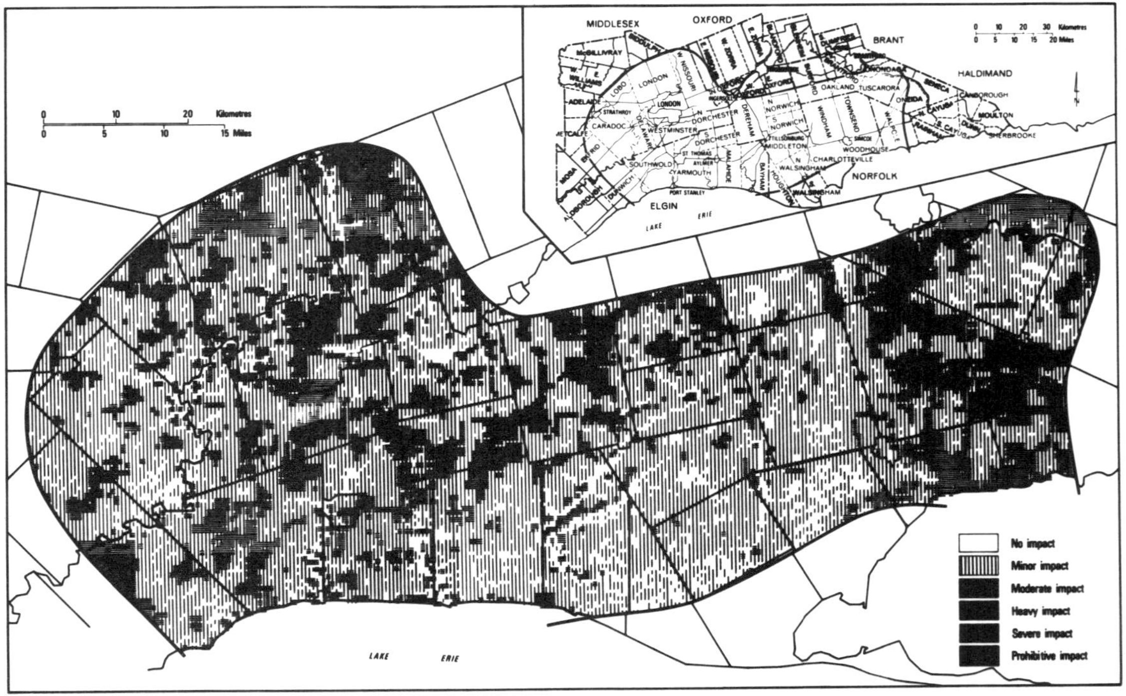

FIGURE 10.4
Viewable General Diversity

the highly viewable areas of important general diversity. The resulting synthesis is shown in Figure 10.4.

VISTAS, NEIGHBORHOOD SCANS, AND RATINGS

An important part of visual analysis is to identify and evaluate significant vistas. As part of a pre-developmental screening method, it is important to direct development away from high value vistas. The synthesis analysis can be examined on a grid cell by grid cell basis to successively determine, for each cell, to which vista it belongs. This is achieved by an additional neighborhood search. Simultaneously, the computer assembles a quantitative tabulation of the "character" of each vista group. This includes determining for each vista: area, average and maximum general diversity, count of existence of (arbitrarily defined) significant features, tabulation of different landform and/or vegetation types, etc. The result is the development of a tabulation which lists the number, character, and extent of each vista. This does not provide a visual value rating for vistas but rather an objective quantitative identification and description of their characteristics. It remains an interpretive professional exercise to develop relative ratings based on such quantitative descriptions.

An important contribution of the computer based approach is to support the consistent application of an assessment algorithm across a study area making assessments both on a detailed area basis and in the context of neighborhood characteristics. The neighborhood scanning capability is currently being further developed. It appears that it has good potential, using a transition matrix approach, to detect the key transition zones and edge conditions essential to a more complete visual evaluation. It also shows important potential for identifying potential resource or natural area management problems.

BIBLIOGRAPHY

Aerospace Corporation. 1975. Technical Compatibility Factors for Joint-Use Rights of Way. Bureau of Land Management, U.S. Department of the Interior, Washington, D.C.

Alberta, Highways and Transportation. 1974. Secondary Road 967 Proposal Primary Highway S. of Lily CR Location Study Report.

Battelle Corp. Ltd and Jones Landscape Architects. 1974. "Quantified Social and Aesthetic Values in Environmental Decision Making" Symposium on Siting of Nuclear Facilities. Battelle, Seattle, WA.

Canada, Environmental Protection Service. 1978. Guide to Environmental Screening. Dept. of the Environment, Queens Printer, Ottawa.

Fittipaldi, J.J., and E.W. Novak. 1980. Guidelines for Review of EA/EIS Documents. Construction Research Laboratory, U.S Army Corps of Engineers, Champaign, IL.

Leopold, L. B. 1969. "Quantative Comparisons of Some Aesthetic Factors Among Rivers." USGS Circular 620. Washington, D.C.

Lewis, P.F., D. Lowenthall, Y. Tuan. 1973. Visual Blight in America. Resource paper 23, Commission on College Geography, Association of American Geographers, Washington, D.C.

Linton, D. L. 1968. "The Assessment of Scenery as a National Resource." Scottish Geographic Magazine. 84(3):219-238.

MacLaren, J. F., Ltd. 1979. The State of the Art of the Environmental Effects of Transmission Lines. Canadian Electrical Association, Montreal.

McHarg, I.L. 1971. Design with Nature American Museum of Natural History. Doubleday/Natural History Press.

MITRE Co., Ltd. 1975. Resource and Land Investigations (RALI) Program: Considerations for Evaluating Utility Lines Proposals. Report to the Bureau of Land Management, U.S. Department of the Interior, Washington, D.C.

Moss, M.R., and W.G. Nickling. 1980. "Landscape Evaluation in Environmental Assessment and Land Use Planning." Environmental Management 4(1):57-72.

Newkirk, R. T. 1979. Environmental Planning for Utility Corridors. Ann Arbor Science Publishers Ltd., Ann Arbor, MI.

Newkirk, R.T. 1982. "Objective Components for Measuring Visual Values." Mimeo, University of Waterloo, Ontario Canada.

Newkirk, R.T., S.H. Janes, and I. Moncrieff. 1974. "Quantifying Landcape Analysis." Geographical Inter-University Resource Management Seminars 15, Wilfrid Laurier University, Waterloo, Ontario Canada.

Riggins, R., and E. Novak 1976. Computer Aided Environmental Impact Analysis for Mission Change, Operations and Maintenance, and Training Activities: User Manual. Construction Research Laboratory, U.S. Army Corps of Engineers, Champaign, IL.

Saskatchewan Environment. 1980. Guidelines for the Development of Electrical Tranmission Lines. Environmental Assessment Secretariat.

Saskatchewan Research Council. 1979. Guidelines for the Preparation of Environmental Impact Statements for Roads, Pipelines, and Electrical Energy Transmission Lines. SRC Report C79-11.

Schlesinger, B., and D. Daetz. 1975. Development of a Procedure for Forecasting Long Range Environmental Impacts, Report to the Resource and Land Investigations Program (RALI). Technical Department 75-3, Stanford University, Palo Alto, CA.

U.S. Department of the Interior. 1975. The Need for a National System of Transportation and Utility Corridors. Washington, D.C.

Vaughn, Alan V. 1974. A Visual Analysis System to Assist in Locating Transmission Corridors. Forestry Department, Ontario Hydro, Toronto.

Waiten, C.M. 1981. A Guide to Social Impact Assessment, Indian and Northern Affairs, Canada.

PART IV
Social and Psychological Approaches to Value Assessment

Social and psychological value assessment tools and methodologies for measuring how attitudes and behaviors are affected by changes in scenic quality may be used to provide the link between perceived changes in visual quality and the impacts of these changes on the affected experience, thus determining whether an impact is adverse. A variety of potential methodologies have been suggested and applications of some are discussed in this chapter, but some authors concentrate on the problems that have arisen and questions that are still to be addressed. This is a topic area where perspectives differ widely and much development in research methodologies, and their theoretical underpinning, is still needed. The complexities of considering the contribution of visual resources in the recreation experience, the cultural and historic importance of visual resources, whether observable behavior accurately reflects values, and the appropriate forum to elicit values concerning visual resource protection make this a different task.

The first two papers set a tone of caution about measuring social values due to the many inherent limitations. Through the use of role theoretical analysis Craik identifies three roles of scenic observers: the landscape assessor, the scenic tourist, and the citizen role. He cautions that all three observers should be given an opportunity to register their views concerning scenic quality levels in national parks and Class I areas.

Stewart focuses upon the formidable difficulties in making visual air quality judgments, which require fact competence, value competence, and cognitive competence of technically complex and often incomplete information. He concludes that public input is of limited use in making informed choice.

The next four papers examine the design and implementation of approaches to measure social values through analysis of behavior and attitudes. Haas establishes that better data will be needed for these purposes and, under the premise that national parks personnel can be effectively utilized to collect useful information with minimal effort and cost, presents data collection and design procedures that might be followed to assist in visual values analyses.

MacFarland, Malm, and Molenar present the methodology and results for visibility value assessments undertaken at Grand Canyon and Mesa Verde National Parks. This research shows clean air and scenic views are among the most highly valued resources at these parks. The authors find a strong link between perceived visual air quality indices, as presented in

Part II, and willingness to pay and length of stay measures used in visual value assessments of adverseness.

In two experiments, Loomis and Greene examine issues concerning the transition from perceived visual air quality to the measurement of social values. The first shows that changes in instructional sets may affect the assessments and the second illustrates one method for identifying recreation behaviors that are affected by visibility changes and can be measured in visibility assessments.

Malm, Shaver, and McGlothin propose a new methodology for visual value assessments. This procedure would formally integrate measures used in visual air quality perception work, as presented in Part II, with changes in time spent at a recreation site as a measure of value. An ordered logit probability choice technique is proposed to examine how time expenditures change as visibility conditions change.

The last two papers present specific social research methodologies used for related problems that could be usefully employed in visual value assessments. Buchanan, Hayter, and Buchanan demonstrate how canonical correlation analysis can be employed to search for critical relationships between perceived psychological benefits from recreation activities and changes in characteristics of visual resources as a link in social values research concerning the adverseness of an impact. Burdge, McAvoy, Absher, and Gramann demonstrate research aimed at determining how perceptions of changes in a physical recreation resource will produce changes in recreation behavior. From this work one could proceed to determine the adverseness of these visual resource changes.

11. A Role Theoretic Analysis of Scenic Quality Judgments

Kenneth H. Craik

This volume follows an earlier workshop on visibility values held in Fort Collins, CO (Fox et al. 1979), but broadens the range of issues being addressed. For the sake of coherence, I want to indicate at the outset how my contribution here extends and builds upon my earlier comments (Craik 1979).

The major question at hand is how to assess scenic quality (including visibility conditions). In that connection, the development and application of Perceived Environmental Quality Indices (PEQIs) (Craik and Zube 1976) are pertinent. In the earlier report, I recommended ways to assess visibility conditions within this framework and also recommended caution about other possible approaches to this task. I will not pass up the opportunity to review those recommendations here.

This volume extends the inquiry to how scenic quality bears upon (1) the experience of visitors to national parks and Class I areas and (2) how these considerations might be incorporated into management policies for the federal lands. I will use role theoretical analysis to suggest avenues of research on the first question and to derive yet another set of cautionary remarks about the second.

VISIBILITY CONDITIONS, SCENIC QUALITY AND PEQIs

Previously, I argued that the assessment of visibility conditions (e.g., clarity of detail in vistas, range of view, absence of discoloration) is a specific task within the assessment of scenic quality and that measures of scenic quality are special instances of PEQIs (Craik 1979). What is required therefore are reliable, systematically gathered, on-site context-pertinent field judgments of visibility quality and impairment. These measures of visibility conditions as experienced on site seemed to me to be what Section 169A of the Clean Air Act Amendments of 1977 called for. If properly developed and used in monitoring operations, they would serve two valuable purposes. First, on-site, observer-based judgments of this kind constitute a basis for monitoring the effectiveness over time of the federal agencies re-

---

Kenneth Craik is Professor of Psychology and Research Psychologist, University of California, Berkeley, and Editor of the Journal of Environmental Psychology. Presentation of this paper was partially supported by The Electric Power Research Institute.

sponsible for maintaining the overall quality of national parks and Class I areas. Second, these on-site observer-based measures provide a basis for communicating trends in visibility quality and impairment to the general public, elected officials and other decisionmakers.

Even with good intentions and agency effectiveness, it is still possible that due to scientific ignorance, technological limitations and considerations of economic and other societal factors, degradations in overall scenic quality, and visibility conditions specifically, could occur. If so, we must be able to gauge the extent of the decline in visibility quality in national parks and Class I areas, for use in long-range policy or political action.

Technical Issues in Developing and Applying Appropriate PEQIs.

I will not go into the details again of the steps to be taken in generating pertinent PEQIs for assessing and monitoring visibility quality and impairments. The general procedure has been presented fully elsewhere (Craik and Zube 1976) and the specific issues were outlined in Craik (1979). The major requirements include: (1) conceptual analysis and instrument development; (2) methodological studies; (3) studies of instructional set; (4) establishing cutting points and standards; and (5) long-term place oriented monitoring operations. Step 1 is the critical phase, and requires a fresh and imaginative approach to recording the varied and subtle experiences afforded by atmospheric conditions and an exploration of the aesthetics of visibility, including both positive and negative attributes. From this exercise, appropriate Visibility Quality Adjective Check Lists (VQACLs) and Visibility Quality Rating Scales (VQRSs) could be constructed, providing criterion measures for appraising the validity of telephometric and related instruments and for evaluating the predictive success of physical modeling efforts.

I recommended that special attention be given to context-pertinent instructional sets, suggesting that on-site observers be fully informed (1) that the scenes are from Class I areas and what that designation entails, (2) of what the Clean Air Act Amendments of 1977 intended, and (3) that their judgments are part of the broad process of implementing that legislation. Similar instructions should be given to panels of the general public in establishing breaking points between conditions that are acceptable and not acceptable in light of the pertinent legislation. These cutting points could then be compared with those derived more indirectly, from bidding games, agency staff analysis, legal rulings, etc.

At the Fort Collins session, Frederica Perera pointed out that in contrast to health-related aspects of air pollution, which entail imperceptible components, in the case of visibility impairment "we have seen it with our own eyes." My major recommendation followed directly from that recognition: that on-site, context-pertinent field judgments of visibility quality and impairment form the fundamental and enduring long-term basis for monitoring visibility quality in national parks and Class I areas.

Cautionary Remarks: Set A

At the Fort Collins session, my major cautionary remarks dealt with the temptation to abandon on-site context-pertinent field judgments of visibility quality and impairment for various imperfect but somehow appealing substitutes. I will re-state these cautions in the briefest possible form.

1. Instrumentation. Proposed optical and electrical surrogates for on-site observer evaluations may not fully capture our ordinary construct of visibility quality and impairment. Research gauging the extent to which they do so is essential.

2. Modeling techniques. The amount of variance in on-site observer evaluations of visibility quality that can be accounted for by predictive models is likely to be severely constrained by limits to current scientific knowledge and technology. Validational research on these models should justify their application before they are used in decisionmaking.

3. Overt and hypothetical behaviors. Monitoring what people do rather than what they report they see and experience may lead to misleading and invalid indices. For example, attendance at national parks may remain high over a decade of steadily degraded visibility conditions because relatively, but not absolutely, the settings are still the best available. Yet visitors in the tenth year, if shown adequate photo-slide records of the highest past levels of visibility quality and the current levels, would immediately recognize the magnitude of their loss and ours. Similarly, hypothetical behaviors such as statements of willingness to pay in bidding games appear to be remote from the intentions of Section 169A and remarkably indirect bases for suggesting cutting points, standards or thresholds of acceptable visibility quality.

4. Psychophysiological measures. An understanding of the psychophysiological concomitants of the experience of high or low scenic quality is an important item on our scientific agenda in environmental perception, but its present bearing on assessing levels of visibility quality is unclear to me. Certainly, levels of visibility impairment or degradations in scenic quality are validly and directly assessed by on-site observer reports, not by phasic alpha levels.

I now want to turn to the extended inquiry for this present volume and earn my way to a second set of cautionary remarks.

VISIBILITY CONDITIONS, SCENIC QUALITY AND VISITOR EXPERIENCES

The present volume extends the inquiry to include the full range of scenic quality issues (including visibility conditions) and addresses two important questions. The first is how scenic quality affects visitors' experiences of national parks and Class I areas, including the impact of degradation in quality. The second asks how these considerations can be incorporated into resource management of the federal lands.

My intention is to show how role theoretical analysis can help to delineate the relation of scenic quality to visitors' experiences, to identify needed research, and to highlight the complexities entailed in arriving at management decisions guided by these considerations.

A ROLE THEORETICAL ANALYSIS OF SCENIC JUDGMENTS AND OF THE SCENIC TOURIST

On-site Impressions, Psychological Sets, Roles, Personality and Culture.

Our primary concern is with the immediate, on-site impressions of place experienced by visitors to national parks and Class I areas (e.g., is the setting majestic, tranquil, eroded, unspoiled). However, research has shown that ordinary perception is an active process, importantly influenced by psychological sets and other cognitive processes (see Figure 11.1). That is, individuals encounter settings with something like a plan to attend to and process certain kinds of information about it. Thus, Leff and his associates (Leff et al. 1974 Leff and Gordon 1980) have instructed their research participants to take certain sets intentionally when presented with an environmental scene (e.g., to attend to the shapes, lines colors and textures of the scene, to attend to imagining changes that would make the scene better, to evaluate the scene and figure out the human values that might be represented in it). But in everyday life, these psychological sets are rarely so explicit or intentional and are typically organized into larger units, often around the role the person is enacting at the time.

FIGURE 11.1
Influences in the Perception Process

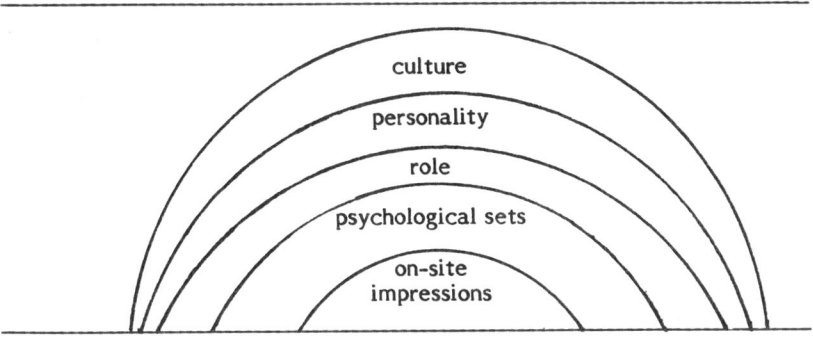

Within the framework of role theory (Sarbin 1954, Sarbin and Allen 1968), the psychological sets of the perceiver of landscapes are structured by the particular role of scenic observer which the individual enacts when touring the countryside (including national parks or Class I areas). Using historical materials, I have elsewhere analyzed the observer of picturesque landscape (Craik 1970) and I will use that worked example at a later point to go into the details of role analysis.

At this point, for the sake of completeness, I wish to note that the roles an individual elects to enact are themselves influenced by or reflect and give expression to his or her enduring personality characteristics, as the research of McKechnie (1974, 1975), Driver and Knopf (1977) and others have shown. Furthermore, the array of scenic roles available to individuals at any given time and place in history represents important cultural forces that are little understood but now receiving some attention (Cohen 1972, MacCannell 1973, Britton 1978, Pizam et al. 1978). Indeed, the way in which these tourist roles are structured and the demands they make upon host cultures and upon the often fragile physical environments that accommodate their enactment have become major international issues in environmental management (Turner and Ash 1975, Smith 1977, Cohen 1978, De Kadt 1979, Hawkins et al. 1980). Thus, the on-site impressions of place formed by visitors to national parks and Class I areas can be fully understood and interpreted only through appropriate use of concepts drawn from the fields of perception, cognition, social psychology, personality, anthropology and sociology.

Application to the Role of Scenic Observer.

Many specific roles can be delineated within the category of scenic observer. Historical sources clearly document the currency in England during the period between 1770 and 1820, and surely beyond, of the role of observer of picturesque landscapes (Manwaring 1925, Hussey 1927, Merchant 1951). The analysis of this scenic role illustrates the questions raised by the role theoretical orientation. First, what are the role expectations, that is, the conduct and the psychological sets deemed appropriate to the role. In touring the countryside, the observer was directed to scan the landscape for the kind of beauty that would look good in a picture, locating the viewpoints from which prospects meet the criteria of good composition employed in pictorial art. The scenic observers were also to sketch and to acquire the terminology of landscape painting, in order to describe scenes to themselves and to others. Special devices aided them in their search for the picturesque. Claude-glasses—small, rectangular sheets of colored glass—served to frame a landscape scene and endow it with the mellow hues of Claude Lorrain's popular landscape paintings. Small mirrors were also used in role enactment. The Reverend Mr. William Gilpin, whose tour books were well read (Gilpin 1792), often toured the District of the Lakes in a carriage, with his head turned away from the landscape but toward its reflection framed in his mirror, all the while marveling at the prices which paintings of nature's well-composed scenes would fetch in the city.

How does one acquire the role of picturesque landscape observer? No problem here—for the means of role acquistion were readily available in the many widely sought guidebooks to the Lake District by Rev. Gilpin, Arthur Young, Thomas Pennant, and William Wordsworth. The most popular, Richard West's Guide to the Lakes, went through ten editions between 1778 and 1812, giving the novice meticulous directions to "Stations" throughout the region, from which the authenticated picturesque scenes might be viewed.

How might the role skill of the enactor be evaluated? In the case of the observer of picturesque landscapes, proper terminology was important, as satirically illustrated in a passage from Jane Austin's Sense and Sensibility. The character Edward disclaims his skill in properly describing what he has seen in a walk through the countryside:

> You must not inquire too far, Marianne—remember, I have no knowledge in the picturesque, and I shall offend you by my ignorance and want of taste, if we come to particulars. I shall call hills steep which ought to be bold; surfaces strange and uncouth, which ought to be irregular and rugged; and distant objects out of sight, which ought only to be indistinct through the soft medium of a hazy atmosphere (Quoted in Merchant 1951, p.9).

A taxonomy of scenic roles would have to include the role of observer of romantic landscape, which places more emphasis upon the impact of scenery on the emotions and imagination (Nicholson 1959). Ever alert to the terrible joy and delightful horror, the sense of the sublime and the intimation of mystery and unattainability afforded by crags, cascades, pools, caves, and deep forest, this role demanded somewhat more adventurous touring into wilder, less pastoral settings. A more recent emergent is the role of observer of wilderness landscape (Lucas 1966), whose role expectations include informed analysis originating in ecological knowledge, spiritual appreciation of holistic patterning, moral imperative and guardian vigilance. What is clearly needed is an ambitious and full analysis of contemporary roles of scenic observer, bringing into use as well the additional concepts of role conflict, environmental props, role location and cues, etc.

A very recent appearance is the scenic role of landscape assessor. Under the impetus of legislative and administrative mandates of the 1970s concerning the analysis and management of the landscape, governmental agencies are being called upon to adopt or develop visual resource and impact assessment (VRIA) systems. In this case, agency professionals and technicians, as well as members of the general public, may enact the role under institutional supervision and for decision-making purposes. The primary requirements entail the making of judgments of landscape quality under conditions and with procedures that enhance their reliability, validity, generality and utility (Craik and Feimer 1979, Feimer et al. 1981). The composite panel judgments are then used in appraising the potential impact of projects proposed for various landscape settings and for post-construction monitoring of impacts upon visual resources. Role enactment requires complete focus upon landscape quality in making such judgments, e.g., modifying the scenic quality ratings to convey a negative judgment about ecological, economic or other impacts of a proposed project is inappropriate conduct for this role. Smardon (1982) is preparing an overview of the emergence of this institutional form of the role of scenic observer.

When properly enacted, judgments generated by the role of landscape assessor yield perceived environmental quality indices that establish prevailing landscape quality levels for specific settings. In this sense, they provide the independent variable for the current research topic, which is examining the impact of variations in scenic quality, so assessed, upon the ordinary conduct and experience of visitors to national parks and Class I areas when enacting scenic roles whose nature we do not yet adequately understand or appreciate. A related question is how often and under what conditions do visitors spontaneously take on the role of landscape assessor, and with what consequences?

The Role of Scenic Observer: Issues of Duration and Breadth.

Visits to national parks and Class I areas typically entail a journey. The scenic journey and the enactment of the role of scenic observer constitute psychological events with a temporal course. The experience and meaning of the journey and visit to these cherished federal lands are difficult to delimit temporally, involving as they do anticipation and post-journey residues. Furthermore, repeated journeys interweave, forming a temporal fabric of experience and meaning that is obviously very important to the question of the impact of degradations in scenic quality of the settings visited.

The scenic journey as a unit must receive much more extensive and subtle analysis than it has enjoyed heretofore. What components of the observer's motivation for undertaking the journey are scenic and which are extrascenic? If the observer is traveling with others, what is the interpersonal significance of the journey? What are the salient themes of the journey itself and to what extent do extrascenic concerns and preoccupations overwhelm the impact of scenery? If the intent of the journey is predominantly scenic, what are the travelers' scenic goals and what strategy is adopted to attain them? To what extent is the journey a form of ritual respect for the scenic beauty of the national parks and Class I areas?

What are the aftereffects of the journey? Are they social? The individuals have new narrative material for presenting themselves to others. Are they personal--enhancing a sense of themselves as individuals close to nature? Are they psychological--recharging their imagery of natural landscape? The intense experiences at vista points may be quickly, if temporarily, lost in the hustle and bustle of the journey back home, but may be beneficially evident later, in revivified daydreams at the workshop. People other than poets may heed advice such as Bishop Berkeley's admonition to Alexander Pope to travel more, in order "to store his mind with strong images of nature" (quoted in Hussey, 1927, p. 88).

This call for longitudinal analysis of scenic journeys over fairly long periods of visitors' lives is not intended to detract unduly from analysis of the immediate situational context of the on-site visit. Certainly, it is possible to examine their deployment of attention (Wagner et al. 1981), to identify the cues they use to size up conditions of the environment (Craik and Appleyard 1980), and to consider the on-site situational factors in their visits (Magnusson 1981). However, the part played by role, personality, and culture in shaping the experience and lasting meaning of these visits over time seems all too easily neglected in this field of research.

The issue of breadth is equally as important as the issue of duration in understanding the role of scenic observer. That is, how does this role interact with other roles of the individual? While other linkages could be considered, e.g., with the role of parent, with occupational role, I want to consider the relation between the roles of scenic tourist and citizen. To do so, I will draw on research we are conducting at Berkeley on perceptions of technology (Craik 1982) for an analogy. In that study, we are focusing upon two issues: first, how do individuals perceive and judge the risks and benefits of various technologies, and second, how would they prefer that society approach the decision of establishing how safe is safe enough for specific technologies. The first kind of research is itself part of one societal decision approach, namely, gathering the expressed preferences of the general public about the perceived risks of and safety standards for technologies. But an array of other societal decision approaches are available, including: (1) cost-benefit analysis, (2) decision analysis, (3) implied preferences, (4) natural standards, (5) political judgment, (6) professional judgment, (7) revealed preferences, and (8) risk comparisons. Our second phase of research thus addresses the issue at a meta-level of analysis by using the expressed preferences approach to examine the general public's relative evaluations of the several approaches to societal decisions themselves, including the expressed preference method.

Analogously, I see two kinds of research that are called for by the issues addressed in this volume. First, research is being conducted on the experiences and conduct of visitors to the national parks and Class I areas, with an eye to how those findings might bear upon decisions regarding the management of these federal lands. Second, research should address explicitly the ordinary citizen's views on these matters: (1) how they evaluate this method of considering land management policies, (2) how they evaluate alternative approaches to setting acceptable levels of scenic quality in these lands, and (3) to what extent they approve the levels of scenic quality established for these lands by means of the considerations we have been addressing at this conference.

CONCLUSIONS

Scenic Quality Judgments and Three Roles of Scenic Observer.

This analysis has revealed three roles of scenic observer that generate judgments regarding scenic quality. First, the recently emerging role of landscape assessor is structured institutionally to yield reliable, valid, generalizable, and useful composite indices of prevailing levels of scenic quality. These indices can monitor the level of scenic quality but do not provide an answer to the policy question of what level of scenic quality is acceptable, e.g., for national parks and Class I areas. Second, the multifarious roles of scenic tourist offer a useful conceptual device for studying the experiences and conduct of visitors to national parks and Class I areas and an approach to understanding the enduring meaning of these visits for the individuals making them. The ways in which these experiences and meanings are affected by given levels of scenic quality and by degradations in scenic quality can also be examined as a specific research question. These latter research findings might then be used to inform and guide decisions regarding the management of these federal lands, including the policy question of what levels of scenic quality are acceptable. Finally, the role of citizen in a democratic society entails indirectly and perhaps sometimes directly the making of scenic quality judgments at the policy level, in forming and expressing views about how society should go about reaching decisions regarding acceptable levels of scenic quality in federal lands and in judging the adequacy of the decisions that have been or might be reached on this question through the various approaches to policy formation, including the approaches discussed in this volume.

Cautionary Remarks: Set B

My closing cautionary remarks can be associated with the three roles that generate scenic quality judgments.

1. *The role of landscape assessor.* In my view, this role facilitates the gathering of on-site, context-pertinent judgments of prevailing scenic quality that are at the heart of the matter. My main caution is that these fundamental judgments of scenic quality not be replaced by any of the various surrogate measures unless full documentation of the adequate validity of the substitute methods is available. I do not consider that such justification has been demonstrated at this time.

2. *The role of the scenic tourist.* In enacting any of the roles of scenic tourist to National Parks and Class I Areas, visitors presumably do sometimes make judgements akin to those reached within the role of landscape assessor--that is, the roles merge. But visitors experience and do much else besides that, which is little understood at this time. The enduring meaning of visits to these federal lands and the influence of levels of scenic quality upon it are worthy topics of research, formidable in their complexity. These research findings might usefully inform and guide such policy issues as establishing acceptable levels of scenic quality in these federal lands. My major caution is that available research findings are quite meager at present and misinterpretation of them is a clear danger.

3. *The role of citizen.* In the role of citizen, individuals can form and express their views (1) about how policy is reached in the management of federal lands, (2) about the acceptable levels of scenic quality yielded by manager attending to research on the experiences and values of visitors to

these lands, and (3) about how these decisions should be reached and about what they themselves deem to be acceptable levels of scenic quality. My caution is that this kind of analysis at present seems to be slighted in favor of the more indirect analysis of visitors to the federal lands. Both kinds of analysis are needed. Indeed, perhaps all research participants in the study of visitor experiences and conduct should also be given an opportunity to register more directly their views concerning scenic quality levels in national parks and Class I areas.

BIBLIOGRAPHY

Britton, R. A. 1978. "The Image of the Third World in Tourism Marketing." Annals of Tourism Research 6: 318-329.
Cohen, E. 1972. "Toward a Sociology of International Tourism." Social Research 39: 164-182.
Cohen, E. 1978 "Impact of Tourism on Physical Environment." Annals of Tourism Research 5: 215-237.
Craik, K. H. 1970. "Environmental Psychology." In Craik, K.H., et al. New Directions in Psychology 4. Holt, Rinehart and Winston, New York, NY.
Craik, K. H. 1979. "The Place of Perceived Environmental Quality Indices (PEQIs) in Atmospheric Visibility Monitoring and Preservation." In D. Fox, R. J. Loomis & T. C. Green (Eds.) Tech. Report WO-18. Proceedings of the Workshop in Visibility Values. U.S. Forest Service, Fort Collins, CO.
Craik, K. H. 1982. Fundamental Research Issues in Risk Analysis: Perceptions of Technologies and Their Societal Contexts. Technology Assessment and Risk Analysis Program, Division of Policy Research and Analysis, National Science Foundation, Washington, D.C.
Craik, K. H., and D. Appleyard. 1980. "Streets of San Francisco: Brunswik's Lens Model Applied to Urban Inference and Assessment." Journal of Social Issues 36: 72-85.
Craik, K. H., and N. R. Feimer. 1979. "Setting Technical Standards for Visual Assessment Procedures." In G. Elsner and R. C. Smardon (Eds.) Our National Landscape. U. S. Forest Service, Berkeley, CA.
Craik, K. H., and E. H. Zube (Eds.). 1976. Perceiving Environmental Quality: Research and Application. Plenum Press, New York, NY.
De Kadt, E. (Ed.) 1979. Tourism: Passport to Development? Perspectives on the Social and Cultural Effects in Developing Countries. Oxford University Press, New York, NY.
Driver, B. L., and R. C. Knopf. 1977. "Personality, Outdoor Recreation and Expected Consequences." Environment and Behavior 9: 169-194.
Feimer, N. R., R. C. Smardon, and K. H. Craik. 1981. "Evaluating the Effectiveness of Observer-Based Visual Resource and Impact Assessment Methods." Landscape Research 6: 12-16.
Fox, D., R. J. Loomis, and T. C. Green (Eds.). 1979. Proceedings of the Workshop In Visibility Values. U. S. Forest Service, Tech. Report WO-18. Fort Collins, CO.
Gilpin, W. 1792. Three Essays: On Picturesque Beauty; On Picturesque Travel; and On Landscape Painting. R. Blamire, London.
Hawkins, D. E., E. L. Shafer, and J. M. Rovelstad. (Eds.) 1980. Tourism Planning and Development Issues. George Washington University, Washington University, Washington, D.C.

Hussey, C. 1927. The Picturesque: Studies in a Point of View. Putnam's and Sons, London.
Leff, H. L., L. R. Gordon and J. G. Ferguson. 1974. "Cognitive Set and Environmental Awareness." Environment and Behavior 6: 395-447.
Leff, H. L., and L. R. Gordon. 1980. "Environmental Cognitive Sets: A Longitudinal Study." Environment and Behavior 12: 291-328.
Lucas, R. C. 1966. "The Contribution of Environmental Research to Wilderness Policy Decisions. Journal of Social Issues 22: 116-126.
MacCannell, D. 1973. "Staged Authenticity: Arrangements of Social Space in Tourist Settings." American Journal of Sociology 79: 589-603.
McKechnie, G. E. 1974. Manual for the Environmental Response Inventory. Consulting Psychologists Press, Palo Alto, CA.
McKechnie, G. E. 1975. Manual for the Leisure Activities Blank. Consulting Psychologists Press, Palo Alto, CA.
Magnusson, D. (Ed.) 1981. Toward a Psychology of Situations. Erlbaum, Hillsdale, NJ.
Manwaring, E. 1925. Italian Landscape in Eighteenth Century England. Oxford University Press, New York, NY.
Merchant, W. M. 1951. "Introduction." In W. Wordsworth, A Guide Through the District of the Lakes in the North of England. Rupert Hart-Davis, London.
Nicolson, M. 1959. Mountain Gloom and Mountain Glory: The Development of the Aesthetics of the Infinite. Cornell University Press, Ithaca, NY.
Pizam, A., Y. Newmann, and R. Reichel. 1978. "Dimensions of Tourist Satisfaction with a Destination Area." Annals of Tourism Research 5:314-322.
Sarbin, T. R. 1954. "Role Theory." In G. Lindzey (Ed.) Handbook of Social Psychology, Volume 1. Addison-Wesley, Reading, MA.
Sarbin, T.R., and V. L. Allen. 1968. "Role Theory." In G. Linzey and E. Aronson (Eds.), Handbook of Social Psychology, Volume 1. Addison-Wesley, Reading, MA.
Smardon, R. C. 1982. Ph.D. dissertation in progress. University of California, Berkeley, CA.
Smith, V. (Ed.) 1977. Hosts and Guests: The Anthropology of Tourism. University of Pennsylvania Press, Philadephia, PA.
Turner, L., and J. Ash. 1975. The Golden Hordes: International Tourism and the Pleasure Periphery. Constable, London.
Wagner, M., J. C. Baird, and W. Barbaresi. 1981. "The Locus of Environmental Attention." Journal of Environmental Psychology 1: 195-206.

12. Visual Air Quality Values: Public Input and Informed Choice

Thomas R. Stewart

INTRODUCTION

Suppose there were a meaningful, valid, and generally accepted index for measuring the social impact of visual air quality (VAQ). How would we then place a value on an observed or predicted change in VAQ? This is an especially difficult question to answer because VAQ is an intangible public good whose value depends on subjective experience.

Putting a specific value on VAQ is important because public policy decisions affect VAQ as well as energy, economic development, environmental quality, and wilderness preservation. Even the best possible technology and farsighted planning does not eliminate the need to make trade-off decisions--either to decide to give up some VAQ for gains in something else, or to decide, for example, to increase the cost of electricity in order to protect VAQ. In either case, the responsible policy maker must ask, "Is it worth it?"

The usefulness of any method that assigns value to VAQ depends on its contribution to decision making. Because different methods may be appropriate in different decision contexts, this paper begins not with the problem of assigning dollars or utils to changes in VAQ, but rather with the problem of making good decisions when VAQ is involved. This is a formidable problem, because decisions involving VAQ generally involve alternatives that are technically complex, information that is incomplete, outcomes that are uncertain, and many criteria.

The first section of the paper describes a model of individual decision making and defines the concept of "informed choice." Informed choice requires competence of three kinds: <u>fact</u> competence, <u>value</u> competence, and <u>cognitive</u> competence. Each kind of competence is defined.

The second part of the paper considers just one aspect of making informed choices--the use of public input. How useful are messages from the public that are generated by (1) public hearings, (2) surveys and opinion

---

Thomas Stewart is a psychologist with the Environmental and Societal Impacts Group of the National Center for Atmospheric Research (NCAR), Boulder, CO. NCAR is sponsored by the National Science Foundation. The author wishes to thank Mary Downton, Dan Ely, Michael Glantz, Maria Krenz, and Paulette Middleton for helpful comments on an earlier version of this paper.

polls, (3) studies using multiattribute utility analysis, and (4) studies using bidding methods? Each of these public input approaches is discussed with regard to representativeness, validity, and applicability to policy making. It is concluded that public input is of limited use in making informed choices.

Informed choice in decisions about VAQ requires improved competence on the part of both the citizens who provide input and the planners and policy makers. Improved competence requires not just study of the facts, but increased emphasis on developing coherent value systems and on developing the ability to integrate facts and values for informed choices. The problem of valuing VAQ should not be approached as merely a problem of assessment, which is solved by more sophisticated measuring instruments. The problem is one of development--creating value systems that are consistent and coherent and can reasonably be applied to decisions involving VAQ.

The paper describes individuals making choices. Social choice theory and social welfare functions for groups of individuals are not considered. Public policy decisions are influenced by the social and political context which involves a network of relations among people and groups. I have not discussed the legal, social, political, or governmental context of decisions regarding VAQ. I have not treated competence in interpersonal skills or the role of persuasion, negotiation, mediation, or conflict resolution in decision making. My hope is that the discussion of values and informed choice at the level of the individual will contribute to an understanding and improvement of decisions in the broader social context.

DECISION MAKING AND INFORMED CHOICE

Individual Decision Making

Figure 12.1 illustrates a general, descriptive model of individual choice adapted from a model suggested by Gary McClelland (personal communication, 1981). The elements of the model are described briefly below.

Formulation. This represents the person's perspective on the problem to be solved. Different people may formulate a decision problem in different ways. For example, in the case of the decision to site a coal-fired power plant in a western state, an environmentalist might formulate the problem as: "How best to preserve visual air quality in the national park?" The public utility company representative might formulate the problem as "Where can the plant be sited for most economical operation and ease of obtaining necessary permits?" The National Park Service representative might formulate the problem as: "What is the appropriate balance between needs for energy and economic development and the needs of recreational park users?" Obviously, problem formulation is closely tied to the individual's goals, and how the problem is formulated will have a significant impact on what is considered, or ignored, in the decision making process.

Alternatives. The alternatives are potential actions, policies, or plans which can be chosen. The set of alternatives may be large or small in number and may change over time as new alternatives are discovered. For a particular decision, different individuals may consider different alternatives. The alternatives considered depend on how a person formulates the problem and on his or her knowledge of the potential alternatives. In the case of power plant siting, the alternative set for one person may consist of two alternatives: grant a permit or refuse the permit. Another person may

FIGURE 12.1
Model of Individual Choice

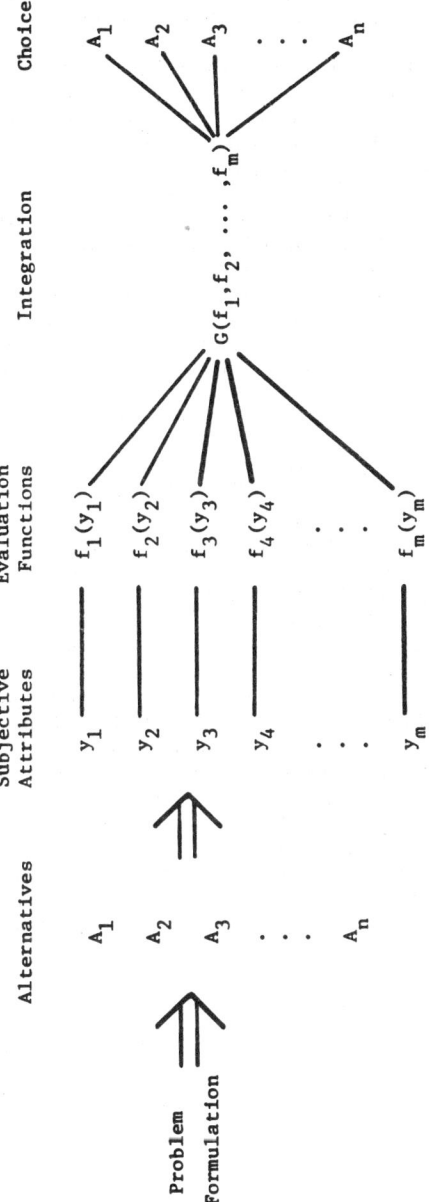

consider a wider range of alternatives including many different combinations of sites, fuels, and emission control technologies.

Subjective Attributes (Beliefs). These represent what the person knows or believes to be true about the alternatives. Beliefs do not necessarily correspond to the known facts, particularly when the alternatives are technically complex and the person does not have the time and/or the background to study them fully. In the case of siting a power plant, subjective attributes might include the estimated cost of construction, the number of construction jobs created, the degree of impact on wildlife, and the change in VAQ created by power plant emissions. Technically, the subjective attributes consist of specific levels or ranges on these factors. For example, the subjective attributes for person A which are associated with the alternative "build the power plant" might include a cost of $1 billion, 500 construction jobs created, negligible wildlife impact, and a 10 percent decrease in average visual range. For person B, subjective attributes for the same alternative might consist of a $1 billion cost, 300 construction jobs, destruction of riparian habitat, and no change in visual air quality. Attributes include both characteristics of an alternative (e.g. location, design) and consequences (e.g. impact on wildlife and VAQ).

Evaluation Functions. Some attributes indicate desirable aspects of alternatives, some indicate undesirable aspects, and some are neutral. The evaluation functions, f_i, assign levels of desirability to attributes. People who share the same beliefs about an alternative may associate different levels of desirability with those beliefs. For example, one person who believes that a power plant will create 500 construction jobs may evaluate that as an extremely desirable attribute in favor of development. Another person may evaluate the same attribute as negative because of the social impact on a neighboring community. Furthermore, a 20 percent decrement in VAQ could be evaluated negatively by one person but considered negligible by another.

Integration. The desirable level for each attribute is integrated into a judgment of overall desirability of the alternative by a process represented by "G" in Figure 12.1. This is the process by which all the positive and negative attributes of these alternatives are combined into a value judgment, i.e., a judgment that some alternatives are more, or less, desirable than others when everything is considered. People with the same subjective attributes, and the same evaluations, may differ with regard to integration function because, for example, one person places more emphasis on VAQ than another person does.

Choice. The overall desirability of the alternatives provides a basis for ranking them and choosing the most preferred alternative. The choice is based on the evaluative judgments of the alternatives.[1]

Although this descriptive model of decision making resembles many prescriptive models of "rational" decision making (Keeney and Raiffa 1976), it can be applied to any case in which a person chooses from a set of alternatives. The model does not assume "rationality," makes no statements about how the person obtains or processses information, and does not specify how decisions should be made. It is equally consistent with a highly intuitive, fragmentary choice process and an analytic process based on the application of an explicit procedure (Hammond 1978).

Since the model is simply a description of a single individual making an isolated decision, it excludes many elements that would be necessary for a complete theory of decision making. In particular, the person's values, experiences, or motives are not represented because they are not required to describe how a person makes a particular decision. Also not represented, for the same reason, is feedback about the results of the decision. Feedback may well influence future decisions, but it is not a factor in the present decision.

It should also be noted that the model is deterministic, i.e., given the same alternatives and subjective attributes, the same choice will result, assuming that the evaluation and integration functions are stable. In fact, choice processes are probabilistic because people are inconsistent. The model applies only to the consistent process which underlies inconsistent judgments.

Competence and Informed Choice. With reference to Figure 12.1, informed choice means (1) the person is aware of all the alternatives and his or her subjective attributes correspond to current knowledge, (2) the evaluation functions and integration function are based on a coherent set of values, and (3) errors, inconsistency and bias in judgment are avoided. Informed choice, therefore, requires three kinds of competence: (1) fact competence—knowledge about all the alternatives and attributes, (2) value competence—development of a coherent value system and the ability to apply it to the decision, and (3) cognitive competence—the ability to integrate information in a consistent, controlled manner.

Different citizens possess these kinds of competence in varying degrees and the same person has different degrees of competence with regard to different decisions at different times. In general, the number of people who are competent to make an informed choice with regard to an issue decreases as (1) the technical complexity of the alternatives and their consequences increases, (2) the difficulty and importance of value issues increase, (3) the amount of information relevant to the decision increases, and (4) the consequences become more removed from most people's everyday experience.

Fact Competence. A person is competent with regard to the facts if her or his knowledge of the alternatives and their attributes matches expert knowledge. In terms of the choice model, fact competence means that subjective attributes are based on the best available information. Note that fact competence does not require perfect knowledge; it requires current knowledge. The continuum of fact competence extends from "complete ignorance" to "current state of knowledge." An informed choice is the best choice, given available current knowledge.

Fact competence is necessary, but not sufficient for informed choice. The person must also have the competence to apply knowledge to the decision, competence which is not necessarily acquired through the process of learning facts. The competence required to apply knowledge to choice includes both value competence and cognitive competence.

Value Competence.[2] A particular person is defined as value competent with regard to a specific problem, at a given time, if two condition hold: (1) The relevant value system is coherent, and (2) evaluation and integration functions are consistent with the value system.

The first condition departs from the belief, adopted by many social scientists, that values are subjective, not empirically verifiable, and therefore not subject to critical examination. According to this view, called "value non-cognitivism," values are to be accepted as given (Fischer 1980). They can be measured, but not questioned.

Many philosophers today believe that values can be questioned. Some hold that the rightness or wrongness of certain values can be derived from universally accepted principles such as duty and justice. Some maintain that as a result of critical examination, a person may reject certain values or may change some values once the implications of holding those values and their consistency with other strongly held values is understood. Coherence of a value system cannot be assumed:

> ...to speak of a "system" of values is usually no more than a courtesy; its unity is not antecedently given but is something to be achieved. There is always a question of the consistency of our values among themselves, of their compatibility in given situations, and of their coherence or mutual support (Kaplan 1964).

This is particularly true with regard to choices involving VAQ. Technology has increased both our range of choices and our ability to predict the consequences of those choices. As a result, challenging new value issues are created (Baier 1969, Kyle and Rabehl 1980), and nothing in our common experience provides an adequate basis for resolving them. We cannot, therefore, let intuition be our guide, but must carefully work out the grounds for value judgments (Kaplan 1964). Difficult choices require a coherent value system which is the result of critical examination of values.

The second condition, that the evaluation and integration functions be consistent with the value system, has not received the same attention in the literature as has the first, but it is an equally important condition for the value competence that is necessary for informed choice. A value system affects choice <u>indirectly</u> through its influence on the selection of criteria and the form of the evaluation and integration functions. Applying a value system to a decision requires the translation of the value system into evaluation and integration functions which can be applied to the subjective attributes. This translation is not trivial, because value systems are abstract and general while subjective attributes are concrete and specific (dollars, stack height, etc.). The difference between having a coherent value system (condition 1) and having applicable evaluation and integration functions consistent with that value system (condition 2) may be likened to the difference between (1) holding the belief that benefits and costs of projects should be distributed equitably, and (2) deciding whether a swimming pool is adequate compensation for a community that has been chosen as a site for a chemical waste dump, or whether a hospital should be provided as well.

Decisions regarding VAQ in pristine areas involve perplexing value questions. What is the proper distribution of costs and benefits among population groups and generations? What are appropriate trade-offs among incommensurate goods (or bads)? To say that informed choice requires value competence is to say that these kinds of issues must be resolved through careful deliberation. Of course, it would be naive to expect that an alternative that is clearly and demonstrably "best" will be discovered through such deliberation, but it is likely that a number of clearly inferior alternatives will be eliminated.

The assertion that value competence is developed through critical examination of values contradicts two views which seem often to influence the public policy formation process. The first view is that technical expertise qualifies one to cope with value issues, either because technical training somehow produces value competence or because value issues are only artifacts created by lay people's misunderstanding of the facts. The second point of view is that everyone is value competent by virtue of being a participating citizen. The roots of this belief are in democratic ideals and the noncognitivist position with regard to verification of values. The view that critical examination of values is a requirement for informed choice is a cognitivist position that rejects the notion that either experts or citizens are naturally value competent.

A value system that is to be used as a guide for action in public policy decisions should be the result of careful thought, critical examination, comparison of alternative value systems, and consideration of the consequences of applying the value system to decisions (Coates 1978, MacRae 1979). Value competence results when a consistent, coherent value system has been developed that can be applied to a particular decision to usefully guide the decision maker toward desirable goals. Developing value competence is, of course, a continuous process with no clear end-point.

Cognitive Competence. There can be little doubt that making informed choices that affect VAQ challenges human cognitive abilities. The literature of judgment and decision research contains many examples of cognitive limitations that can affect decisions that require complex judgments under conditions of uncertainty (Einhorn and Hogarth 1981, Kahneman et al. 1982, Slovic et al. 1977). Although there is some debate regarding the generality of cognitive limitations and the degree to which they affect performance in actual decision making (Hogarth 1981), it seems that decisions that affect VAQ are vulnerable to cognitive limitations because they involve many attributes, trade-offs (which produce negative attribute intercorrelations), uncertainty, and because the opportunity to make adjustments in the case of bad decisions is limited.

Because of cognitive limitations, a person's choice may be inconsistent with his or her beliefs and value system. This may result from forgetting an important attribute at the time of the decision, from errors in the process of organizing and integrating large amounts of information or from use of shortcuts or rules-of-thumb which work well in some situations but not in others.

Cognitive competence is required in order to apply fact and value competence to decision making. As with the other kinds of competence, cognitive competence is not a general characteristic of a person, but applies only to a specific decision problem. A person who is cognitively competent with respect to a given problem can derive the choice that is most consistent with his or her factual beliefs and value system. Cognitive competence requires that a person making a judgment execute the evaluation functions and the integration function in a consistent fashion.

Hammond (1978) discusses the problem of cognitive competence in policy makers, describes dominant modes of thought, and suggests the use of "judgment aids" to improve competence. Many decisions require judgments of such complexity that the use of cognitive aids may be the only way to assure cognitive competence. These aids range from use of pencil and paper as a memory aid, through simple decomposition, rating and ranking methods (Dawes and Corrigan 1974, Gardiner and Edwards 1975, Einhorn and McCoach 1977), to elaborate procedures often requiring the support of computers (Hammond et al. 1975, Keeney and Raiffa 1976, Warfield 1974). The research on judgment and decision making has demonstrated that the cognitive competence required to deal with complex judgments cannot be taken for granted and that cognitive aids can be useful in improving the consistency with which facts and values are integrated to make a choice (Hammond and Adelman 1976).

The process of developing cognitive competence may improve fact and value competence. Judgment aids require formulation, decomposition, definition and analysis of a decision problem that forces one to think carefully about beliefs and values. Gaps or inconsistencies in knowledge may be identified, and the critical examination of values may be facilitated by this process.

PUBLIC INPUT AND INFORMED CHOICE

Public input for decision making is important because, ideally, it is the public's values (or the values of the various publics), not the planners' or the policy makers' values, that are to guide decision making. Knowledge of public values may be considered part of the fact competence required for informed choice. Furthermore, comparison of the decision maker's own values with public values might aid in the development of the decision maker's value competence.

Different kinds of information from and about the public constitute public input that is relevant to decison making. The following discussion concerns only input with regard to citizen's value judgments, as distinguished from value expressions or factual beliefs.[3] Since value judgments suggest answers to the decision maker's question "What shall I do?" (Baier 1958), they are an important component of meaningful and useful citizen input.

Public input is generated through hearings, public comments, direct contact with constituents or lobbyists, media reports of speeches or demonstrations, and surveys or public opinion polls. Not all input about public values is equally valuable for informed choice and not all public input methods are equally useful in obtaining valuable input. I suggest that, in order to contribute to informed choice, input about public values must meet three criteria.

1. <u>Representatives.</u> Public input is representative to the extent that it describes the views of the entire relevant population, rather than those of just a segment of the population.

2. <u>Applicability.</u> Input is applicable to a particular decision if it can be applied directly to the decision without assumptions or judgments on the part of the decision maker. Input that is vague or imprecise or does not address critical issues is not directly applicable to decision making because it does not clearly indicate a choice.

3. <u>Validity.</u> Input from a citizen is valid if it is not misleading, i.e., if acting upon it would really increase the welfare of the citizen. Validity requires more than just the ability to use the input to predict what choice a person would make (Sinden and Worrell 1979); it requires the ability to predict what choice a person would make if he were competent to make an <u>informed</u> choice.

Validity defined in this way is more rigorous than the concept of validity that is traditional in the social sciences. Traditionally validity refers to the accurate measurement of some existing psychological characteristic. The existence or validity of the characteristic itself is not questioned.

In the discussion of value competence above, the belief that citizens are naturally value competent with respect to complex issues was rejected. In many cases, the citizen does not have values that are applicable to a decision, or the values exist in a poorly developed form. The problem of validity is then not just a problem of measuring existing values, but a problem of development of a coherent value system worth measuring. The point has been concisely stated by Henry Rowen (1976):

...preferences are generally built through experience and through learning about facts, about relationships, and about consequences. It is not that values are latent and only need to be "discovered" or "revealed." There is a potentially infinite number of values; they are not equally useful or valid, and part of the task of analysis is to develop ones that seem especially "right" and useful and that might become widely shared.

If one accepts the cognitivist view with regard to the verification of values which is implied by Rowen, then the validity of public input about values requires both (1) that the citizen be value competent and (2) that the description of the citizen's value system be accurate. If either of these conditions is not met, the description of a citizen's values can be misleading, in the sense that, if used to guide decision making, it may not lead to a result that is really desired by the individual (Rubenstein 1975).

The importance of the value development component of validity is demonstrated in a paper by Fischhoff et al. (1980) which suggests that often people do not "know what they want" and contains many examples of the ways that the process of measuring or eliciting values can influence the values themselves. They conclude:

> If one is interested in what people really feel about a value issue, there may be no substitute for an interactive, dialectical elicitation procedure, one that acknowledges the elicitor's role in helping the respondent to create and enunciate values. That help would include a conceptual analysis of the problem and of the personal, social, ethical value issues to which the respondent might wish to relate.

In the following section, four methods for obtaining public input about the value that citizens place on VAQ will be evaluated with regard to representativeness, applicability and validity.

Public Hearing and Comments. Public hearings and comments are required by many public laws, notably the National Environmental Policy Act. Public input is only one purpose of hearings; they also serve educational and political functions. With regard to providing useful public input, hearings succeed in one respect: they greatly reduce the chance that something will be completely overlooked. Given the volumes of testimony and public comment generated by an important project, it is difficult to imagine how an alternative, an impact, or a point of view could be inadvertently omitted.

While public hearings succeed in generating large amounts of information, they fail to provide the information needed for informed choice for several reasons. First, the people who testify are typically not representative of the general population. Participants are self selected and may differ from the general population in a number of aspects, including their stake in the outcome, political activity, knowledge, speaking ability, income, education, ethnic background, and party affiliation. Opening a hearing to all does not assure that those who attend adequately represent those who do not. Observing the number of people expressing a particular view at a hearing may help the policy maker make a politically wise choice, but does not necessarily help him or her to make an informed choice.

Second, the information provided at public hearings may not provide a valid basis for informed choice. Testimony at public hearings may deal with any aspect of the decision, including alternatives, beliefs, evaluation, integration, and choice. Often people testify in support of one alternative, stating their choice and giving reasons for the choice. Their support for an alternative is a political fact, but may not reflect an informed choice. The person testifying may not have considered all the alternatives, may have erroneous beliefs about the alternatives, or may evaluate beliefs in a way that is not coherent. Therefore, the person's choice may not be the one that will lead to what the person desires. Furthermore, the reasons given for the choice are selected to persuade, and may not be the person's real reasons for the choice.

Third, the information provided at public hearings is rarely directly applicable to the decision that must be made. The information can be too general or too specific. It is too general when only broad goals or values are discussed, e.g., "We want to preserve scenic values." It is too specific when it deals only with one alternative, e.g., "We want the power plant here." Neither type of communication addresses trade-offs, which are a central problem for the decision makers. Public hearings seldom directly address the questions that decision makers face such as "How many dollars is it worth to protect this vista?" or "How much weight should be given to the need for electricity versus the quality of the recreational experience?" yet informed choice requires answers.

Public hearings are an important part of the political process, but they cannot be justified solely on the grounds of the quality and usefulness of the information they provide to help a decision maker make an informed choice. If public hearing input is not carefully evaluated, it can even result in less well informed policy choices. Meaningful participatory democracy would seem to require that public input be sought from other sources.

Surveys and Opinion Polls. One of the other sources most frequently used is the sample survey or opinion poll. Typically, a random sample of citizens is polled by telephone, in person, or by a mail questionnaire. Assuming an adequate response rate and a statistically valid sampling procedure, inferences about the views of the general population can be made from the responses of the sample. Sample surveys emphasize representativeness.

Unfortunatley, surveys gain representativeness at the expense of validity and applicability. Representativeness requires that there be no screening of citizens with regard to competence. As a result, the views of citizens surveyed do not necessarily provide a good indicator of what is in the citizens' best interest because the views cannot be assumed to reflect informed choice. This problem might be partially corrected by providing information during the interview and by asking more probing questions, but anything that lengthens the time required of a respondent will decrease the response rate and limit representativeness. The problem of competence then, places a severe constraint on the value of information from surveys as a guide to informed choice.

For similar reasons, surveys fail with regard to applicability of the information obtained (Sharkansky 1982). Like public hearings, surveys rarely deal with trade-off issues. The typical survey question addresses one alternative or one issue at a time. Those that do consider trade-offs generally handle only two dimensions at a time even though decisions involve trade-offs among many dimensions simultaneously.

Multiattribute Utility Theory. Both public hearings and surveys have been criticized on the grounds of applicability. They do not directly address trade-offs, so the decision maker is forced to guess what trade-offs people would be willing to make, and there is little reason to believe those guesses will be accurate (Stewart and Gelbard 1976, Milbrath 1980). Multiattribute utility theory (MAUT) has provided several techniques for directly addressing trade-offs. MAUT methods are intended to assess the evaluation functions and the integration function (Figure 12.1) for an individual with respect to a specific problem. Once expressed explicitly, usually in the form of a mathematical equation, the results of MAUT analysis can be applied directly to the attributes in order to produce a ranking of alternatives with regard to desirability.

MAUT methods can be divided into three groups: (1) those that ask a person to estimate directly the evaluation functions and the integration function (Einhorn and McCoach 1977, Gardiner and Edwards 1975); (2) those that infer the evaluation and integration functions from a series of carefully constructed trade-off judgments (Keeney and Raiffa 1976); and (3) those that infer evaluation and integration functions from global judgments about the desirability of alternatives (Hammond et al. 1975).

The results of MAUT can be directly applied to choosing among alternatives and answering trade-off questions (Edwards 1978). Once evaluation and integration functions for an individual are made explicit in a mathematical form, information about alternatives can be "plugged in" and the alternatives can be ranked from best to worst. Furthermore, the equations can be used to calculate answers to questions such as "How much benefit in electricity production is needed to balance a decrease of 20 percent in the visual air quality index?"

MAUT methods are attractive because, in addition to providing information that is directly relevant to decision making, they appear to solve some of the competence problems in citizen input. The problem of fact competence might be avoided by using the correct facts about alternatives in place of the person's beliefs about them as a basis for evaluating alternatives. Substituting known facts for subjective attributes in the evaluation functions might be considered an estimate of what the person would choose, given his or her current value system, if the person had full knowledge of the correct facts. The problem of value competence is addressed because the evaluation and integration functions which are based on the value system, are made explicit and can be examined and changed. The problem of cognitive competence might be reduced because the integration of information is performed in a controlled, consistent fashion by calculating rather than judgment. Calculation is consistent and guards against inadvertently leaving something out.

Unfortunately, my recent experience with MAUT methods leads me to seriously question their value in improving the use of public input for making informed choices. My pessimism is due primarily to the failure of these methods to adequately address the problem of value competence. MAUT methods <u>assume</u> that a person is capable of making difficult evaluative judgments. In order for the results of MAUT to be a valid guide to informed choice, the person must not only have a coherent value system relevant to the problem, but must be able to use that value system to derive consistent answers to complex questions. Even for a thoughtful person whose value system is well developed, the translation from that value system to answers to the kinds of questions required by MAUT methods is difficult.

The validity of MAUT is further limited by a related problem. Substituting correct facts for erroneous subjective attributes in the evaluation functions in order to predict informed choice <u>assumes</u> that the evaluation and integration functions are independent of the person's level of knowledge. In other words, the assumption is that the person would apply his or her value system to a problem in the same way regardless of his or her state of knowledge. This assumption is questionable. Even if one's value system were independent of knowledge, it is likely that the application of that value system to a particular set of alternatives would change as new knowledge is acquired. For example, the trade-offs a person is willing to make between the cost of automobiles and their emissions of pollutants might well change depending on whether the person believed that effective emission controls were available.

Because MAUT is time consuming, representativeness will suffer. The likelihood of getting a random sample of citizens to cooperate for even a limited MAUT analysis (30 minutes to 1 hour) is extremely low. The use of MAUT increases applicability of results at the expense of representativeness, and it does not assure validity.[4]

Bidding Methods. Bidding methods, based on the economic theory of consumer valuation, have been proposed for valuing VAQ (Rowe and Chestnut 1981). These methods are designed to measure a person's willingness to pay (WTP) or willingness to accept payment (WTA) for specified changes in air quality. The methods are well grounded in economic theory, and assume that citizens are competent to make informed choices. "Individuals are perceived as having the ablility to make well defined, rational decisions regarding trade-offs among alternatives in order to maximize their own utility." (Rowe and Chestnut 1981). This assumption justifies the "criterion of individual choice" which is the basis for most economic valuation techniques. This criterion states that "the individual's own preferences are the standard by which we judge his well being." (Stokey and Zechhauser 1978).

Bidding methods are not as time consuming for the respondent as MAUT methods, but are less straightforward than standard survey procedures. As a result, bidding methods can be used with representative samples of citizens, though not as readily as the standard survey.

The research on the validity of bidding games assumes that valuing VAQ is a measurement problem, i.e., that a "true value" exists, and the only problem is to find it. Therefore, bidding methods are carefully designed to eliminate certain kinds of bias in detecting the true values. The diligence of economists in rooting out sources of bias is impressive. In addition to properly accounting for differences in willingness to pay introduced by property rights and income differences, economists have identified hypothetical biases, strategic biases, information bias, contingent market rejection and problem bids (Rowe and Chestnut 1981). One has the impression that bidding methods, properly used, are accurate measuring instruments.

The validity and applicability of bidding methods rest on three assumptions: (1) the citizens are competent to make informed choices, (2) VAQ can be meaningfully valued in dollars and those dollars are equivalent to the dollars that measure the value of other attributes, such as jobs or construction materials or wildlife habitat, and (3) the value of VAQ does not depend on the levels of other attributes that enter into the decision, i.e., preferential independence (Keeney and Raiffa 1976). The first assumption has been questioned above. As is the case for MAUT, bidding methods assume competence, they do not help assure it. The second and third assumptions need to be tested empirically, although as far as I know this is not typically done in the use of bidding methods.

As a method for obtaining useful public input for informed choice, bidding methods do not solve problems of competence. This failing also limits the usefulness of the other approaches reviewed.

CONCLUSION

This analysis leads to the pessimistic conclusion that no social value assessment procedure provides input for informed choice that is simultaneously representative, valid, and applicable. The usefulness of every method is severely limited by problems in meeting one or more of these criteria. The following points support this conclusion.

1. Truly informed choice requires not only the best available information, but a coherent value system and the cognitive ability to apply that value system to a complex information base in order to make a choice.

2. With regard to any complex technical policy decision, most citizens lack the fact competence, value competence, and the cognitive competence to make an informed choice. This is not a result of apathy, but is simply a consequence of the number and complexity of the technical and value issues created.

3. Representativeness in social value assessment is gained by increasing sample size which requires decreasing the time spent questioning each person. Validity and applicability require increasing the amount of time spent with each person, but that will decrease representativeness.

4. In order to obtain input that is applicable to a decision, the right questions must be asked. The right questions involve trade-offs, but supplying adequate answers to these questions places a heavy cognitive burden on the citizen. This produces a conflict between validity and applicability--a choice between good answers to bad questions and bad answers to good questions.

Are the goals of informed choice and participatory democracy achievable in decisions involving VAQ? Perhaps, but the answer does not lie in the use of sophisticated methods for assessing public values. The key problem is the lack of competence required to make informed choices, and methods for eliciting and assessing public input do not address that problem.

Informed choice and participatory democracy seem to require increasing the level of competence, both of citizens and policy makers. Since informed choice requires fact, value and cognitive competence, technical education alone does not create "informed citizens." Examination of value systems, grounded in philosophy, the humanities, and the social sciences, and practice in applying value systems to specific decision problems is required. Obviously, all citizens cannot be competent on all issues, but perhaps the kind of specialization in issues (not special interests) described by Wildavsky (1979) is part of the solution.

Policy makers and their staff spend large amounts of time studying complex issues. Most of that time is spent developing fact competence. This activity, while it contributes to informed choice, is not sufficient. To make informed choices, policy makers must learn to use technical information in conjunction with an understanding of value systems and information about public values.

The problem of valuing visual air quality should not be considered simply a problem of measurement or assessment, to be treated by psychometric or economic techniques. To act as if visual air quality values exist out there in the public waiting to be measured with the right instrument is to ignore the gulf between the kinds of evaluative judgments people are accustomed to making and the kinds of value judgments required in order to assign a specific value to visual air quality. The problem is not assessment of value but development of a consistent, coherent, reasonable value system that can be applied to informed choices regarding VAQ.

NOTES

1. See, for example, Hammond et al. (1980) for a discussion of the distinction between judgment and choice.
2. For purposes of this discussion I will use the terms values and value system as defined by Rokeach (1973). "A value is an enduring belief that a specific mode of conduct or end-state of existence is personally or socially preferable to an opposite or converse mode of conduct or end-state of existence. A value system is an enduring organization of beliefs concerning preferable modes of conduct or end-states of existence along a continuum of relative importance."
3. It is important to distinguish between value expressions and value judgments. A value expression is simply an expression that something seems good or bad, desirable or undesirable, at a particular time in a particular context. It is a person's subjective report of his feelings at a given time. Its truth or validity depends only on how aware the person is of his feelings and on his sincerity.
    A value judgment is a judgment that something is, or will be, good or desirable. If a person pays a fee to enter a national park or expresses pleasure while observing a scenic vista, she has made a value expression. If the same person says that a vista should be preserved at a specific cost, she has made a value judgment.
4. These methods apply to MAUT as an alternative to traditional survey methods for public input. MAUT can be usefully employed by policy makers themselves as a decision aid and it is in that context that it is generally proposed by its primary advocates.

BIBLIOGRAPHY

Ayer, A. J. 1946. Language, Truth and Logic. Dover, New York, NY.
Baier, K. 1958. The Moral Point of View: A Rational Basis of Ethics. Cornell Univeristy Press, Ithaca, NY.
Baier, K. 1969. "What Is Value? An Analysis of the Concept." in K. Baier and N. Rescher (Eds.), Values and the Future. The Free Press, New York, NY.
Coates, J. F. 1978. "What Is a Public Policy Issue?" in K. R. Hammond (Ed.), Judgment and Decision in Public Policy Formation. Westview Press, Boulder, CO.
Dawes, R. M., and B. Corrigan. 1974. "Linear Models in Decision Making." Psychological Bulletin 81: 95-106.
Edwards, W. 1978. "Technology for Director Dubious: Evaluation and Decision in Public Contexts." in K. R. Hammond (Ed.), Judgment and Decision in Public Policy Formation. Westview Press, Boulder, CO.
Einhorn, H. J., and R. M. Hogarth. 1981. "Behavioral Decision Theory: Processes of Judgment and Choice." Annual Review of Psychology 32: 53-88.

Einhorn, H. J., and W. McCoach. 1977. "A Simple Multiattribute Utility Procedure for Evaluation." Behavioral Science 22: 270-282.
Fischer, F. 1980. Politics, Values, and Public Policy: The Problem of Methodology. Westview Press, Boulder, CO.
Fischhoff, B., P. Slovic, and S. Lichtenstein. 1980. "Knowing What You Want: Measuring Labile Values." in T. Wallsten (Ed.), Cognitive Processes in Choice and Decision Behavior. Erlbaum, Hillsdale, NJ.
Gardiner, P. C., and W. Edwards. 1975. "Public Values: Multiattribute--Utility Measurement for Social Decision Making." in M.F. Kaplan and S. Schwartz, Human Judgment and Decision Processes. Academic Press, New York, NY.
Hammond, K. R., and L. Adelman. 1976. "Science, Values, and Human Judgment." Science 194: 389-396.
Hammond, K. R. 1978. "Toward Increasing Competence of Thought in Public Policy Formation." in K. R. Hammond (Ed.), Judgment and Decision in Public Policy Formation. Westview Press, Boulder, CO.
Hammond, K. R., T. R. Stewart, B. Brehmer, and D.O. Steinmann. 1975. "Social Judgment Theory." in M. F. Kaplan & S. Schwartz (Eds.), Human Judgment and Decision Processes. Academic Press, New York, NY.
Hogarth, R. M. 1981. "Beyond Discrete Biases: Functional and Dysfunctional Aspects of Judgmental Heuristics." Psychological Bulletin 90: 197-217.
Kahneman, D., P. Slovic, and A. Tversky (Eds.). 1982. Judgment Under Uncertainty: Heuristics and Biases. Cambridge University Press, New York, NY.
Kaplan, A. 1964. The Conduct of Inquiry. Chandler, San Francisco, CA.
Keeney, R. L., and H. Raiffa. 1976. Decisions with Multiple Objectives: Preferences and Value Tradeoffs. Wiley, New York, NY.
Kyle, F., and A. Rabehl. 1980. "Structure and Use of the Integrated World Model." Technological Forecasting and Social Change 17: 73-87.
Milbrath, L. W. 1980. "Using Environmental Beliefs and Perception to Predict Trade-offs and Choices among Water Quality Plan Alternatives." Socio-economic Planning Sciences 14: 129-136.
Mackie, J. L. 1977. Ethics: Inventing Right and Wrong. Penguin, New York, NY.
MacRae, D., Jr. 1979. "Concepts and Methods of Policy Analysis." Society (Sept./Oct.): 17-23.
Nagel, T. 1979. Mortal Questions. Cambridge University Press, London.
Rokeach, M. 1973. The Nature of Human Values. The Free Press, New York, NY.
Rowe, R. D., and L. G. Chestnut. 1981. "Visibility Benefits Assessment Guidebook." U.S. Environmental Protection Protection Agency Publication No. 450/5-81-001, Research Triangle Park, North Carolina.
Rowen, H. S. 1976. "Policy Analysis as Heuristic Aid: The Design of Means, Ends, and Institutions." in L. H. Tribe, C. S. Schelling and J. Voss (Eds.), When Values Conflict: Essays on Environmental Analysis Discourse and Decision. Ballinger, Cambridge, MA.
Rubenstein, M. F. 1975. Patterns of Problem Solving. Prentice-Hall, Englewood Cliffs, NJ.
Sharkansky, I. 1982. Public Administration: Agencies, Policies, and Politics. Freeman, San Francisco, CA.
Sinden, J. A., and A. C. Worrell. 1979. Unpriced Values. Wiley, New York, NY.

Slovic, P., B. Fischhoff, and S. Lichtenstein. 1977. "Behavior Decision Theory." Annual Review of Psychology 28.

Slovic, P., and S. Lichtenstein. 1973. "Comparison of Bayesian and Regression Approaches to the Study of Information Processing in Judgment." in L. Rappoport and D. A. Summers, Human Judgment and Social Interaction. Holt, Tinehart and Winston, Inc., New York, NY.

Stevenson, C. L. 1944. Ethics and Language. Yale, New Haven, CT.

Stewart, T. R., and L. Gelbard. 1976. "Analysis of Judgment Policy: A New Approach for Citizen participation in Planning." American Institute of Planners Journal (January) :33-41.

Stokey, E., and R. Zechhauser. 1978. A Primer for Policy Analysis. Norton, New York, NY.

Warfield, J. N. 1974. Structuring Complex Systems. Battelle Monograph No. 4. Battelle Memorial Institute, Columbus, OH.

Wildavsky, A. 1979. Speaking Truth to Power: The Art and Craft of Policy Analysis. Little, Brown and Co., Boston, MA.

13. Social Research Methods for Public Land Managers

Glen E. Haas
David M. Ross

Research related to atmospheric visibility impairment in Class I areas has generally focused on three interrelated questions (Fox et al. 1979): (1) What is the likely extent and nature of the emissions from a proposed development; (2) Can recreationists detect the visibility impairment from such emissions; and (3) What value judgment do recreationists place on such visibility impairment? The answers to these questions are necessary, but not sufficient for reaching the decision to permit the development of a particular industrial facility or not.

Additional information relating to the behavior and characteristics of the visitors is needed to realistically assess the impact of visibility impairment in a Class I area. How much visitation occurs during this time? What type of recreation activities and experiences are visitors seeking? What are the travel patterns of the visitors? What kind and how many recreation facilities (e.g., trails, campsites, visitor centers, overlooks) are in the impacted area? How many visitors will be exposed to the visibility impairment? How many visitors will consciously perceive the impairment and feel a sense of dissatisfaction? Answers to these types of questions could significantly influence management decision-making.

The premise of this paper is that National Park Service personnel (e.g, park naturalists, rangers, backcountry guards, resource inventory specialists) can be effectively utilized to systematically collect visibility-related information with minimal impact on their primary job responsibilities (Haas and Nachtman 1979). This is not to say that park naturalists and rangers will become park researchers, but rather that with careful planning, work scheduling, and training, park personnel can systematically collect information which could help to answer questions related to the human detection and valuation of visibility impairment and the behavior and characteristics of visitors.

Glenn E. Haas is an Assistant Professor and David M. Ross is a Research Associate in the Department of Recreation Resources, College of Forestry and Natural Resources, Colorado State University, Fort Collins, CO. Also contributing to this paper were students in a graduate research methods class: Doug Deppe, Pat Devlin, Willene Hendon, Roberta Hilbruner, Bill Miller and Tom Rhodes. The authors acknowledge and appreciate the travel expense support from the Electric Power Research Institute.

The purpose of this paper is to acquaint park personnel to several options available to them at three interrelated decision points in the scientific inquiry of the visibility impairment issue: (1) data collection methods, (2) research designs, and (3) sampling designs. A detailed discussion of associated strengths and weaknesses of the various options is not the intent of this paper.

DATA COLLECTION METHODS

Park personnel can implement several data collection methods that have been labeled as naturalistic or unobtrusive research. Tunnell (1977) refers to naturalistic research as the systematic selection, recording, and encoding of natural behavior, in natural settings and involving natural treatments. The procedural goal of naturalistic research is that there be no investigator (e.g., park naturalist or ranger) intervention which "unnaturally" influences the people being studied. Since the presence of park personnel is expected and desired in national parks, there are many situations where information could be collected without unnaturally influencing the visitor being studied. The following section discusses archival research, structured observation, and structured conversation as naturalistic methods of data collection.

Archival Research

Archival research is also referred to as historical or retrospective research. It is a form of naturalistic research which relies on the content analysis or synthesis of information from various documents currently available. Selltiz et al. (1981) defines archival research as the systematic analysis of existing documents in order to reconstruct or explain real world events. Documents available to park personnel can take many forms:

1. Statistical Records:
   - Park visitation records
   - Activity participation records
   - Park programming records
   - Park maintenance records
   - Industry emission records
   - Related agency statistics (EPA, U.S. Weather Bureau)
2. Government Documents (local, state, federal):
   - Environmental impact statements
   - Legislative histories
   - Technical reports and publications
   - Management policies, handbooks, and plans
   - Historical photos
3. Research Reports:
   - University reports and publications
   - Industry reports and publications
   - Symposium and conference proceedings
4. Mass media:
   - Newspapers
   - Radio
   - Television
   - Speeches

5. Public Opinion Documents:
   - Letters and notes
   - Comments in suggestion or registration boxes

Park visitation records may suggest how many visitors visit the vistas or a portion of the park that will be visually impaired to some degree by industrial emissions. Legislative histories may suggest the relative value of a park's visibility attributes at the time of congressional designation. Historical photos may suggest a baseline, a limit of acceptable change standard, or provide a comparison between two points in time. Mass media and public opinion documents may suggest the sentiments (e.g., perceptions, preferences, attitudes) of local communities and visitors toward visibility attributes.

Systematically analyzing these types of documents can provide insight and, in some cases, answers to many of the questions related to visibility impairment. This information can become increasingly meaningful when it is related to information gathered from more scientifically rigorous studies.

Stuctured Observation

Structured observation is a form of naturalistic research that relies upon seeing or observing particular behaviors or characteristics of visitors. Because of the amount and social desirability of interaction between park personnel and visitors (Haas 1977), there are many naturalistic opportunities to collect observation-based information. The contribution that observation research can make to recreation planning and management has been recognized since the early 1960's (Burch 1964).

Friedrichs and Ludtke (1975) identified four types of observation research: unstructured non-participant, unstructured participant, structured non-participant, and structured participant. The major difference between structured and unstructured methods is that with the latter there are often no clearly defined objectives, explicit plan of action, or standardized method of recording information. In essence, all park personnel are unstructured participant and non-participant observers. By providing more structure (e.g., standardized recording forms, training, supervision), park personnel could not only provide insights and answers to many visibility related questions but also answers which are more reliable and valid.

An example of a structured non-participant observer could be a park ranger or backcountry guard who systematically and unobtrusively observes behaviors and characteristics of visitors. A structured participant observer could be a park naturalist on a guided nature walk or backcountry guard camping at a lake area who systematically and unobtrusively observes behaviors and characteristics of visitors. Below are examples of behaviors and characteristics of visitors that could be observed and recorded:

- number of visitors
- size of party
- party composition
- mode of travel
- travel pattern
- time of visitation
- type of activity participation
- length of time viewing from vista

- direction of viewing from vista
- percentage stopping at vistas
- number of pictures taken
- direction of picture taking

Structured participant and non-participant observation can be very useful in describing the "type" of visitor to a particular vista and also in providing insight into the fundamental questions of visitor detection and value judgments related to visibility impairment. This information can become increasingly meaningful when it is related to information obtained from telephotometers, meteorological instruments, or other visitor research studies.

Structured Conversation

A form of interviewing exists that has synonymously been referred to as structured conversation, unobtrusive interview, casual interview, and informal interview. Structured conversation is a form of naturalistic research that relies on conversing with people. It can be defined as the purposive and systematic collection of information by "unobtrusively steering" conversations to particular topic areas. Moeller et al. (1980) found that results obtained from structured conversations more accurately reflected visitors' actual behavior than did formal (obtrusive) interviews.

Similar to structured observation, there are many unobtrusive opportunities for park personnel to verbally interact with visitors. Park personnel with proper training and supervision can serve as unbiased and unobtrusive catalysts in steering conversations to particular aspects of visibility impairment and, subsequently, recording relevant information on a standardized form. Examples of these aspects might include:

- travel patterns
- place of residence
- number of previous visits
- type of recreation experience being sought
- detection of visibility impairment
- attributes of impairment detection
- value judgments related to impairment
- level of satisfaction
- perceived changes over time
- plans for future visitation

Structured conversation can not only provide a description of visitor's behaviors and characteristics, but also can provide information about visitor perceptions, preferences, attitudes, and satisfactions. Park personnel can get a better understanding of both the behavioral and cognitive characteristics of visitors. For example, a park naturalist guiding a nature walk to several vistas can collect information by both observing and/or conversing with visitors about their travel patterns, length of stay, place of residence, number of pictures taken, direction of viewing, and comments about visibility impairment detection or valuation. This type of information collected by structured converstion in a naturalistic setting can be valuable in helping to validate information from other visitor research studies or in the correlation with such information as from telephotometers, meteorological instruments, or EPA air quality records.

RESEARCH DESIGNS

The selection of a research design is a very important decision point in scientific inquiry. Considering various research designs helps (1) in conceptualizing what people and what treatments effects will be studied and when, (2) in analyzing and interpreting collected information, and (3) in recognizing the strengths and weaknesses associated with any particular research design. There are a variety of research designs applicable to the visibility impairment issue in Class I areas.

Two characteristics of true experimental research designs is the random assignment of people to different treatments and the control of extraneous variables which may confound the interpretation of results. While this type of design is probably not practical for national park personnel, there are a variety of quasi-experimental research designs that may be practical. The following section illustrates several research designs which could be implemented by park personnel to collect data via archival research, structured observation, or structured conversation.

One-Shot Case Study

The one-shot case study involves a group of people being given a particular treatment (X_1), after which some type of measurement (0_1) is taken. Much of the research presented in this volume utilizes this design. For example, a group of people is shown a set of pictures (X), after which some measurement of their reaction is taken (0); or after a group of people have visited a particular national park or vista (X), a mail survey (0) is sent to them to assess their willingness-to-pay for the visibility attribute of their experience.

In the context of this paper, park personnel can conceptualize the treatment to be a particular vista, guided nature walk or interpretive slide show (X), and the measurement tool being structured observation and/or structured conversation (0). Or, given the operation of particular industrial facility since 1975 (X), park personnel may utilize archival research to detect visitation changes (0).

Non-Equivalent Group Comparison

The non-equivalent group comparison design is defined as the measurement of some reaction of two different groups of people $(0_1, 0_2)$ to two different treatments (X_1, X_2). Relative to the one-shot research design, the non-equivalent group comparison design provides for comparing reactions to two different treatments.

Park personnel may be interested in comparing visitors' reaction $(0_1, 0_2)$ to different vistas (X_1, X_2), in which one vista is known or anticipated to have higher levels of visibility impairment. Similarly, the same design may be implemented to compare two different Class I areas, different geographic areas within a Class I area, different weeks or months of the year, or to compare visitors participating in different activities or seeking different recreational experiences.

Non-Equivalent Group Time Series

Non-equivalent group time series design is defined as the comparison of different people's reactions $(0_1, ..., 0_6)$ before and after a particular treatment (X_1). Park personnel may want to compare visitors' perceptions of impairment detection before and after (or during) those weeks of the year when emissions are anticipated to be most visible or to compare visitors' perceptions of impairment before and after an industrial facility becomes operable. Another possibility of this design is to compare the relative success of various mitigating actions taken by management when impairment is a problem to recreationists. Such actions might include re-scheduling of guided nature walks, various interpretive devices, or different verbal explanations of the situation.

Non-Equivalent Group Time Series Comparison

Non-equivalent group time series comparison is defined as the comparison of different people's reactions (0_1, ..., 0_{12}) before and after some treatment (X_1, X_2) and a comparison of the two different treatment effects. The research design is similar to the previous design but also provides for the comparison of treatments. Examples of these treatments include different vistas, parks, times of year, activity participation, guided nature walks, levels of emissions, or different mitigating actions.

SAMPLING TECHNIQUES

Another important decision point in scientific inquiry relates to identifying the people to be studied. In most social research studies, it is not practical or feasible to study all the people possibly involved (i.e., the population). How one defines the people to be studied directly influences how the information is interpreted and how generalizable it is to other people and park areas. For example, collecting data from visitors to a park visitor center during the first week of July and generalizing the findings to all visitors to that park would be tenuous.

There are a variety of basic sampling techniques (Selltiz et al. 1981): accidental, quota, purposive, simple random, stratified random, and cluster sampling. Each sampling technique has asssociated strengths and weaknesses ranging from time and travel costs to generalizability of results.

A rigorous, yet practical, sampling technique for park personnel to implement is cluster sampling, sometimes referred to as multi-stage sampling (Oderwald et al. 1980). The cluster sample procedure identifies the people to be studied through a series of decision steps. The following visibility-related example illustrates this process of identifying visitors to be studied.

Park personnel could identify all the vistas that are or will have the visibility impairment problem based on computer simulations of emission patterns. From this array, several vistas are systematically (perhaps randomly) selected based on clearly defined critera (e.g., availability of archival telephotometer data, amount of visitation, visual attributes from vista, feasibility of scheduling park naturalist or ranger patrol). From this less inclusive array of vistas, park personnel may identify several particular types of visitors to purposively include in the sample (e.g., day versus overnight visitors, hiker versus horseback riding visitors, morning versus afternoon visitors, older versus younger visitors). The results of these decision steps is a logical, traceable, and criteria-based selection of vistas and visitors to be studied.

Park personnel interested in collecting visitor information do not need to rely on a first-visitor first-studied approach. Cluster sampling, and in some instances, other sampling designs, provides a procedure for systematically identifying the visitors to be studied. The result will be a more reliable and valid information base for management decision making.

CONCLUSION

The purpose of this paper is to acquaint National Park Service personnel to available options for various data collection methods, research designs, and sample designs. The fundamental premise is that park personnel have the time, tools available, and aptitude to systematically and unobtrusively collect visibility-related information with minimal impact on their primary job responsibilities. The benefits of park personnel becoming more actively involved in collecting visitor information are numerous and extend beyond the visibility issue in Class I areas. Several of these benefits are overviewed below.

First, naturalistic research information can help to answer questions related to whether visitors detect visibility impairment and, if so, what value is attached to such impairment. It can also help to identify those visitors (e.g., who, where, when, how many) who will be impacted from an industrial development.

Second, naturalistic research information can supplement and help to validate findings from other studies. Given that the visibility impairment issue in Class I areas is significant and complex enough to warrant a multi-method research approach, naturalistic research information can and should be collected in conjunction with and related to information obtained from telephotometers, metereological instruments, visitor perception studies, and others.

Third, naturalistic research information will not only help determine whether a proposed development should be approved or not, this information may also help (1) to evaluate the effects of proposed developments adjacent to other parks, (2) to monitor changing visibility impairment values and visitor behaviors over time, and (3) in the selection of mitigating actions that may be implemented in response to visibility impairment.

Fourth, naturalistic research information may suggest to researchers particular variables, relationships among variables, or hypotheses that could be included in subsequent studies. This information may not only suggest what information should be collected, but also what variables should be explicitly considered in developing the research and sampling designs of these more scientifically controlled studies.

Fifth, naturalistic research information will result in park personnel having a better understanding of the behavior and characteristics of the national park visitor. Relative to the visibility impairment issue, park personnel will become more involved with the decision to permit an industrial development and be in a more informed position to justify to local communities and visitors why a particular decision was made. Relative to many other management issues confronting national parks, understanding the behavior and characteristics of the visitors may provide a better information base to make informed, equitable, and responsive decisions.

Lastly, recreation planning, management, and research is rapidly becoming more analytical, systematic, and quantitative. Such concepts as "monitoring" and "evaluation" are now commonplace among public land managers. This change will result in many of the traditional recreation "job descriptions" being redefined. One change we believe is inevitable, and which will contribute to an improved professionalism, relates to park personnel becoming more involved with the systematic collection of naturalistic research information and the subsequent improved understanding of the recreation clientele being served.

BIBLIOGRPAHY

Burch, W.R. 1964. Observation as a Technique for Recreation Research. USDA Forest Service Research Paper, Pacific Northwest Forest and Range Experiment Station, Portland, OR.

Fox, D., R.J. Loomis, and T.C. Green (Eds.). 1979. Proceedings of the Workshop in Visibility Values. U.S. Forest Service, Technical Report WO-18, Fort Collins, CO.

Friedrichs, J., and H. Ludtke. 1975. Participant Observation: Theory and Practice. Saxon House, England.

Haas, G.E. 1977. "Recreation and Parks: A Social Study at Shenandoah National Park." National Park Service Scientific Monograph Series, Number 10, Superintendent of Documents, Washington, D.C.

Haas, G.E., and S.C. Nachtman. 1979. A More Efficient and Effective Utilization of a Forest Service Seasonal Employee. Report to Fishlake National Forest, Utah. Department of Recreation Resources, Colorado State University, Fort Collins, CO.

Moeller, G.H., M.A. Mescher, T.A. More, and E.L. Shafer. 1980. "The Informal Interview as a Technique for Recreation Research." Journal of Leisure Research 12(2):174-182.

Oderwald, R.G., J.D. Wellman, and G.J. Buhyoff. 1980. "Multi-Stage Sampling of Recreationists: A Methodological Note on Unequal Probability Sampling." Leisure Science 3(2):213-217.

Selltiz, Wrightsman, and Cooks. 1981. Research Methods in Social Relations. Holt, Rinehart, and Winston, New York.

Tunnell, G.B. 1977. "Three Dimensions of Naturalness: An Expanded Definition of Field Research." Psychological Bulletin 84:426-437.

14. An Examination of Methodologies for Assessing the Value of Visibility

Karen Kelley MacFarland
William Malm
John Molenar

INTRODUCTION

Under the Clean Air Act and the Environmental Protection Agency's Visibility Regulations, the manager of Class I federal lands must make a decision if a new major facility (such as a power plant) would have an unacceptable, adverse impact on the air quality related values (including visibility) of the subject lands. If emissions from a source are predicted to be perceptible [any humanly perceptible change in visibility (visual range, contrast, coloration) from what would have existed under natural conditions] the federal land manager must then determine if the perceptible visibility impact would diminish the national significance of the area or impair the quality of the visitor's visual experience, thus possibly constituting an unacceptable, adverse impact. Once this "adverse" determination has been made by the federal land manager, the permitting authority (in many cases the state) may weigh cost, energy and other relevant factors with the land manager's recommendation in determining whether to permit construction of a new major source.

Several studies have been conducted in the past few years that examined the relationship between human perception of visibility and physical measures of visual air quality in an attempt to quantify visibility impairment (see Malm et al. 1980, 1981, Chapter 4 in this volume, and Latimer et al. 1980). Some research has also focused on the cost-benefit aspects of visibility protection (see Schulze et al. 1981 and Rae 1981a,b and Chapter 19 of this volume). However, little, if any, research has concentrated on quantification of what constitutes an unacceptable, adverse impact on visibility. Studies over the past decade have examined visitors' activities and enjoyment of parks and wilderness areas, but did not specifically focus on air quality (Arnold et al. 1981, Brown and Haas 1980, Loomis and Greene in Chapter 15 of this volume, and Haas et al. 1980). It is interesting to note, however, that enjoyment of scenic views and clean

---

Karen Kelley MacFarland and William Malm are with the Air Quality Division of the National Park Service, Fort Collins, CO, and Washington, D.C. John Molenar is with the John Muir Research Institute, Fort Collins, CO. The assumptions, findings, conclusions, judgments, and views presented herein are those of the authors and should not be interpreted as representing official National Park Service policies.

air did surface as important values in national parks and wilderness areas (Loomis and Greene, Chapter 15).

This paper reports preliminary results of pilot studies addressing effects of visual air quality on visitor enjoyment of National Park Service areas conducted by the National Park Service and Environmental Protection Agency in conjunction with visibility perception studies in Grand Canyon and Mesa Verde National Parks in 1980.

STUDY DESIGN

The studies were designed to test two methodologies for measuring the value park visitors place on good visibility. A willingness to pay methodology was used to measure "economic value" placed on good visibility, and an allocation of time methodology was used as a measure of "social value." The willingness to pay (WTP) methodology used in these studies was of similar design to that discussed in Randall et al. (1974) and Brookshire et al. (1976). For the bidding procedure, survey participants were shown various sets of five color slides ranging from poor to good visual air quality, labeled A through E. Approximate visual ranges (VR) for each slide were 100, 160, 210, 270, and 350 kilometers (km) (see Figure 14.1).

The Grand Canyon study utilized two sets of five composite slides to assess WTP to improve visual air quality associated with uniform haze. Composite slides were made by photographically placing three separate scenes onto one 35 mm slide format. One set of composite slides consisted of three representative Grand Canyon vistas while the second set consisted of three vistas representative of Grand Canyon, Zion, and Mesa Verde National Parks. The second set of composite slides was used to suggest how visual air quality changed across the Southwest region. A third set of two slides was used to assess WTP for not having a thin coherent dark plume visible over a typical Grand Canyon vista.

At Mesa Verde, three sets of five slides representing Grand Canyon, Mesa Verde and Zion National Parks were used. However, in this study each set of five slides contained only one representative scene under various air quality conditions.

After all five slides were displayed on the screen, slides B through E were blocked out, leaving only slide A, representing poor visual air quality. Survey participants were instructed to assume that the visual air quality was as shown in slide A on the day of their visit to the park. Slide B was then projected next to slide A and participants were asked how much above the current $2.00 daily park entrance fee they were willing to pay per day to improve the visual air quality to that shown in slide B. The same bidding procedure was used for slides C, D, and E of each series.

The bidding procedure varied only slightly for the plume slides at the Grand Canyon. Survey participants were shown a slide with a thin coherent dark plume crossing the entire vista just above the sky-canyon interface. They were then shown the same scene without the plume and were asked how much above the current $2.00 daily park entrance fee they were willing to pay per day to have the improved visual air quality condition occur on the day of their visit to the park.

For both uniform haze and plume bidding sequences, respondents were provided options for each bid ranging from $0.00 per day to $100.00 per day. The last section of the willingness to pay response sheet also asked participants who bid $0.00 on any of the questions to indicate their reason for bidding zero dollars.

FIGURE 14.1
Slides Used in the Willingness to Pay Methodology

POOR		AVERAGE		GOOD
A	B	C	D	E
7.02 µg/m^3	3.78 µg/m^3	2.43 µg/m^3	1.08 µg/m^3	0.27 µg/m^3
110 km	160 km	210 km	270 km	350 km

It is important to note that all slides used in the willingness to pay studies were a subset of those used in the visibility perception studies. Therefore, electro-optical measures of air quality were available for each slide as well as the Perceived Visual Air Quality (PVAQ). See Malm et al. (1981) for a discussion of how the PVAQs were calculated.

A separate questionnaire was used to test the allocation of time methodology, with different groups of survey participants. Participants were first given a list of ten items and asked how important each item was to their enjoyment of the park. The items were scenic views, geological formations (ruins at Mesa Verde), museums, natural forest, wildlife, solitude, clean air, interpretive programs, scientific study and park rangers. The response options were not important, somewhat important, important, very important, and extremely important.

The same set of five slides as used in the willingness to pay studies were then displayed on the screen and participants were then asked if visual air quality would affect the length of time they looked at vistas in the park. They were then shown slide C, with an explanation that it represented the average visual air quality at the park, and asked if the length of time they looked at the vistas would change if the visual air quality was improved to that shown in slide A (good visual air quality). Participants recorded how much longer or shorter they would look at the vistas in either minutes, hours or days. The same procedure was then followed using slide C, comparing it to slide E, with poorer visual air quality.

Survey participants were then asked if visual air quality would affect their length of stay at the park. The same allocation of time questions were asked as explained above. Figure 14.2 shows the allocation of time procedure.

A socio-economic/demographic questionnaire was completed by participants in both surveys. Information gathered included home zip code, rural/suburban/urban residence, age, education, sex, length of stay at the park and in the region, activities while in the park, annual household income, number of previous visits to the park, and number of previous visits to all other NPS areas.

STUDY RESULTS

Willingness-to-Pay Methodology

Figure 14.3 shows the mean bids for each WTP sequence (A to B, A to C, A to D, and A to E) for the Grand Canyon study. The y-axis is plotted in

FIGURE 14.2
Slides Corresponding to Air Quality Levels Used in the Allocation of Time Methodology

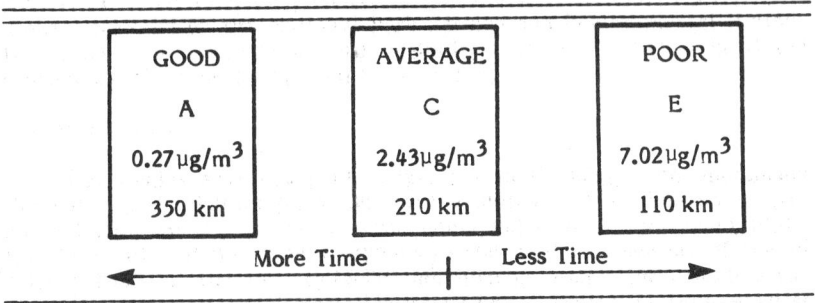

dollars and the x-axis shows the fine particulate concentration and visual range that correspond to each slide. For example, visitors would be willing to pay an average of 98 cents per household over the current $2.00 daily entrance fee to improve the visual air quality at Grand Canyon from how the canyon appears with 7.02 µg/m^3 atmospheric fine particulate concentration (VR = 110 km) to how it appears with a 3.78 µg/m^3 fine particulate concentration (VR = 160 km). The solid line represents bids for the Grand Canyon composites while the broken line represents bids for the regional composites. The dotted line shows the mean bid that individuals would pay to not have a thin dark coherent plume present over a Grand Canyon vista.

Even though the standard deviations associated with mean bids were large, the 90 percent confidence intervals around each mean are quite small. Because of the large sample population (approximately 1,000) the confidence intervals were typically less than 10 cents. These confidence intervals are shown in Figure 14.3 as the shaded area around the heavy dashed or solid lines that represent the mean bids. Note that the differences between bids on the Grand Canyon composites and regional composites are statistically significant.

The mean bid for improving visual air quality by not having a thin dark coherent plume present in a Grand Canyon vista ($1.98) was less than the mean bid for improving visual air quality from the poor to good case ($2.74) but higher than the mean bid for improving visual air quality from poor to average ($1.61). Other studies have shown that visual air quality is not judged to be worse when plumes are present in a scene if they do not obscure the vista (Malm et al. 1982). The results of this study also seem to show that visitors place a higher value on being able to clearly see the vista itself and place less importance on the presence of plume if it does not obscure any of the vista's scenic features.

Figure 14.4 shows the results for a similar study carried out at Mesa Verde. In this case, single scenes rather than composites were used to represent vistas of various national parks. The mean bids for improving visual air quality at Grand Canyon and Mesa Verde vistas are shown. Again, bids for improving visual air quality are higher for the Grand Canyon vista than for the Mesa Verde vista. The differences in mean bids are statistically significant. Bids for improving visual air quality at Zion National Park were also obtained. In every case the Zion bids fell in

FIGURE 14.3
Mean Bids for Willingness to Pay Sequence for Grand Canyon National Park

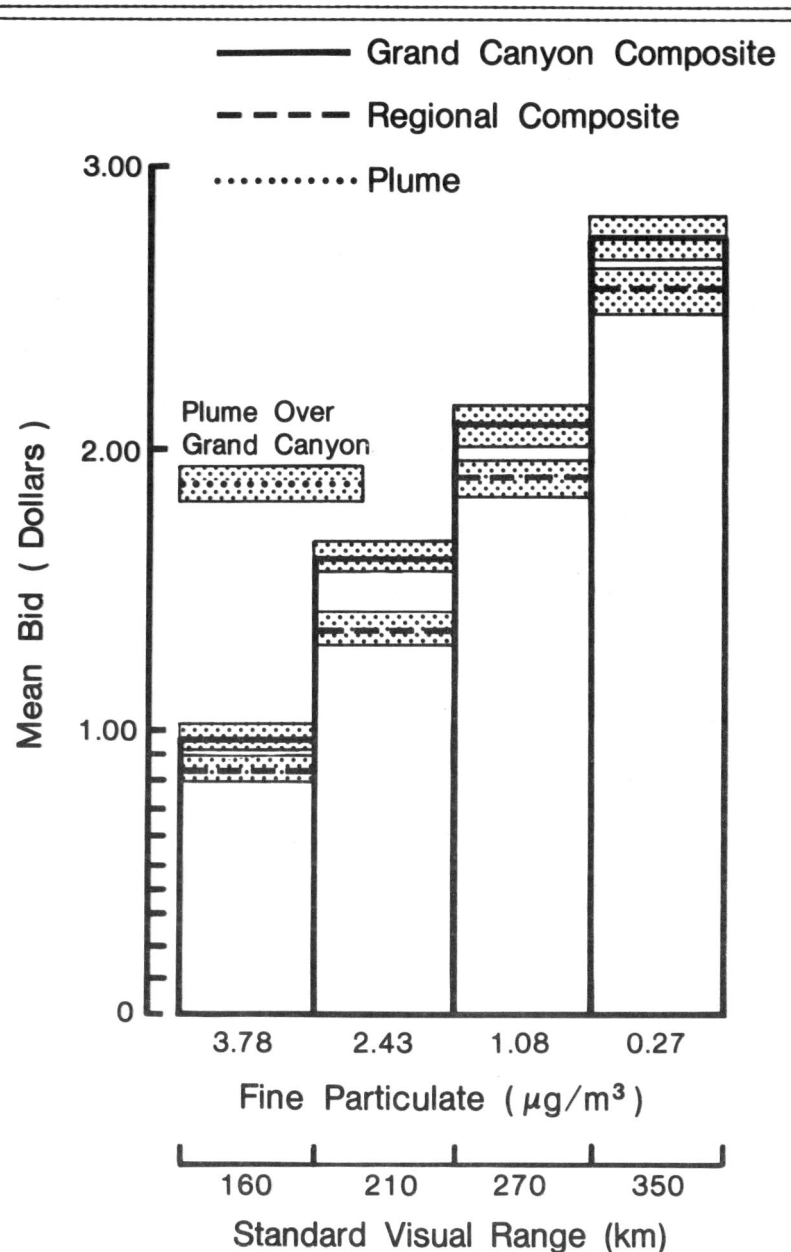

FIGURE 14.4
Mean Bids for Willingness to Pay Sequence for Mesa Verde National Park

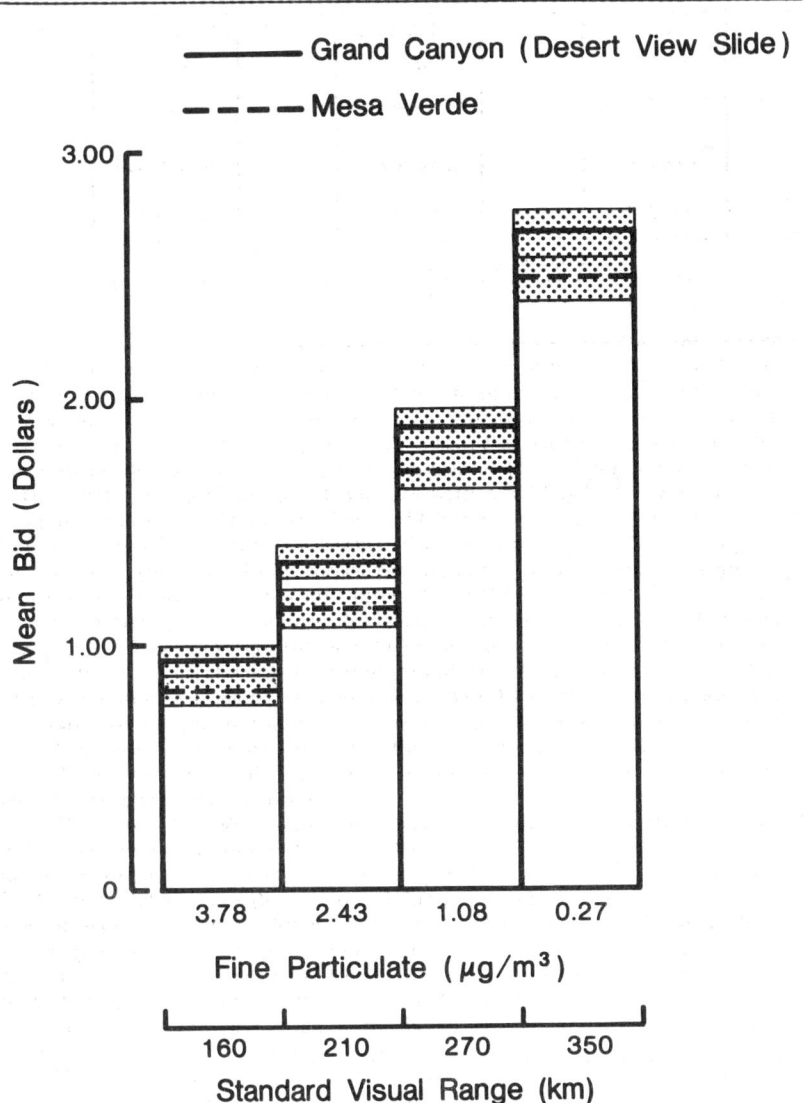

between the Grand Canyon and Mesa Verde bids. The differences between Zion and Grand Canyon or Zion and Mesa Verde bids were not statistically significant.

Visitors were willing to pay more to improve visual air quality at the Grand Canyon than they were for the region. This suggests that the Grand Canyon in and of itself is valued as a scenic resource more than the region as a whole. It should be pointed out, however, that: (1) this conclusion can only be reached if the composite slides were truly indicative of the scenery both within the canyon and for the region, and (2) the mean bids for the Grand Canyon are only slightly higher than mean bids for the region. If visitors value visual air quality at the Grand Canyon more than at other representative regional vistas it is not by large amounts.

It should be pointed out that the slides representing vistas in each park were chosen in such a way that differences in judged visual air quality between each slide were nearly the same. That is, PVAQ increased monotonically from slide to slide. It is interesting to note that mean bids also seem to increase monotonically from slide to slide; as PVAQ increases so does WTP for improved visual air quality. There seems to be, for the series of air quality levels examined, a linear relationship between WTP for improved visual air quality and PVAQ.

Allocation of Time Methodology

Whereas in the WTP questionnaire visitors were given specific bid responses to select from, this portion of the study allowed visitors to choose amounts of time without any guidance from the survey instrument. Therefore, when analyzing the data, length of stay (LOS) responses were grouped into cells. Responses between 0 and 1 minute were coded as 1, those between 1 and 3 minutes were coded 2, those between 3 and 7 minutes were coded 3, and so forth. Except for the predominance of LOS greater than 120 minutes, the distribution of LOS tends to be log normal. This was also the case for WTP. However, the WTP survey yielded standard deviations that were nearly equal to the mean, while standard deviations for LOS tended to be three to five times greater than the mean. The distribution has a standard deviation of 76 minutes greater than and 18.3 minutes less than the mean of 23.5 minutes. (The unsymmetric standard deviation is a result of a distribution that is not normal.)

This greater variance in LOS than in WTP responses may be a result of allowing the visitor to arbitrarily indicate amounts of time without guidance from the survey instrument, or it may indicate that visitors are less "standardized" in their allocation of time.

Figure 14.5 shows the mean time visitors would either increase or decrease their LOS at a vista depending on whether the visual air quality was increased from average to good or decreased from average to poor. The average day slide showed how the vistas appeared when there was an atmospheric loading of 2.43 $\mu g/m^3$ of fine particulate (VR=210 km). The x-axis in Figure 14.5 shows the corresponding fine particulate concentrations and visual ranges for good and poor days. The shaded areas around the means are the 90 percent confidence levels.

First note that the differences between increased and decreased LOS are not statistically significant. As in the relationship between WTP for improved air quality and PVAQ, this suggests a linear relationship between LOS and PVAQ. However, the data shown in Figure 14.5 suggests a further generalization. Equal changes in visual air quality, whether from average

FIGURE 14.5
Mean Amount of Time Visitors Would Increase or Decrease Their Length of Stay at a Grand Canyon Vista with Varying Visual Air Quality

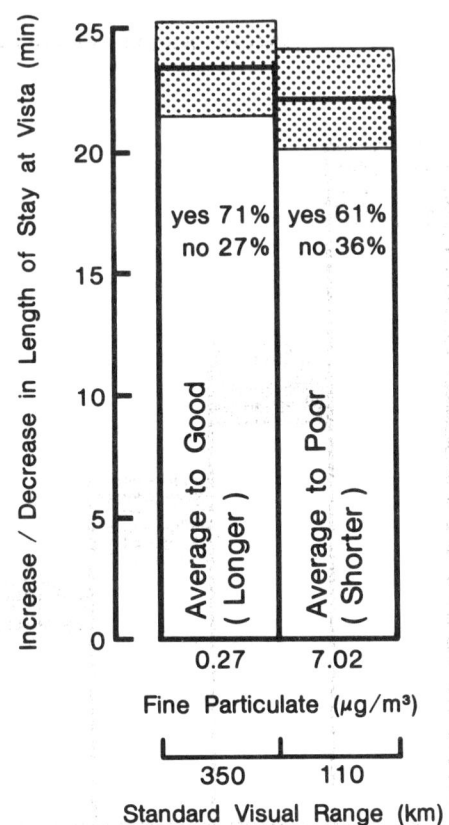

to good or average to poor, elicit equal LOS responses. This suggests that the LOS increments would be the same if the visitors were asked to indicate the increase in LOS when the visual air quality went from poor to average or from average to good. These results suggest that surveys that ask the visitors WTP for preserving an increment of visual air quality might yield the same answers as those where the visitor is asked WTP associated with improving visual air quality.

Finally, Figure 14.5 also shows the percentage of visitors who would not increase or decrease their LOS at a vista as visual air quality varied. Twenty-seven percent of the respondents would not increase their LOS at a vista if the visual air quality improved, while 36 percent would not decrease their LOS with declining visual air quality.

It might be suggested that those individuals who said they would not vary their LOS at a vista as a function of air quality are insensitive to impacts of air pollution. This is not necessarily the case. Rather, the study results might suggest that certain individuals have a specific time schedule that they are trying to meet or that once they arrive at the Grand Canyon they will spend x amount of time viewing the vistas no matter how impaired they may be. However, this does not mean that these individuals are necessarily enjoying the vistas as much on a poor day as they would on a good day.

This suggests a major weakness of willingness to pay and allocation of time methodologies. They do not allow a visitor to specifically express whether his enjoyment is being affected by variations in visual air quality.

Figure 14.6 shows the results of asking visitors to indicate how they might modify their LOS in the park instead of at a vista. Again, the results are generally the same except that average times are increased from minutes to hours. An increase or decrease in visual air quality from average to good or from average to poor yielded similar LOS responses of about twelve to fifteen hours.

The one striking difference between LOS at vistas and LOS in the park is the number of people who said their LOS in the park would not be affected by improved visual air quality. Seventy-one percent of the respondents said they would increase their length of viewing a vista but only 56 percent of the respondents indicated that they would increase their LOS in the park if the visual air quality improved from average to good. However, 80 percent of the visitors sampled said they would shorten their LOS in the park if the visual air quality decreased from average to poor but only 61 percent would decrease the amount of time viewing a vista.

An interesting question arises. Was the smaller number (56 percent) of the respondents who would increase their LOS in the park if the visual air quality increased from average to good versus the 80 percent who said they would decrease their LOS if the visual air quality declined a result of more individuals concerned about the visual air quality becoming poorer than average, or was it a result of the way the question was asked? The lower response (56 percent) of visitors who would increase their LOS as a consequence of improved visual air quality might just be due to constrained vacation schedules. Most visitors do not have the leeway of increasing their vacation time irregardless of how much visual air quality is improved. However, if visual air quality decreases, they have the option of shortening their stay in the park significantly, and visiting an alternate area.

FIGURE 14.6
Mean Amount of Time Visitors Would Increase or Decrease Their Length of Stay in Grand Canyon NP with Varying Visual Air Quality

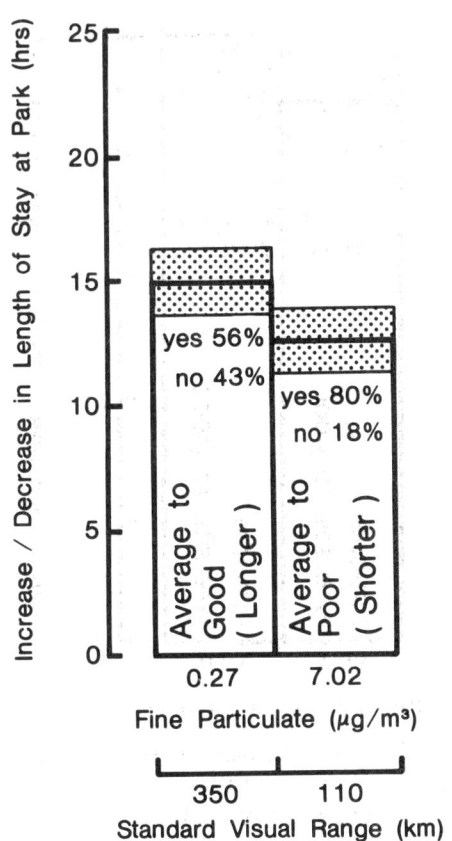

COMPARISON OF METHODOLOGIES

In the remaining sections of the paper the willingness to pay and allocation of the time methodologies are compared. Tables 14.1 and 14.2 are correlation matrices for the Grand Canyon and Mesa Verde data. Each matrix includes WTP and LOS responses and PVAQ. For purposes of correlating WTP and PVAQ with LOS it is necessary to translate LOS into either a continually increasing or decreasing scale. Therefore, it was assumed that Slide C was zero time from which either an increase or decrease in LOS could be referenced against. An increase in LOS was treated as a positive number while a decrease in LOS became a negative number. Thus, as PVAQ increased so did the LOS scale, initially from a negative time to zero and then from zero to a positive time. In this way LOS can be compared and correlated with PVAQ as well as WTP. WTP responses were chosen such that they could be compared to the same PVAQ increments used in the LOS questions. Only bids from poor to average and poor to good were used in order for these numbers to be comparable to the WTP bids. Numbers in the tables without parentheses are correlation coefficients while those within parentheses show the level of significance associated with each. Note that all correlations are significant at less than the 1 percent level.

Correlations between various WTP bids are generally high, suggesting that the survey participants were very consistent in their use of the bidding scale. Once a participant established a scale for the Grand Canyon bids he tended to be consistent in using the same scale for bids for other park areas. It should be emphasized that the correlations apply to each individual's consistency, not consistency within the entire survey sample. Standard deviations for mean bids of sample groups were large, indicating that the sample as a whole was not consistent in its use of the bidding scale. Conversely, this same individual consistency did not show up in LOS responses.

Correlations between WTP and PVAQ are around 0.45, while correlations between LOS and PVAQ are about 0.20. Again, although these correlations are low, they are significant at less than a .01 percent level. It was expected that these correlations would be higher since participants were asked to respond to changes in PVAQ. These relatively low correlations are indicative of the very large amount of variance in responses. Also, the low correlations may be explained by other variables which were not examined in these studies.

As indicated previously, respondents were given some guidance for establishing their bid criteria in that a range of possible bids was presented to each participant. The respondent indicated his preferred bid by placing a mark next to a dollar amount already on the response sheet. On the other hand, participants were given no criteria for choosing LOS responses. They could indicate LOS in minutes, hours or even days.

It seems that as the survey instrument becomes more structured, the correlations between visual air quality indicators and participant responses are higher. Where the responses are entirely open ended, such as in LOS questions, the correlations are lowest. In previous studies where the respondents were given very structured response criteria (a maximum and minimum range) and judged visual air quality was correlated against vista contrast, extremely high correlations of around 0.90 were achieved (Malm et al. 1980, 1981).

TABLE 14.1
Correlation Matrix for Grand Canyon

	WTP for Grand Canyon Composite	WTP for Regional Composite	LOS at Vista	LOS in Park	Perceived Visual Air Quality (PVAQ)
WTP for Grand Canyon Composite	1.00[1]	0.88 (0.001)[2]	0.19 (0.001)	0.23 (0.001)	0.40 (0.001)
WTP for Regional Composite	0.88 (0.001)	1.00	0.20 (0.001)	0.23 (0.001)	0.40 (0.001)
LOS at Vista	0.19 (0.001)	0.20 (0.001)	1.00	0.34 (0.001)	0.22 (0.001)
LOS at Park	0.23 (0.001)	0.23 (0.001)	0.34 (0.001)	1.00	0.16 (0.002)
Perceived Visual Air Quality (PVAQ)	0.40 (0.001)	0.40 (0.001)	0.22 (0.001)	0.16 (0.002)	1.00

[1] Correlation coefficients.
[2] Level of significance associated with correlation coefficient.

TABLE 14.2
Correlation Matrix for Mesa Verde

	WTP for Mesa Verde Slides	WTP for Grand Canyon Slides	WTP for Zion Slides	LOS at Vista	LOS in Park	Perceived Visual Air Quality (PVAQ)
WTP for Mesa Verde Slides	1.00[1]	0.89 (0.001)[2]	0.88 (0.001)	0.14 (0.016)	0.32 (0.001)	0.44 (0.001)
WTP for Grand Canyon Slides	0.89 (0.001)	1.00	0.94 (0.001)	0.16 (0.007)	0.34 (0.001)	0.49 (0.001)
WTP for Zion Slides	0.88 (0.001)	0.94 (0.001)	1.00	0.19 (0.001)	0.36 (0.001)	0.46 (0.001)
LOS at Vista	0.14 (0.001)	0.16 (0.001)	0.19 (0.001)	1.00	0.28 (0.001)	0.23 (0.001)
LOS at Park	0.32 (0.001)	0.34 (0.001)	0.36 (0.001)	0.28 (0.001)	1.00	0.18 (0.002)
Perceived Visual Air Quality (PVAQ)	0.44 (0.001)	0.49 (0.001)	0.46 (0.001)	0.23 (0.001)	0.18 (0.002)	1.00

[1] Correlation coefficient.
[2] Level of significance associated with correlation coefficient.

A principal component analysis was carried out in an attempt to reveal trends between the dependent variables of LOS and WTP and independent socio economic variables such as home, education, and income. Determination of principal components is a statistical approach to assess whether observed correlations between variables in a large variable set can be explained by a few hypothetical variables or factors. The analysis essentially classifies those variables which tend to cluster as part of a group or factor. The variables grouped in a factor often suggest an underlying physical interpretation.

Table 14.3 shows results of the principal component analysis for the Grand Canyon data. If factor loadings are less than 0.25 they are not shown. First of all, notice that WTP and PVAQ group together under Factor 1 while LOS and PVAQ group under Factor 4. It is not surprising that PVAQ groups with WTP and LOS since it is visual air quality that participants were asked to respond to. What is significant that LOS and WTP responses fall into separate factors. This suggests that those individuals who bid high for WTP were not the same individuals with high LOS responses.

It might be expected that WTP and/or LOS would group with education, income, or residence. However, there does not appear to be any relationship between WTP or LOS with these socio economic variables. The only other grouping between independent and dependent variables occurred under Factor 2. Those individuals who stayed the longest in either the park or region were most sensitive to changing their LOS in the park as a function of changing visual air quality.

Table 14.4 is a similar analysis for the Mesa Verde data. The results are generally the same except that WTP, LOS, and PVAQ all fall into the same factor. This difference may be partially explained by the fact that visitors to the two parks have different socio economic profiles, with mean education and income higher for the Mesa Verde visitor sample. The one important relationship between length of stay in park and LOS responses still shows up. Those individuals who are staying longest in the park are most sensitive to changing their length of stay as a function of visual air quality.

These clusterings were further examined with a more detailed analysis of variance (ANOVA) procedure. Table 14.5 shows the results for the Grand Canyon survey data. The dependent variables are listed across the top of the table and the independent variables are listed in the left hand column. Bids that correspond to changes in atmospheric fine particulate loadings of 7.02 $\mu g/m^3$ to 3.78 $\mu g/m^3$ to 0.27 $\mu g/m^3$ were examined (A to B and A to E). The first number in each set is the variance explained by that specific independent variable; the second number is the total variance; the third number is the percentage of variance explained by the independent variable and the last number is the level of significance.

Again, education and household income had no relationship to WTP and LOS. The one common variable to both WTP and LOS is the actual length of stay in the park. Those individuals who stayed the longest in the park were most apt to be willing to pay more for good visual air quality or stay longer if the visual air quality were improved. The reason that the WTP and LOS were grouped into different factors in the principal component analysis can be seen in the variance explained under these two categories. Length of stay at the park explains only 1 percent of WTP variance but about 20 to 25 percent of the LOS variance. It is important to point out that the ANOVA analysis reemphasizes that those individuals who are

TABLE 14.3
Principal Component Analysis (Grand Canyon)

	Factor 1	Factor 2	Factor 3	Factor 4
WTP for Grand Canyon Composites	0.94	NO	NO	NO
WTP for Regional Composites	0.90	NO	NO	NO
LOS at Vista	NO	NO	NO	0.65
LOS at Park	NO	0.43	NO	0.43
Perceived Visual Air Quality (PVAQ)	0.43	NO	NO	0.27
Length of Stay in Park	NO	0.64	NO	NO
Length of Stay in Region	NO	0.55	NO	NO
Previous Number of Visits to Grand Canyon	NO	NO	0.65	NO
Previous Number of Visits to Other Parks	NO	NO	0.38	NO
Home (Rural/Suburban/Urban)	NO	NO	NO	NO
Education Level	NO	NO	0.31	NO
Annual Household Income	NO	NO	NO	NO

TABLE 14.4
Principal Component Analysis (Mesa Verde)

	Factor 1	Factor 2	Factor 3
WTP for Mesa Verde Slides	0.91	NO	NO
WTP for Grand Canyon Slides	0.97	NO	NO
WTP for Zion Slides	0.95	NO	NO
LOS at Vista	NO	NO	0.38
LOS at Park	0.25	NO	0.64
Perceived Visual Air Quality (PVAQ)	0.48	NO	NO
Length of Stay in Mesa Verde	NO	0.85	0.34
Length of Stay in Region	NO	0.44	NO
Previous Number of Visits to Mesa Verde	NO	0.25	NO
Previous Number of Visits to Other Parks	NO	NO	NO
Home (Rural/Suburban/Urban)	NO	NO	NO
Education Level	NO	NO	NO
Annual Household Income	NO	NO	NO

TABLE 14.5
Analysis of Variance (Grand Canyon)

	WTP for Grand Canyon Composite		WTP for Regional Composite		LOS at Vista		LOS in Park	
	A - E	A - B	A - E	A - B	Longer	Shorter	Longer	Shorter
Home (Rural/Urban/Suburban)	56.1[1] 4333.2[2] 0.013[3] (0.006)[4]	NO	28.8 3779.4 0.008 (0.046)	NO	NO	NO	NO	NO
Length of Stay in Park	46.3 4319.9 0.009 (0.06)	24.7 2940.9 0.008 (0.082)	40.7 3742.2 0.011 (0.034)	29.6 2833.7 0.01 (0.040)	80.2 1632.1 0.05 (0.001)	96.6 1947.5 0.05 (0.001)	38.2 1259.6 0.21 (0.001)	345.6 1055.7 0.25 (0.001)
Previous Number of Visits to Grand Canyon	NO	NO	NO	NO	NO	NO	36.4 1602.9 0.02 (0.035)	29.2 1438.5 0.02 (0.05)
Previous Number of Visits to Other Parks	NO	NO	NO	NO	NO	NO	32.4 1460.0 0.02 (0.054)	NO
Education Level	NO	NO	NO	NO	NO	NO	NO	NO
Annual Household Income	NO	NO	NO	NO	NO	NO	NO	NO

[1] Variance explained by that specific independent variable.
[2] Total variance.
[3] Percentage of variance explained by the independent variable.
[4] Level of significance.

are staying in the park the longest are the same individuals who are most sensitive to changing their LOS in the park as a function of visual air quality. Table 14.6 shows results of a similar analysis for the Mesa Verde data.

The relationship between length of stay in the park and increase/decrease in LOS in the park as a function of visual air quality again shows up as being strong and significant. Individuals who stayed the longest were most sensitive to changes in visual air quality. Another strong relationship that is unique to the Mesa Verde data is between annual household income and LOS in the park as function of visual air quality. Ten percent of the LOS variance was explained by household income.

Finally, the relative ranking of the ten items visitors considered important to their enjoyment of the parks shows that scenic views and clean air were the two most important items at Grand Canyon. Ruins were rated most important at Mesa Verde, with clean air second, and scenic views third. This indicates that scenic views and clean air are important even in a park which has primary emphasis on cultural features. Figure 14.7 is a histogram of the relative ranking of the ten items for each park.

CONCLUSIONS

This study concentrated on two similar but different methodologies for determining how individuals would allocate their time and money for preserving a given length of stay at a vista and in the park as a function of varying air quality. This question was open ended in that they could respond any amount of time desired. In addition, they were given a choice to either increase or decrease their length of stay depending on whether the visual air quality was judged to be better or poorer. On the other hand, questions concerning willingness to pay for improving visual air quality were structured in such a way as to give participants specific choices for their bids.

Although the willingness to pay and allocation of time methodologies differed substantially, they did yield a number of similar results:

1. Both WTP and LOS response frequency distributions tended to logarithmic in nature.
2. Both responses (WTP and LOS) correlated significantly with PVAQ. However, the correlations were higher between WTP responses and PVAQ (about 0.35) than between LOS and PVAQ (about 0.25). Again this may be attributable to the variation in structure between the two questionnaires, or how people budget their time.
3. Over the visual air quality levels examined, there tended to be a monotonic relationship between WTP and PVAQ and LOS and PVAQ.
4. In general, neither LOS or WTP could be significantly related to other socio economic variables such as household income, education level, residence, age, etc.
5. Most significantly, both LOS and WTP could be statistically linked to how long the individuals stayed in the park. Those individuals who came to stay in the park the longest were most sensitive and placed more value on good visual air quality.

TABLE 14.6
Analysis of Variance (Mesa Verde)

	WTP for Mesa Verde Slides		WTP for Grand Canyon Slides		WTP for Zion Slides		LOS at Vista		LOS in Park	
	A - E	A - B	A - E	A - B	A - E	A - B	Longer	Shorter	Longer	Shorter
Home (Rural/Urban/Suburban)	45.7[1] 3880.0[2] 0.01[3] (0.025)[4]	38.8 3309.6 0.01 (0.025)	32.7 3610.4 0.01 (0.058)	31.6 3128.9 0.01 (0.042)	38.9 3418.8 0.01 (0.030)	35.9 2903.8 0.01 0.02	13.5 1060.3 0.013 (0.029)	NO	NO	NO
Length of Stay in Park	54.4 3984.2 0.01 (0.031)	48.4 3388.8 0.01 (0.025)	48.0 3703.9 0.01 (0.037)	62.0 3185.7 0.02 (0.006)	40.8 3571.8 0.01 (0.059)	41.5 2968.2 0.01 (0.028)	NO	NO	358.7 1102.8 0.25 (0.001)	279.8 1169.5 0.19 (0.001)
Previous Number of Visits to Mesa Verde	84.1 3832.0 0.02 (0.003)	99.1 3198.9 0.03 (0.001)	68.7 3567.1 0.02 (0.006)	89.3 3029.1 0.03 (0.001)	59.6 3433.8 0.02 (0.010)	102.4 2774.6 0.04 (0.001)	NO	NO	NO	NO
Previous Number of Visits to Other Parks	NO	NO	NO	NO	NO	NO	NO	NO	58.4 1354.0 0.041 (0.004)	
Education Level	NO	NO	NO	NO	NO	NO	NO	NO	NO 171.9 1216.4 0.12 (0.001)	NO 138.4 1240.5 0.10 (0.012)
Annual Household Income	NO	NO	NO	NO	NO	NO	NO 76.2 1046.5 0.07 (0.001)	NO 53.6 785.3 0.06 (0.061)		

[1] Variance explained by that specific independent variable.
[2] Total variance.
[3] Percentage of variance explained by the independent variable.
[4] Level of significance.

FIGURE 14.7
Relative Ranking of Ten Items Visitors Considered Important to Their Enjoyment of the Parks

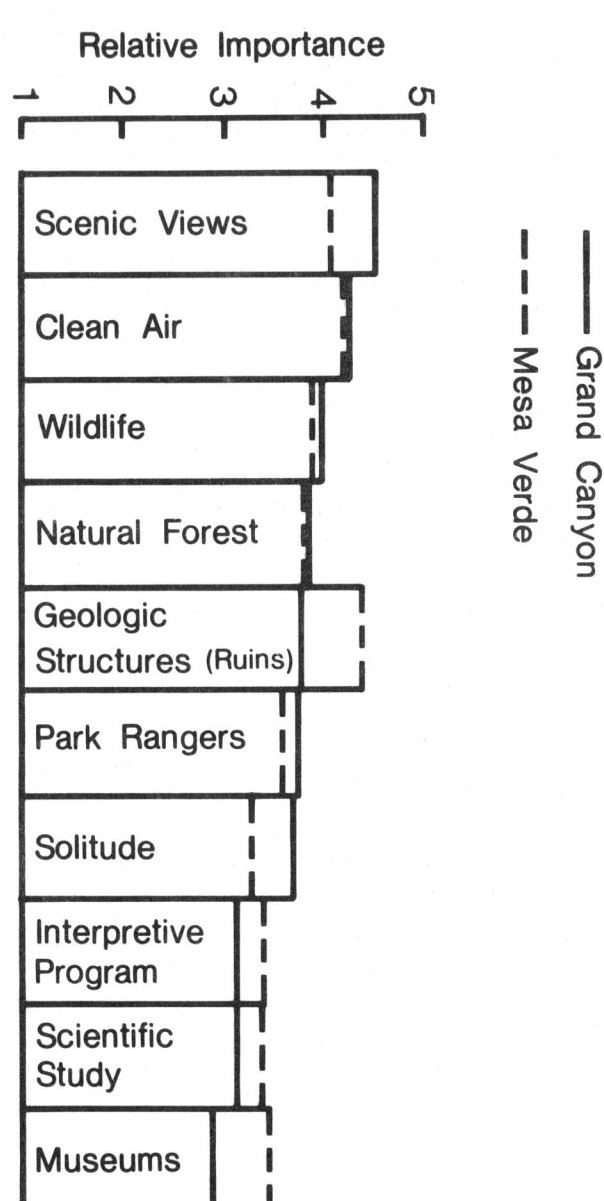

There also were some interesting differences in results between the two methodologies. The variance in responses associated with the more open ended questionnaire (LOS) was much higher than for the WTP questionnaire. WTP standard deviations were about equal to the mean bid where LOS standard deviations often exceeded the mean response by as much as a factor of five.

Individuals tended to develop consistent economic value scales for the WTP questions. They tended to respond to changes in visual air quality for Grand Canyon in much the same way as they did for Mesa Verde. This was not the case for LOS responses. Individuals tended to respond differently to one vista as compared to another. In addition, those individuals who placed a high economic value on good visual air quality were not the same individuals who would vary their length of stay either in the park or at the vista as a function of varying air quality. In fact, principal component analysis for the Grand Canyon survey showed LOS and PVAQ grouped under one factor and WTP and PVAQ under another. This suggests that there may be two distinct socio economic cohorts who visit Grand Canyon; one group "economically" sensitive to visual air quality changes while the other tends to be more sensitive from a "behavior" standpoint, altering their allocation of time as a function of visual air quality.

Finally, it should be reemphasized that visitors identified clean air and scenic views as among the three most important factors that contributed to their enjoyment of the parks. Future research efforts in this area should be directed at quantifying how visitor enjoyment is affected by changes in air quality.

BIBLIOGRAPHY

Arnold, J.R., B.L. Driver, P.J. Brown, and L. Udick. 1981. "Measuring Dispersed Use and Visitor Preferences on the Bureau of Land Management's National Resource Lands: King Range Study." Report to Bureau of Land Management. USDA Forest Service Rocky Mountain Forest and Range Experiment Station, Fort Collins, CO.

Brookshire, D., B. Ives and W. Schulze. 1976. The Valuation of Aesthetic Preferences. Journal of Environmental Economics and Management 3(4):325-346.

Brown, J., and G.E. Haas. 1980. "Wilderness Recreation Experiences." Journal of Leisure Research 12(3):229-241.

Haas, G.E., B.L. Driver, and P.J. Brown. 1980. "A Study of Ski Touring Experiences on the White River National Forest." In Proceedings of North American Symposium on Dispersed Winter Recreation. College of Forestry, University of Minnesota, St. Paul, MN.

Latimer, D., T. Daniel, and H. Hugo, 1980. "Relationships Between Air Quality and Human Perception of Scenic Areas." Publication 4323, Systems Applications Incorporated, San Rafael, California.

Malm, W.C., K. Leiker, and J. Molenar. 1980. "Human Perception of Visual Air Quality." Air Pollution Control Association Journal 30(2):122-131.

Malm, W.C., K. Kelley, J. Molenar, and T. Daniel. 1981. "Human Perception of Visual Air Quality (Uniform Haze)." Atmospheric Environment 15(10/11):1875-1890.

Rae, Douglas A. 1981a. "Benefits of Improving Visibility at Mesa Verde National Park." Research Project 1742, Draft Report. Prepared by

Charles River Associates, Inc. for the Electric Power Research Institute. November. Boston, MA.

_____. "Benefits of Improving Visibility at Great Smoky National Park." Research Project 1742, Draft Report. Prepared by Charles River Associates, Inc. for the Electric Power Research Institute. December. Boston, MA.

Randall, A., B. Ives, and E. Eastman. 1974. "Bidding Games for Valuation of Aesthetic Environmental Improvements." Journal of Environmental Economics and Management 1:132-149.

Schulze, W.D., D. Brookshire, E. Walther, K. Kelley, M. Thayer, R. Whitworth, S. Ben-David, W. Malm, and J. Molenar. 1981. Methods Development for Environmental Control Benefits Assessment. Volume VIII. The Benefits of Preserving Visibility in the National Parklands of the Southwest. Prepared for the U.S. Environmental Protection Agency. Office of Research and Development. Washington, D.C.

15. Two Examples of Psychological Assessment of Visual Values

Ross J. Loomis
Thomas C. Greene

Will people change their assessment of the visual air quality of scenic slides as a function of specific instructions used in the assessment? Are there identifiable patterns to the way recreationists utilize visual resources on a hiking trail and could these patterns be used as behavioral indicators of changing levels of visual quality? These two questions were part of two assessment strategies intentionally designed to reflect the need to create a variety of indicators of how people value visual quality. The two examples are from a series of pilot studies conducted by the authors and intended to suggest alternate ways of evaluating the value of visibility or visual quality in general (Loomis and Greene 1982).

Both of the examples discussed in this paper are based on ideas suggested at a Fort Collins, CO workshop concerning visibility values (Fox et al. 1979). In that report, Craik (1979) suggested that instructional sets (i.e. information given to subjects about the task they are asked to do, reasons for the research and the precise nature of the stimuli used) should be directly studied to determine their influence on assessments of visual air quality. Driver et al. (1979) emphasized the importance of identifying specific behaviors related to visual resources and assessing the psychological benefit of visibility-related behaviors and experience. The second assessment example reported in this paper deals with the first part of the Driver et al. suggestion that visibility-related behaviors be identified.

INSTRUCTIONAL SETS AND ASSESSMENT OF VISUAL QUALITY

Three experiments were conducted using college students as subjects. A variety of instructional sets were given prior to viewing scenic

---

Ross J. Loomis is a Professor of Psychology, Colorado State University, Fort Collins, CO. Thomas C. Greene is an Assistant Professor of Psychology, Saint Lawrence University, Canton, NY. Support for portions of the work described in this paper was from Project No. 16-952GR, Rocky Mountain Forest and Range Experiment Stations. Participation at the workshop was made possible, in part, by travel funds from the Electric Power Research Institute and Saint Lawrence University. The authors wish to thank Linda Kline and Beth Etlinger for their assistance in data collection, and Doug Fox and Henry Cross for their assistance and review of the manuscript.

vistas with moderate levels of visual impairment present. In two of the exeriments, the dependent variable of interest was the rating of visual air quality for the scenes shown in the slides. In the other experiment, the dependent variable was a rating of how desirable a pictured scene was as a place to visit.

The first study (Greene et al. 1982) used intentionally broad instructional sets consisting of a neutral condition (only basic instructions as to how to make a slide quality rating) and, a Clean Air Act condition in which the subject was informed that the purpose of the study was prompted by the visibility section of the Clean Air Act. The requirements of that section of the act were explained to subjects. The independent groups design experiment also had a third instructional set condition in which subjects were told that trade-offs were necessary to meet national energy needs and one specific trade-off was some visual impairment of scenic vistas. It was expected that the Clean Air Act instructions would produce more critical evaluations (lower quality ratings) while the trade-off instruction subjects would be more lenient (higher quality ratings) than subjects receiving neutral instructions.

Results of a multivariate analysis of variance indicated that a marginal instructional set effect was present ($\underline{F}(16,30) = 1.7$, $\underline{p} < .06$) between the three conditions.[1] The most critical ratings, however, were for the trade-off condition, not the instructions sensitizing subjects to the Clean Air Act. Inspection of univariate comparisons across the three instruction conditions for each of the eight slides showed a consistent pattern of lower ratings for trade-off instructions with statistically significant differences appearing on three of the eight comparisons.

It is important to note that little difference in ratings was found between the Clean Air Act and neutral instruction conditions. The Clean Air Act instructions were intentionally drawn from existing field research efforts.[2] As mentioned, the instructional sets used in the first experiment were intentionally broad or general which could account for the marginal outcomes. In addition, it was possible that the actual influence of the trade-off condition was to sensitize subjects to the distinction of human versus natural caused visual impairment. In the Fort Collins workshop, some attention was given to the practical problems faced by land managers in determining the importance of visibility protection. Included among those problems was whether or not the public can correctly attribute that some visibility impairment is due to natural causes such as naturally caused fires. Two follow-up experiments were conducted to produce more effective and focused instruction sets and assess the importance of attributed natural versus human origins to visual quality impairment.

In the second experiment, subjects were once again asked to rate the visual air quality of scenic slides containing moderate levels of visibility impairment. Instructional set conditions were prepared that attributed visual impairment to (1) haze from natural water vapor, (2) a wildfire that was being allowed to burn for ecological reasons, and (3) an industrial source. Results of this independent groups experiment revealed that subjects who were told that the haze was due to the industrial (human) cause rated slides significantly lower in quality than did subjects in either of the conditions attributing impairment to natural causes ($\underline{F}(2,33) = 4.89$, $\underline{p} < .05$).

The third study limited the instructional sets to overall attributions of natural or human causes and asked a sample of student subjects to rate three slides of hazy mountain scenes as to how desirable the placed pictured would be to visit. All three scenes were rated as less desirable for visitation in the human than natural caused attribution condition and statistical significance was reached on two of the three slides ($\underline{F}(1,65) = 15.33$, $\underline{p} < .01$ for slide two, $\underline{F}(1,65) = 4.95$, $\underline{p} < .05$ for slide two).

In conclusion, some evidence for instructional set influence was found across the three experiments. In his presentation at the Visual Values Workshop, Terry Daniel discussed a caveat about the limitations of laboratory psychology research that certainly applies to the reported experiments. Only a small portion of the variance was accounted for by the independent variables. In addition, the experiments were limited to student subjects. The first and third experiments were performed at a Rocky Mountain University and the second completed at a Northeastern school.

The results, however, are consistent with long-standing research findings that instructional sets can influence judgments. Textbook (Dember and Warm 1979) discussions of psychological or perceptual sets, including the influence of different instructions, have often distinguished between response bias versus perceptual sensitivity explanations of perceptual set research. The first explanation really has little to do with perception and suggests that instructional sets merely make some responses more probable than others. For instance, subjects in the three experiments just summarized may have responded with lower ratings in human attributed visual impairment conditions, because they are familiar with the idea that modern lifestyles cause undesirable air pollution. Their ratings would be somewhat analogous to a "top of the head" response that pollution is bad which continually shows up in public opinion polls.

However, it may be that instructional sets create different perceptual sensitivities to scenes containing visual impairment. That is, people may really perceive impairment differently as a function of whether they think it is natural or human caused. Thresholds or sensitivity levels may actually be different and the effects of instructional sets serve as indicators of how people value visual quality. If this explanation of perceptual set is valid, then efforts at manipulating instructional sets could be one indicator of the underlying social values people hold toward scenic beauty and the impact of human caused impairment.

Two suggestions about assessment of how the public values visual air quality can be made based on a consideration of instructional sets. First, any assessment of sensitivity or quality of visual resources should check for bias being introduced through instructions as an <u>unintentional</u> source of variance. This consideration of instructional set places an emphasis on methodological control, such as controlling for slide order effects (see Latimer et al. in this volume).

A second assessment strategy would emphasize the intentional use of instructional sets as a means of informing subjects about critical information necessary to make intelligent judgments. Alan Randall noted in some of his comments at the Visual Values Workshop that outcomes in willingness to pay assessments can change as a function of information (instructional sets) given subjects.[3] The impact of specific information contained in instructions can be to elicit different underlying social values. This strategy might also be one way of speaking to the issues raised by Stewart (in this volume) concerning the need for informed choice in public responses to the visual air quality issue. It certainly underscores his point that more than just passive psycho-metric and economic assessment procedures are needed to understand public reactions to complex social issues.

It may be possible to come up with an assessment strategy that would allow for both detection of sensitivity to visual quality and the level of response bias. Dember and Warm (1979) point out that a signal-detection-theory approach considers bias as a variable of the subject's decision criterion and allows for independent assessments of sensitivity and bias. Modifying the type of experiments reported in this paper to include a signal detection response format could be a major improvement to the research method and worthwhile pursuing.

INSTRUMENTING A HIKING TRAIL FOR VISIBILITY-RELATED BEHAVIORS

At first glance, it might seem like an overly abrupt transition to move from a discussion of instructional set effects to field observation of visibility-related behaviors. However, behaviors related to visual resources can be seen as the action or expressive components of expectancies people have about the environment and its features. The emphasis on behavior stems around the question of what do people do with visual resource opportunities and will they behave differently when visual resources change?

Driver et al. (1979) based their thinking on expectancy-value research from psychology.[4] The idea of expectancy-value is a close conceptual cousin to perceptual sets mentioned in the first assessment example described in this paper. Just as perceptual sets may evoke underlying ideas people have about visual quality, expectancy can refer to their perceptions of what a given setting is like and what kinds of behaviors would be possible and desirable. Ultimately, a link must be established between expectancies and actual behavior as well as how people evaluate the importance of being able to engage in expected acts and experiences. Driver et al. (1979) suggested that a first step in understanding the psychological benefits and values associated with expectancies about visual quality would be to identify specific visibility-related behaviors.

The importance of identifying visitor or public behaviors related to visual quality is a theme in a number of papers in this volume. Brown underscores the importance of understanding recreationist behavior in the context of visual resources and Randall presents a fascinating possibility for using a time-series analysis of potential relationships between visual quality and activity choices in the Chicago area. Malm et al. report on a behavioral approach to visual resource demand analysis and Haas and Ross discuss some practical ways resource managers could learn about visitor visual resource-related behaviors.

In the present example, a pilot study effort at instrumenting a specific hiking trail (Mount Audubon Trail in the Colorado Indian Peaks Wilderness Area) was completed. The first part of the study asked a sample of twenty-five hiker groups to carry a Polaroid camera with them and take pictures as they might normally do if they had a camera of their own. The hikers were encouraged to take pictures of interest and to disregard the limited quality of snap shot type photographs. One-half of the sample was contacted at the top of the trail and took pictures while moving downward; the other half took pictures while going up the trail.

Two-hundred and twenty-nine usable pictures were taken and it was possible to organize the photos around nine specific trail points. The content of the photographs suggested that there were indeed several preferred locations for taking photographs, with remarkable agreement in both location and composition in several instances. This agreement was particularly striking for one trail point at the base of a mountain and at two trail switchbacks. The larger number of photos taken at these sites (including a number that were virtually identical) suggested that these views may be particularly striking. Additionally, the data are consistent with what Gustke and Hodgson (1980) have called "discontinuity." This term refers to the observation that information rate and pleasure increase dramatically as one moves from one continuous and predictable landscape into a new environment. For instance, the first trail point in the study marked a transition from a relatively level forest to a rocky (and later barren) hillside. Similarly, the two switchbacks offered vistas to hikers who have been traveling through relatively enclosed forest, and further signaled a change in both directions of travel and dominant view.

The 229 pictures were sorted into a few general categories of picture content as shown in Table 15.1. Vista pictures as such were limited in number although some pictures of the trail, people, and scenery had medium or long-range vistas as background. If vista only and vista background pictures were considered together, about 43 percent of the photographs contained at least some vista orientation.

TABLE 15.1
Photograph Content for Polaroid Task in Trail Instrumentation of Visibility-Related Behavior

Content		Number	Percent
Scenery	(tree groupings, lakes, etc.)	87	38
Foreground	(logs, flowers, small animals, etc.)	65	28
People		30	13
Trail		29	13
Vista		9	4
Clouds		2	1
Unclassified		7	3
		229	100

It is possible that people avoided vista pictures because a snap shot photograph does not represent a long-range view very well. However, several subjects volunteered when turning their cameras in that they did not take pictures of the eastern three vista points because they were "too smoggy." Refinement of this photograph technique could document just how often such an avoidance of picture taking activity occurs.

A second part of the trail instrumentation assessment consisted of having four observers look for visibility-related behaviors at nine different trail points. Six of the trail points had been suggested by the Polaroid task. Observations were spread across six days and completed in one hour sessions. All observations were unobtrusive in nature. Table 15.2 summarizes the kinds of behavior observed, based on ninety-four incidents classified as visibility-related as sampled across 144 hikers. A coding system had been worked out based, in part, on some preliminary work done by having recreationists suggest behaviors they would typically engage in if at a setting containing vista viewing points.

TABLE 15.2
Observed Incidents of Visibility-Related Behaviors

Description of Activity	Frequency Observed
Verbal statements and discussions with other hikers	33
Body and head postured in apparent long-distance viewing	26
Pointing to vista	17
Shading eyes or obvious movement to improve view	11
Using a camera	7
	94

Results of the behavior observations suggested that there were site-specific behavioral tendencies. For example, three of the sites shared characteristics that appeared to make them important viewing points. In each case, the trail made a sharp curve or switchback and offered a relatively unobstructed view. Three other trail points all offered panoramas across the eastern foothills to the Great Plains, and were frequently selected for stops of up to one-half hour, which sometimes included lunch. It is probably safe to say that these sites would be the most likely to show adverse effects from impaired visibility because of the long periods some hikers spent at these sites. In addition, the locations offered by far the most distant views along the observed section of the trail and their horizon views contained the Front Range population corridor stretching from Fort Collins to the Denver area.

Very little of what could be termed visibility-related behavior was observed at either of the wooded sites or at a mountainside location. Clearly, the wooded sites offered little opportunity for more than close or middle-range viewing. In the case of the mountainside location, distant views were available, but could easily be enjoyed while hiking downhill, whereas uphill hikers were apparently not prompted to stop and turn around to view what lay behind them. The differences in behaviors observed at this site and those seen at the switchback locations demonstrated that a switchback curve represents a final opportunity to look in the direction one has been traveling before turning in a new direction. Furthermore, the curves were placed in a manner which made them a "window" at the end of an enclosed path.

Those trail points that displayed visibility-related behaviors could be sampled over time to plot the frequency of behaviors under different conditions of visual impairment. Such an expansion of the trial instrumentation task would provide an indicator of any visual resource lost because of impairment.

The two field instrumentation tasks obviously represent a purely descriptive and pilot effort at identifying some ways hikers utilize visual resources, including vista viewing that could involve air quality impairments. Museum visitation also involves visitor use of highly visually oriented resources and some parallels can be made between the trail study and existing museum visitor research (Elliott and Loomis 1975).

1. As mentioned in connection with the Polaroid picture task, transition points in the visual properties of a trail are apt to stimulate visual exploration. These points of discontinuity have also been documented in museum settings (Robinson 1928). Visitors typically slow their pace of movement, look around more and may stop to view specific points of interest. Visual quality and interpretation are apt to be of greater value at points of discontinuity.

2. Visual quality may also be valued less if a trail area serves primarily as a passageway to other destinations. It became apparent to the four observers that many hikers on the trail section studied were on their way to other points which may account for the lack of more visibility-related behaviors not being observed. Any behavioral indicators of visibility value must take into account this kind of factor of visitor activity choices. A similar (Wolf and Tymitz 1979) observation has been made in museums where some galleries have a large degree of visitor passageway behavior which accounts for less attention being given to exhibits.

3. Both the photograph and observation tasks confirmed that use of recreational environments is often a highly social experience. A picture of a valued vista is apt to have a person in the foreground and people may share their reactions to visual points on a trail and quite likely their awareness of any visual impairment. Museum visitor surveys (Elliott and Loomis 1975) have also confirmed the social nature of leisure time visitation activities.

4. Finally, a long-standing outcome of museum visitor research (Melton 1972) has shown that people sample or highlight a given area of visual opportunities. There is selective attention to specific features of an environment and it should be possible to use the tasks described in this paper to identify those most important vista points and other visual resources in a specific park or recreation setting. Both the Polaroid and observation measures revealed that some specific trail points were most important to visual resource utilization.

It should be noted in passing that when subjects in the present study were debriefed while turning their cameras in, they were given to express concerns about air quality. While some of these reactions may simply have reflected a response bias of it being socially acceptable to compain about air pollution, it is also likely that interviews of visiting groups could yield significant assessments of visitor concerns about visual quality.

As a final note about this second example of psychological assessment of visual quality, the four points just mentioned could serve as partial guidelines in setting up the kind of research model proposed by Malm et al. (in this volume) in which any visitor loss in enjoyment due to visual quality degradation could be assessed over time. A key feature to that model, and any assessment effort, is that multiple indicators of visual values be organized in such a way as to maximize the evidence of real effects. These indicators need to be integrated into a decision-making framework such as that proposed by Fox and Rosenthal (1981) to assist managers charged with making decisions about protection of visual quality in parks and other natural resource settings.

Fox and Rosenthal (1981) developed a decision tree model that includes, as a major decision point, the determination if any impairment is detectable. Of significance to the present paper, Fox and Rosenthal noted the importance of distinguishing whether or not the source of visual impairment is human or natural caused and implied in their model that source attribution could be a factor in sensitivity. Another critical decision point consists of determining just how much impact impairment will have. The trail instrumentation task reported in this paper could evolve, with refinement, into an information source for impact assessment.

The major implication, however, of a management decision tree model for visual air quality is that a variety of assessment tasks are needed to provide natural resource managers with the information they need to decide what degree of visibility protection is appropriate. A multiple indicator research strategy is called for with specific indicators utilized at different decision points. Future research on visual air quality should be more integrative in nature and combine the various kinds of assessment tasks being developed.

SUMMARY

In the first assessment example described, a series of three laboratory experiments demonstrated that an influence of instruction set can occur at least for some stimuli depicting impaired visual quality. Assessment of how the public values visibility should take into account the methodological implications of perceptual set research as well as the use of such research paradigms to suggest underlying social values. The second assessment strategy involved a first effort at instrumenting a hiking trail for visibility-related behaviors. It was possible to identify critical trail points for visual resource use and also identify some use patterns that could be included in a more extensive assessment effort.

NOTES

1. Note that the degrees of freedom reported include a third variable not described. See Green et al. (1982) for a complete summary of the experiment.
2. Instructions were similar to those used by Latimer et al. described in Chapter 5. In addition, Malm, Keiker and Molenar (1979) referred to Clean Air Act legislation in their instructions used in a National Park Service Visibility Perception Study.
3. Comments made at 1982 Visual Values Workshop, Keystone, Colorado.
4. See, for example, Fishbein and Ajzen (1974) and Fishbein (1979).

BIBLIOGRAPHY

Craik, K. H. 1979. "The Place of Perceived Environmental Quality Indices (PEQUIS) in Atmospheric Visibility Monitoring." In D. Fox, R. J. Loomis, and T. C. Greene (Eds.). Proceedings of the Workshop on Visibility Values. U.S. Forest Service Technical Report WO-18, Fort Collins, CO.

Dember, W. N., and J. S. Warm. 1979. Psychology of Perception. Second Edition. Holt, Rinehart and Winston, New York, NY.

Driver, B. L., D. Rosenthal, and L. Johnson. 1979. "Air Pollution, Values, and Environmental Behavior." In D. Fox, R. J. Loomis, and T. C. Greene (Eds.). Proceedings of the Workshop on Visibility Values. U.S. Forest Service Technical Report WO-18, Fort Collins, CO.

Elliott, P., and R. J. Loomis. 1975. "Studies of Visitor Behavior in Museums and Exhibitions: An Annotated Bibliography." Office of Museum Programs, Smithsonian Institution, Washington, D.C.

Fox, D., R. J. Loomis, and T. C. Greene (Eds.). 1979. <u>Proceedings of the Workshop on Visibility Values.</u> U.S. Forest Service Technical Report WO-18, Fort Collins, CO.

Fox, D. G., and D. Rosenthal. 1981. "On the Use of Decision Analysis in Determining Significant Visibility Impairment." <u>Atmospheric Environment</u> 15:2503-2510.

Greene, T. C., M. Breeding, and J. Grant. 1982. "The Effects of Instructional Set and Slide Composition on Ratings of Air Quality." Manuscript, St. Lawrence University, Canton, NY.

Gustke, L. D., and R. W. Hodgson. 1980. "Rate of Travel Along an Interpretive Trail: The Effect of Environmental Discontinuity." <u>Environment and Behavior</u> 12:53-63.

Loomis, R. J., and T. C. Greene. 1982. <u>Pilot Study of Class I Area Visibility-Related Behaviors and of the Perceived Psychological Benefits of Those Behaviors.</u> Final Report, Project No. 16-952-GR, Rocky Mountain Forest and Range Experiment Station, Fort Collins, Colorado.

Melton, A. W. 1972. "Visitor Behavior in Museums: Some Early Research in Environmental Design." <u>Human Factors</u> 14:393-403.

Robinson, E. S. 1928. <u>The Behavior of the Museum Visitor.</u> No. 5 in Publications of the American Association of Museums, New Series, Washington, D.C.

Wolf, R. L., and B. L. Tymitz. 1979. "Eastside, Westside, Straight Down the Middle: A Study of Visitor Perceptions of 'Our Changing Land'." The Bicentennial Exhibit. Office of Museum Programs, Smithsonian Institution, Washington, D.C.

16. Assessing the Effect of Visual Air Quality Degradation on Visitor Enjoyment

William C. Malm
David Shaver
Gerald E. McGlothin

INTRODUCTION

Good visibility has long been recognized as an important value in our national parks. In the National Park Service Act of 1916, Congress directed the Service to manage the parks to "conserve the scenery and the natural and historic objects...and to provide for the enjoyment of the same in such a manner and by such means as will leave them unimpaired for the enjoyment" of visitors (NPS 1916). In order to effectively carry out its Congressional mandate the National Park Service (NPS) needs to be able to determine how scenic landscape features effect a visitor's experience and enjoyment of a park and must assess how these scenic values will be affected by changes in air quality. With this information the decision maker can better judge the adversity of potential changes in air quality.

Increasing emissions of air pollution that result from continued urbanization and industrialization, both within established pollutant centers and in remote wilderness areas, have already compromised the good visual air quality that once existed in the national park system (Trijonis 1978). Has this ongoing degradation in our ability to see the natural wonders affected the ability of the park visitor to enjoy the scenery? Currently, the federal land manager (FLM) is inadequately equipped to deal with this question. He cannot determine with certainty how air pollution will affect public enjoyment. He cannot predict the effect of alternate control strategies or siting criteria on the visitor experience. This paper outlines the issues involved in analyzing the effect of air quality on visitor enjoyment and presents a methodology for assessing and valuing the visitor's visual experience.

The answer to the FLM's question requires analysis of three key areas of the complex concept of visitor enjoyment. The first step is to establish and quantify how a visitor perceives changes in visual air quality as a function of a measurable increment of air pollution. Conventional perception studies are not designed to extract from the park visitor the value he or she

---

William Malm and David Shaver are with the Air Quality Division of the National Park Service, Fort Collins and Denver, CO. Gerald McGlothin is with the Electrical Engineering Department at Northern Arizona University, Flagstaff, AZ. The assumptions, findings, conclusions, judgments, and views presented herein are those of the authors and should not be interpreted as necessarily representing official National Park Service policies.

attaches to specific park attributes or whether visual impairment affects the enjoyment of the scenic resources in a natural area. That is, is the visual air quality degradation adverse? Second, therefore, it is necessary to establish the value, whether it be social, psychological or economic, that a person places on a perceptual change in the visual resource. Finally, it is necessary to establish a link between visitor values and visitor enjoyment. For instance, if a decrease in visibility can be linked to increased stress or behavior modification it can be argued that visitor enjoyment has been altered.

Quantification of visitor perceptions first requires a determination of the amount of pollution that causes just noticeable difference in scenic quality. This quantification is a complex problem for both uniform and layered haze (Henry 1977). If the atmosphere is free of haze layers (uniform spectral radiance) the eye-brain system is quite sensitive to an increment of pollution that manifests itself as a spectral discontinuity between itself and some background. The eye-brain system is also very sensitive to a change in uniform haze which causes an object that is just visible to disappear. However, if the haze layer were already present and all landscape features were impaired but clearly visible, a just noticeable difference in visual air quality is more difficult to define. Since a change in the brightness or color of a haze layer or the density of a uniform haze takes place over time periods of hours or days, a just noticeable difference in visual air quality under these circumstances would require a person to "remember" what the scene looked like before a given change in air pollution took place.

A second important component of the quantificaiton of visual air quality perception is the establishment of the relationship between perceived visual air quality (PVAQ) and air pollution at an aggregate level (Malm et al. 1981). Only after this relationship is understood, is it possible to then continue on to link, in a general way, the value of the visual experience to perceptions of visual air quality and to pollution control levels.

Determining the shape of the aggregate benefits curve as a function of visibility and scenic quality is important in establishing the level of air pollution control that results in maximum utility. To do so, it is necessary to differentiate between perceived changes in the visual resource and the value, whether it be social, psychological, or economic, that people place on a perceived change. For instance, a person may place a high value on preventing a decrease in visual range of 100 km when viewing a particular vista while that person may not place any value on the same visual range preservation when viewing a different scene. Did this individual indeed value one vista more than the other or was one vista just perceptually less sensitive to visual range reduction?

The valuation analysis must be carried out in the context of the overall visitor experience in an area. This accounts for the variation from park to park in the relative significance of the visual experience compared to other park attributes. Although the visual experience is a significant component of visits to Grand Canyon and Mesa Verde National Parks, protection of that experience might require different magnitudes of air pollution protection for each area. In cases where perceptions of visibility impairment are less acute or visibility values have less importance than other park attributes some visibility impairment may be tolerable if the impairment does not affect user enjoyment of a natural area.

Assessment of the benefits derived from maintaining a certain level of visual air quality requires that the FLM have the available "tools" to accurately forecast visitor response to changes in the visual resource system and measure the importance of the visual resource compared to other park attributes. The following sections address four questions that are formulated to address this visual resource demand analysis: (1) What is the scope of a visual resource demand analysis? (2) What are the objectives of visual resource demand analysis? (3) How does the air pollution impact to a visual resource affect the behavior of a park user? and (4) How does a knowledge of behavior affect the judgment of whether the parks visitor enjoyment of an area has been compromised?

SCOPE OF VISUAL RESOURCE DEMAND ANALYSIS

Any management plan should reflect the demand for a visual resource in conjunction with other park attributes. When visiting a park, a user makes a set of activity decisions based on the availability of various park attributes such as camping, hiking, picture taking, attending lectures interpreting the park, and enjoying the natural scenic beauty of the area. A visual resource demand analysis then must take into account the demand for these activities, specifically the visual resources, as a function of visual air pollution impacts.

OBJECTIVE OF VISUAL RESOURCE DEMAND ANALYSIS

The FLM must concentrate on preserving the total experience of visitors to a natural area. A procedure for assessing the effectiveness of alternate visual resource management plans should include four steps.

1. Address the comparative importance of a visual resource to other resource attributes. For example, since most visitors to the Grand Canyon National Park expect to "see" the visual attributes of the canyon, visibility in this park is probably extremely important. On the other hand, a visitor to Carlsbad Caverns National Park would have as his first priority a tour of the caverns and the visual resource may have secondary importance.

2. Address the relative importance of one vista to another. Even though one vista may be more sensitive to incremental changes in air pollution than another vista, the more sensitive vista may not be as important to the visitor experience and hence may be valued less. It is therefore important to determine just which vista or vistas are important to the visitor's experience. Mt. Trumbull, located approximately 100 km from the Grand Canyon National Park subtends only a small percentage of the total viewing area and appears to be dark or black in contrast to the brightly colored geologic structures of the canyon. This mountain may not be an important element of the Grand Canyon visual resource. However, Navajo Mountain as viewed from Bryce Canyon is 180 km away from the observation point and tends to dominate the easterly view from Bryce Canyon and may be an important element of the visual resource at Bryce Canyon.

3. Address the value sensitivity of each vista deemed important to the visitor experience to the air pollution impact. Vista value sensitivity can be thought of as the value associated with perceiving a change in a given level of visual air quality. Each vista has its own sensitivity to an air pollution impact depending on whether it is highly colored or uniformly gray, far or near, etc. Each vista that is important to the visitor's experience will have to be examined in terms of its value sensitivity to increases in atmospheric pollutants. However, the key phrase is "deemed important to the visitor's experience." Not all possible vistas or scenic elements

viewed from a given observation point will have equal importance. Consequently not all vistas need to undergo a sensitivity analysis--only those vistas that are important for the public enjoyment of the area.

4. <u>Determine whether degradation of a visual resource affects the public enjoyment of an area.</u> When is degradation so adverse that a visitor alters his behavior pattern and chooses to spend less time viewing vistas and remain at a campsite reading a book; or when is the degradation of the visual resource so great that the visitor wants to change his whole recreation itinerary and go water skiing rather than visit another national park?

The procedure should be general enough to allow for a variety of resource problems, it should be transferrable to various geographic areas and it should be efficient in terms of yielding maximum forecasting credibility per dollar spent on data collection.

A BEHAVIOR APPROACH TO VISUAL RESOURCE DEMAND ANALYSIS

A behavior model is one which represents the decisions that a visitor will make when confronted with a number of alternative park attributes. The park visitor will choose between how much time he will spend camping, hiking, picture taking, etc. The concern here lies in how air pollution will effect the use of competing alternatives within the park system. It is hypothesized that the visitor will choose between these various park attributes in such a way as to yield the highest state of psychological and physical well being. This implies several assumptions which need to be examined.

Assumption 1: Rationality

It is assumed that the park visitor will try to maximize his enjoyment of the park resources and will allocate accordingly his time spent in various park activities. It is assumed that the visitor would not rationally choose an activity that results in increased psychological stress or in greater discomfort.

Assumption 2: Limited Resource

In order for a meaningful choice problem to exist, it is necessary to assume that the time available for a person to spend in the park is limited. With a fixed time base the park visitor must trade off the amount of time he spends at one park activity versus another.

Assumption 3: Perfect Knowledge

It is assumed that the traveler knows the characteristics of all the park activities available to him; he knows the time requirements, costs, and comforts associated with all the choices available to him.

Assumption 4: Preference Reflected Behavior

It is assumed that the choice of a given activity implies that he has considered the characteristics of all activities and has reached a conclusion that one set of activities is preferable; that is, <u>it is assumed that the amount of enjoyment that a park visitor extracts from an activity is reflected in the way he acts.</u>

This last assumption makes the link between behavior and public enjoyment of a scenic resource. If the "use" of a park involves viewing natural scenery and impaired visibility interferes with that use, then visitor behavior will change. People will do less viewing at an observation point and spend time doing something else because their visual enjoyment has been compromised.

ASSESSING THE DEMAND FOR GOOD VISIBILITY USING THE ORDER LOGIT PROBABILITY CHOICE TECHNIQUE

A number of alternative methods of measuring the value of atmospheric visibility have been suggested (Peterson 1979). Brookshire et al. (1981) and Rae (1981) have carried on a number of studies assessing economic value while Driver et al. (1979) has attempted to measure psychological benefits. This paper examines a method for assessing the public enjoyment of a park through the use of the ordered logit probability choice technique (Watson 1974). This approach is used to determine the value, measured in terms of time expended to see a vista, placed on good visual air quality.

When vacationing in a region, an individual makes a set of complex decisions with respect to how he wants to maximize attributes associated with a recreation area. Each visitor can be thought of as having a utility function, U, which represents the desirability of an option that consists of a number of recreation area attributes.

More specifically, consider a population of park visitors, S, and suppose it is possible to subdivide the visitors into groups, S_n, with each group described by a measured socioeconomic vector s_n. An individual, i, selected from this group can be described by a socioeconomic characteristics vector s^i where $s^i = s_n$. A set of alternatives, X, is available to the population of visitors. A subset of alternatives, X_i, is presented to individual i. Each alternative, j_s, is described by a vector of measured attributes x^{ji}. Let $E_{ji} = E(x_{ji}, s_i)$ be an unobserved vector, containing all the attributes of the alternative and characteristics of the individual which cannot be measured. The utility of the alternative j to individual i is a real number

$$U_{ji} = U(x^{ji}, s^i, E^{ji}). \tag{1}$$

U_{ji} is a utility function depending on each alternative in X and each individual in S. Now if it is assumed that individual i is picked randomly from the population S then he may be considered as being picked randomly from the group S_n. Therefore, the values of E^{ji} will be random and U will be a random variable. Thus, given the observed data (x^{ji}, s^i) consisting of the observed attribute of alternative j presented to individual i and the observed socioeconomic characteristics of individual i, a stochastic distribution of utility is induced by unobserved data in each trial of the experiment.

A convenient parameterization of random utility of alternative j to individual i is

$$U_{ji} = V(x^{ji}, s^i) + N(x^{ji}, s^i) = V_{ji} + N_{ji}. \tag{2}$$

N is a scalar valued random variable defined in a statistical sense for each alternative x^{ji} and each set S_n of individuals with measured socioeconomic vector s_n. V is a nonrandom variable defined over the same set as N. The number $V(x^{ji}, s^i)$ represents the average worth of the alternative j to the group of individuals described by socioeconomic characteristic vector $s_n = s^i$ and is called the utility of x^{ji} to the representative individual in S_n. Associated with the function V is a set of explanatory functions Z^k, $k = 1, ..., K$, where each Z^k is a scalar valued deterministic function defined over X and S. Since x^{ji} and s^i are measured quantities then $Z^k(x^{ji}, s^i)$ is a well defined real number for each alternative described by x^{ji} and faced by individual i in S_n. The functions Z^i convert the possibly subjective descriptions x^{ji} of the alternatives and socioeconomic descriptions s^i of the individuals into numbers $Z^k(x^{ji}, s^i)$.

It is assumed that representative utility can be approximated as the following linear function on R^K (K dimensional Euclidean space):

$$V(x^{ji}, s^i) = B_1 Z^1(x^{ji}, s^i) + ... + B_K Z^K(x^{ji}, s^i) = B'Z^{ji} \qquad (3)$$

where B' is the $B_1, B_2, ..., B_k$ parameter vector in R^K, and Z^{ji} is the $Z^1(x^{ji}, s^i)...Z^k(x^{ji}, s^i)$ vector of explanatory variables in R_K and where prime denotes the vector transpose operation.

The random term $N(x^{ji}, s^i) = U(x^{ji}, s^i, E^{ji}) - V(x^{ji}, s^i)$ essentially reflects the idiosyncracies of individual i in tastes for the jth alternative in X_i with measured attributes x^{ji}. We may assume that the mean of $N(x^{ji}, s^i)$ is zero since any non-zero term would be absorbed in representative utility $V(x^{ji}, s^i)$. Once a functional form for N_{ji} is chosen then the model probability of individual i choosing alternative x^{ji} can be determined to within the value of B.

The unknown parameter vector B must be estimated from sample data. Once it is determined to within a specified degree of statistical accuracy Equation (3) can be used to perform a trade-off analysis of the attributes of alternatives by holding utility constant while varying the numbers $Z^k(x^{ji}, s^i)$, $k = 1, ..., K$.

The sampling procedure considered here requires that I individuals be sampled where the ith individual rank orders the alternatives in X_i. The sample data therefore consist of a description of the set of alternatives X_i presented to individual i and his socioeconomic characteristics s^i in the form of the numbers: $Z^k(x^{ji}, s^i)$, where $i = 1, ..., I$; $j = 1, ..., J_i$; $k = 1, ..., K$ and the ranking assigned to each alternative j in X_i, R_{ji}.

In forming a probabilistic model of choice it is assumed that individual i will rank order alternatives presented to him in the order of how the individual perceives their relative value.

Let x_i^{rji} represent the description of the alternative ranked in jth place by individual i where i equals x^{mi} and where R_{mi} equals j,

$$Z^{rji} = \begin{bmatrix} Z^1(x^{rji}, s^i) \\ \cdot \\ \cdot \\ \cdot \\ Z^K(x^{rji}, s^i) \end{bmatrix} \quad \text{and}$$

$$U_{r_j i} = V(x^{rji}, s^i) + N(x^{rji}, s^i) = V_{r_j i} + N_{r_j i} \qquad (4)$$

The probability P_i that individual i ranks the alternatives in the order $x^{r_1 i}$ ... $x^{r_j i}$ is $\text{pr}\{U_{r_1 i} > U_{r_2 i} > \cdots > U_{r_{j} i}\}$. Using Equation (4) this probability can be written as

$$P_i = \text{pr}\left[N_{r_j i} < V_{r_{j-1} i} - V_{r_j i} + N_{r_{j-1} i}, \, j = 2, \ldots, J_i \right]. \quad (5)$$

Assuming a joint density function for the $N_{r_j i}$'s of

$$f(N_{r_1 i}, \ldots, N_{r_{J_i} i}) = \prod_{j=1}^{J_i} e^{-(N_{r_j i} + \alpha)} e^{-\left[e^{-(N_{r_j i} + \alpha)}\right]} \quad (6)$$

Equation 5 can be written as

$$P_i = \int_{-\infty}^{\infty} \int_{-\infty}^{V_{r_1 i} - V_{r_2 i} + N_{r_1 i}} \cdots \int_{-\infty}^{V_{r_{j-1} i} - V_{r_j i} + N_{r_j i}} f(N_{r_1 i} \ldots N_{r_{J_i} i}) \cdot dN \ldots dN_{r_j i} \quad (7)$$

Substituting Equations (3) and (6) into Equation (7) yields

$$P_i(B) = \prod_{j=1}^{J_i - 1} P_{jji}(B), \quad i=1, \ldots I \quad (8)$$

$$P_{i\ell j}(B) = \frac{e^{Z^{r_\ell i'} B}}{\sum_{m=j}^{J_i} e^{Z^{r_m i'} B}} \quad (9)$$

Equations (8) and (9) constitute the ordered multinomial logit model of consumer choice. The only unknown quantity in the model is the parameter vector B and it must estimated from sample data.

Sampling procedures provide data on the behavior of individuals in a hypothetical choice situation. The empirical estimates of B derived from the data depend greatly on the degree to which the hypothetical choice situation reflects individual choices in the real world. Let $R_i = (R_{1i}, \ldots, R_{J_i i})$ denote the rank ordering of individual i in the sample of I individuals. The vector R_i can be viewed as an independent drawing from a multinomial distribution where the probability of a given rank ordering is given by Equation (9). The probability of the group of I individuals choosing the sample $R = (R_1, \ldots, R_I)$ is therefore

$$P(R,B) = \prod_{i=1}^{I} P_i(B) \tag{10}$$

The method of maximum likelihood estimation argues that the calculated probability, Equation (10), of observing the given sample should be highest when the unknown B is near its true value and hence that a satisfactory estimate of B is a value B which maximizes $P(R,B)$ or, equivalently maximizes the log likelihood function $\log_e P(R,B)$. The function $L(R,B)$ is a nonlinear function on R^k possessing first and second order Frechet derivatives. The maximizing vector B, if it exists, is a value of B for which the first Frechet derivative is zero. Uniqueness of B is established if the k x k matrix, which is the second order Frechet derivative of $L(R,B)$, is positive definite. Under this condition B may be found by the Newton-Raphson method or modifications thereof.

By designing a survey that asks repondents to rank order desirability of alternate activities within the park context, it is possible to obtain data on the relative ordering of visitor choices that will allow estimation of B's and thus the utility functions. The maximum likelihood technique is usually used to extract these attribute parameters.

STUDY DESIGN

In order to assess the value that a park visitor places on various park attributes it is necessary to formulate a utility function which sufficiently addresses each variable of concern. The utility function could take the form

$$V = \sum_{n=1}^{m} Vis_n B_n + B_{m+1} t_v + \sum_{Q=1}^{l} Con_Q B_Q. \tag{11}$$

For purposes of analytic clarity the utility function of a park is constructed at the aggregate level. Thus the subscript i, which indexed individual utilities, has been dropped in Equation 3. The number of vistas is represented by m, Vis refers to a continuous variable which characterizes an existing physical state of the visual resource, and t_v is the amount of available time the park visitor is willing to give up to travel to a set of vista observation points. The model assumes a linear relationship between the B coefficients, Vis and t_v. Con is a congestion factor expressed in terms of waiting time. This may be waiting time to visit an archeological site, waiting time to hear a presentation by a ranger, or waiting time to enter the visitor center.

The visibility index Vis could be a discrete dummy index such as 0 for "clean" air and 1 for "dirty". The difficulty with building a model which relies on dummy variables is that it can never be universally applied to other situations without repeating a whole new study. A specific air pollution condition could be considered "dirty" in one park and "clean" in another. It would also be difficult to assess the change in utility for a visual air quality condition that is somewhere between the 0 and 1 levels. The point is that one does not know the general relationship between visual air quality and the dummy indices. Other choices for the visibility variable could be any of the various visibility related variables such as visual range, or extinction. As with the case of the dummy indices, the use of these variables is restricted by their nonlinearity with perceived changes in visual air quality. For instance, a change in visual range of 10 km in an atmosphere that is already limited to a 15 km visual range would be perceived to be different than that same change in a rayleigh atmosphere. If the visibility variable, Vis, is to be treated as a continuous variable a scale must be chosen such that a given change is perceived to be the same at either the high or low end of the scale. Only in this way is it possible to differentiate between changes in utility that take place as a result of changes in value as opposed to those that are perceptual in nature.

On-going research has shown that there exists a quantitative relationship between perceptions of visual air quality and air pollution (Malm et al. 1981). This relationship can be expressed as:

$$P = \sum_i f_i (P_{oi} - C) e^{-\Delta b_{ext} R_i} + C. \tag{12}$$

P is perceived visual air quality (PVAQ) measured on a 1 to 10 scale, f_i is the fraction of the total area subtended by the ith scenic element, P_{oi} is the inherent perceived visual air quality of the ith scenic element, Δb_{ext} is the increment of atmospheric extinction above rayleigh, R_i is the distance to the i_{th} scenic element and C is a constant.

In general, P_{oi} and C are constants that relate the past experience of an individual observer. That is, an observer's past experience determines the origin and set the "span" on his rating scale. If an individual is accustomed to seeing only polluted or haze scenes he will tend to be less sensitive to the effect of air pollution on a Grand Canyon vista, than a person living in the Grand Canyon area.

In previous NPS perception studies (Malm et al. 1981) observers were essentially calibrated by showing them ten "preview" slides which sensitized them to the worst and best conditions typically found in a given area. The preview slides form the boundary conditions that determine the maximum and minimum rating an observer will assign to best and worst air quality conditions. In most cases, once C and P_{oi} are determined, Equation 12 describes the relationship between PVAQ and extinction coefficient. However, it does not allow for a determination of the value that an observer places on a given landscape feature. In the example above, the observer rated the same scene differently depending on how he was "calibrated." Does this mean that he valued that scene less in the one circumstance and if so, what sort of commodity is he using to set his scale? The point is that Equation 12 describes the perceptual relationship between air pollution and PVAQ but does not necessarily assign a value to a given change.

Equation 12 can be used in conjunction with Equation 3 to describe the utility function associated with a scenic resource as a function of the continuous perception variable P.

$$V = \sum_{n=1}^{m} P_n B_n + B_{m+1} t_v + \sum_{Q=1}^{\ell} Con_Q B_Q \tag{13}$$

If for example there are three separate vistas and one congestion factor, i.e., waiting time to see an archaeological site, Equation 13 becomes:

$$V = P_1B_1 + P_2B_2 + P_3B_3 + B_4t_v + B_5C_A \tag{14}$$

where:

$$P_k = \sum_{i_k=1}^{j_k} f_{i_k}(P_{oi_k} - C_k) e^{-\Delta b_{ext}R_{i_k} + C_k}, \quad k=1,2,3 \tag{15}$$

j_k, k=1,2,3 refer to the number of unique landscape features in each of the three vistas. The constants C_1, C_2, and C_3 are adjusted so that P_1, P_2, and P_3 will span the whole PVAQ interval for a given extinction coefficient increment. For southwestern vistas P_1 might be chosen to vary from 1 to 10 for aerosol extinction coefficients varying between 0.0 km^{-1} and 0.05 km$_{-1}$. Since the P_i's all vary from 1 to 10 over the same air pollution increment the magnitude of the B coefficients are a direct measure of the relative value that a visitor places on a perceptual change in one vista as compared to another.

In general it is possible to define a Value Index of Sensitivity (VIS) as

$$VIS_n = \frac{\partial V}{\partial P_n} \frac{\partial P_n}{\partial \Delta b_{ext}} = \sum_{m=1}^{m} B_n \frac{\partial P_n}{\partial \Delta b_{ext}} . \tag{16}$$

Specifically for Equation 15

$$VIS_k = B_k \sum_{i_{l=k}}^{j_k} f_{i_k} R_{i_k} (P_{oi_k} - C_k) e^{-\Delta b_{ext}R_{i_k}}, \quad k=1,2,3 \tag{17}$$

The VIS_n's are dependent on the amount of background air pollution, distance to scenic elements, inherent scenic beauty of various landscape features, and size of each landscape feature. Most importantly these indices are true value sensitivities in that they not only account for the observer's perceptual sensitivity to changes in visual air quality but also take into account the value that an observer places on a given perceptual change.

It is possible to go a step further and calculate a park value sensitivity:

$$VIS_p = \frac{\partial V}{\partial \Delta b_{ext}} = \frac{\partial V}{\partial P_1} \cdot \frac{\partial P_1}{\partial \Delta b_{ext}} + \frac{\partial V}{\partial P_2} \cdot \frac{\partial P_2}{\partial \Delta b_{ext}} + \frac{\partial V}{\partial P_3} \cdot \frac{\partial P_3}{\partial \Delta b_{ext}} \tag{18}$$

where the subscript P on VIS_p stands for park rather than an individual vista sensitivity. Through the use of VIS_p it would be possible to compare parks and determine their relative sensitivity to degradation in the visual resource.

This formulation of a park utility function also allows for assessment of the actual value, in terms of time, that a visitor places on a given visual resource as a function of an increase in air pollution.

One can ask the question, "How much less time will a visitor be willing to spend traveling to see some vista under increased air pollution conditions and still maintain an unchanged utility?" That is

$$t_{v_k} = \frac{B_1}{B_4} \left[\sum_{i_k=1}^{j_k} f_{i_k}(P_{oi_k} - C) \left[e^{-\Delta b_{ext} R_{i_k}} - e^{-\Delta b'_{ext} R_{i_k}} \right] \right] \quad (19)$$

$$k = 1,2,3$$

t_{v_1}, t_{v_2}, and t_{v_3} are the decreases in time an individual would be willing to expend traveling to a specific scenic overlook as a function of a change in air pollution level from Δb_{ext} to $\Delta b'_{ext}$. These times are a direct measure of the value that individuals place on a specific degradation of a visual resource that results from an increase in air pollution. The total park value is just the sum of the individual times:

$$t_v = t_{v_1} + t_{v_2} + t_{v_3}. \quad (20)$$

VIS's and t_v's, when established for a number of parks, would allow for a comparison of the value that visitors place on maintaining a level of air quality (as it pertains to the visual resource) within the boundaries defined by park vistas.

The utility function specified by Equation 14 can also be used to assess the relative importance of the visual resource as compared to other park resources. Instead of trading visual degradation of a scenic resource for travel time to an observation point where the scene can be seen, it is possible to compare travel time and waiting time, expressed as a congestion factor, to visit, say an archeological site. In terms of the example represented by Equation 14.

$$t_v = \frac{B_5}{B_4} C_A. \quad (21)$$

This time can be directly compared to the time expressed in Equation 20 and thereby establish the relative importance of observing the scenic resource as compared to the importance of visiting an archaeological site. This can also be used to formulate a visual resource demand function:

$$VISD = \int_{t_1}^{t_2} \frac{\partial V}{\partial \Delta b_{ext}} N(t) dt \quad (22)$$

where $\partial V/\partial \Delta b_{ext}$ is the park value sensitivity and $N_{(t)}$ is the number of people present in the park as a function of time. V is also a function of time because it is dependent on the amount of background air pollution which in turn is dependent on time. The integral is over a time period of interest.

This visibility demand function factors in the vista physical sensitivity, the values that an observer places on that visual resource, as well as the number of people who might actually see the vista. For instance, a vista could be very sensitive to air pollution and be valued as an important scenic resource, and yet if visitors did not visit the area the demand function would be very low.

The visual resource demand function can be used in the decision making process that determines whether a given increment in air pollution would significantly affect public enjoyment of a natural area. If the demand function is unaltered by air pollution, the impact would certainly be judged not to be adverse. However, the amount of alteration in the demand function that constitutes an adverse impact would still have to be a subjective decision. The demand function of the value sensitivity analysis would provide a common ground for value assessment. It would allow for assessing the physical sensitivities of a resource to an increase in air pollution, the value an observer places on maintaining a certain level of visual air quality, the comparative value that an observer places on the scenic resource as compared to other park attributes and finally the demand for the scenic resource attributes.

CONCLUSION

This methodology, based on the ordered logit probability choice technique, allows for a determination of how a visitor will alter his behavior, under a prescribed set of circumstances, as a function of changes in air pollution. It is argued that a visitor will actively choose between various park attributes in such a way as to yield the highest state of psychological and physical well being. It is assumed that the amount of enjoyment that a park visitor extracts from a set of activities is reflected in the way he acts. If a visitor is willing to spend less time traveling to a scenic overlook because the air is more polluted, his enjoyment of the park has been compromised.

If the survey is carried out in a standardized, prescribed way this approach can be used to 1) determine the relative importance of visual resource as compared to other park attributes, 2) assess the relative importance of one vista as compared to another, 3) determine the value sensitivity of each park vista, and 4) determine whether a change in visual air quality affects public enjoyment of a national park. Finally a visual resource demand function, VISD, is defined. This function encompasses the physical sensitivity of the vista to changes in air pollution, the value placed on a given change in visual air quality, as well as the number of people who might actually see the vista. The VISD index can be employed by the informed decision maker to determine whether a given increase in air pollution is adverse to public enjoyment.

BIBLIOGRAPHY

Barker, M.L. 1976. "Planning for Environmental Indices: Observer Appraisals of Air Quality." In Perceiving Environmental Quality, K. H. Craik and E. H. Zube (Eds.), Plenum Press, New York, NY.

Brookshire, D., R. d'Arge, W. Schulze, and M. Thayer. 1981. "Experiments in Valuing Public Goods." Advances in Applied Microeconomics. V. Kerry Smith (Ed.), JAI Press, Greenwich, Conn., 1: 123-172.

Driver, B.L., D. Rosenthal, and L. Johnson. 1979. "Air Pollution, Values, and Environmental Behavior." In <u>Proceedings of the Workshop on Visibility Values,</u> D. Fox, R.J. Loomis, and T.C. Green (Eds.), USFS Tech. Report WO-18, Fort Collins, CO.

Henry, R.C. 1977. "The Application of the Linear System Theory of Visual Activity to Visibility Reduction by Aerosol." <u>Atmospheric Environment</u> 11: 697-701.

Hummel, C., R. Loomis, and J. Herbert. 1975. "Effects of City Labels and Cue Utilization on Air Pollution." Working Papers in Environmental-Social Psychology <u>1</u> Department of Psychology, Colorado State University, Fort Collins, CO.

Malm, W.C., K. Kelley, J. Molenar, and T. Daniel. 1981. "Human Perception of Visual Air Quality (Uniform Haze)." <u>Atmospheric Environment</u> 15: 1875-1890.

Peterson, G.L. 1979. "Atmospheric Visibility Assessment." In <u>Proceedings of the Workshop on Visibility Values,</u> D. Fox, R.J. Loomis, and T.C. Green (Eds.), USFS Tech. Report WO-18, Fort Collins, CO.

Rae, D. 1981. <u>Visibility Impairment at Mesa Verde National Park: An Analysis of Benefits and Costs of Controlling Emissions in the Four Corners Area.</u> Charles River Associates, Inc. Report to the Electric Power Research Institute. Boston, MA.

Schulze, W., R. d'Arge, and D. Brookshire. 1981. "Valuing Environmental Commodities: Some Recent Experiments." <u>Land Economics</u> 57 (May): 151-172.

Trijonis, J., and K. Yuan. 1978. <u>Visibility in the Southwest.</u> EPA-600/3-78-039, U.S. EPA, Research Triangle Park, NC.

U.S. Congress. 1916. <u>Organic Act.</u> August 25, 1916 39 stat 535, USC title 16. Washington, DC.

Watson, P.L. 1974. <u>The Value of Time: Behavioral Models of Modal Choice.</u> D.C. Heath and Company, Lexington, MA.

17. Potential Contributions of Canonical Analysis to Visual Value Research

Thomas Buchanan
Marcia J. Hayter
Jacquelin P. Buchanan

INTRODUCTION

Recreation resource planners and managers have for many years recognized the importance of aesthetic dimensions in the provision of recreation opportunites (Buyoff and Wellman 1979). While landscape beauty and aesthetics have been the focus of considerable management attention, such factors have often been considered intangible or at the very least, difficult to conceptualize and measure. Early research attempts toward understanding the nature and role of aesthetics in the provision of recreation services examined a variety of issues, including the importance of physical site characteristics, man-made intrusions, visitor induced resource damage, and environmental pollutants. All were examined with specific attention given to how these factors influenced the quality of outdoor recreation.

The need to understand the role of aesthetics and landscape beauty was enhanced by environmental legislation passed during the 1970s. One particularly important piece of legislation, and a major motivating force behind many of the issues being raised in this volume, was the Clean Air Act of 1970. Originally intended to improve air quality in already polluted areas, this legislation did not protect areas characterized by exceptionally high air quality (Flachsbart 1979). Consequently, as a result of successful litigation brought before the U.S. Supreme Court by the Sierra Club, amendments to the original Clean Air Act were added in 1977.[1] The purpose of these amendments is to protect, preserve, and enhance air quality in Class I areas.[2] The rationale for these amendments, at least in part, was that if visual impairment inhibits our ability to enjoy the recreation opportunities provided by Class I areas, then the values which originally motivated their designation would somehow be reduced (Daniel 1979).

Research inspired by this legislation has helped responsible individuals and agencies to understand visual values and their importance to Class I areas. Engineers, chemists, and atmospheric scientists have developed

---

Thomas Buchanan is with the Department of Recreation and Park Administration, University of Wyoming, Marcia J. Hayter is with the Department of Geography, University of Wyoming and the Laramie Energy Technology Center, and Jacquelin P. Buchanan is with the Water Resources Research Institute, University of Wyoming.

increasingly sophisticated technology to measure, monitor, and identify factors impairing visual quality. Environmental psychologists, physiologists and other social researchers have made great strides in understanding environmental perception. Still other social researchers have studied recreation and visual values from the perspective of the influence that visual impairment has on the outdoor recreation experience. It is this latter perspective that we would like to address in this paper.

UNDERSTANDING THE RELATIONSHIPS BETWEEN VISUAL VALUES AND RECREATION OPPORTUNITIES

Our approach to visual value research is based upon the assumption that recreation is not an activity or an environment, but rather a goal-directed type of human behavior. The basis for this approach is an outgrowth of expectancy value theory and a need satisfaction model of leisure behavior, which is well supported in existing research (Fishbein and Ajzen 1975, Driver and Tocher 1970, Tinsley and Kass 1978).

As we see it, one of the main issues facing the assessment and management of visual values is that in order to assess the value of visibility, we must understand human response to a wide variety of visual factors. Although this topic has been the focus of recent research, the quantification of human response to visual values is in general less well developed than other more technical aspects of visibility research, such as pollution assessment. Additionally, the range of issues which must be researched to understand the interface between pollution, environmental characteristics, and human response do not lend themselves well to short concise management applications.

One need only examine the technical literature on visual values and their management to recognize that a complex array of variables are involved (Carls 1974). It is readily apparent that in order to understand the social costs of visual impairment, it will be necessary to understand not only what is lost as a result of visual impairment but also the magnitude of different types of impacts, the effectiveness of alternative mitigation strategies and the extent to which different publics are affected. The complexity of the issue is a result not only of the large number of variables involved, but also of the complex interactions between these variables. As suggested by Shafer (1969), one of the main problems facing perception research is that there may be no simple additive effect of individual visual variables. What is needed, at least in part, is to better integrate such factors as human response to visual values and environmental preference with increased attention to the connections, strengths, and weaknesses of each. While each individual component may require detailed analysis and study, we must also recognize that, in a practical sense, all these forces operate as one general system (Lowenthal 1972).

POSSIBLE RESEARCH APPROACHES

Based upon a review of existing research and as a result of considerable discussion with interested researchers and managers, we have recognized that visual value research should be addressed from two perspectives. The first perspective we label the "micro" approach. "Micro" research is by necessity the initial concern of visual value research. It includes, among other things, the identification and measurement of specific relevant visual variables and their relationship to visual impairment. We equate this approach to some extent with the "theory of little things" suggested by Driver et al. (1979). The theory of little things suggests (in part) that because visual impairment is a gradual and interactive process resulting from the combined effect of pollution variables, attention must be given to each potentially detrimental element. We agree with this approach and suggest that it is the "micro" approach which has currently made the most substantial contributions to visual research.

We would also suggest, however, that there is a need for a "macro" approach to visual value research. This "macro" approach examines the complex interaction of human and environmental variables that influence visibility values. While each individual piece of "micro" research is important, it is extremely difficult (if not impossible) to combine the results from these individual efforts into the multidimensional taxonomic models needed to understand and manage the entire process in which we are interested. In addition to continued work in the "micro" research component, we see a strong need for increased "macro" research that searches for patterns and consistency in relationships between recreation behavior and visual impairments across individuals and locations. One specific analytic technique with which we have been experimenting and which we feel may offer important opportunities for understanding complex interrelationships in visual research is canonical correlation analysis.

MECHANICS OF CANONICAL CORRELATION ANALYSIS

At a very generalized level, canonical correlation analysis can be viewed as a type of regression analysis (Horst 1961). Its value, however, results from its ability to examine more complex interrelationships than can be considered by multiple regression. Specifically, canonical analysis can be used to examine the interrelationships between two sets of measurements made on the same individuals (Cooley and Lohnes 1971).

The basic task of canonical analysis, to determine the pattern of correlations between the x set and y set of variables, is accomplished by replacing the variables in each set with pairs of linear combinations of the original variables. For example, if the x set of variables is represented by $(x_1\ x_2\ x_3 \ldots x_p)$ then a linear combination of those variables $(\underline{x})$ $a_1x_1 + a_2x_2$ can be express as:

$$\underline{x} = a_1x_1 + a_2x_2 + a_3x_3 \ldots a_px_p \tag{1}$$

Where a_p are coefficients that represent the strength of the contribution each x_p makes to the linear composite $\underline{x}$. Therefore, the linear composite for the x set of variables can be expressed as Σa_px_p and the linear composite for the y set of variables can be expressed as Σa_py_p. Canonical analysis produces linear combinations of each variable set in such a way that the linear composite of the x set is maximally correlated with the linear composite of the y set, and the linear composite of the y set is maximally correlated with the linear composite of the x set. The correlation between the linear composite $\underline{x}$ and the linear composite $\underline{y}$ is called a canonical correlation (R_c) and is the highest value attainable.

It is possible, however, that there are additional pairs of linear composites whose canonical correlations are significantly greater than zero. Canonical analysis also produces the second most highly correlated linear composite for both the y and x sets of variables. An additional constraint is imposed on this and all subsequent linear composites, however, which requires that they be uncorrelated with all preceding sets of linear composites. The procedure identifies and allows for the examination of independent patterns of relationship between two sets of variables. The number of possible canonical correlations is equivalent to the number of variables in the smallest set.

ADVANTAGES OVER TRADITIONAL TECHNIQUES

We believe that the ability of canonical correlation analysis to examine the relationship between two sets of variables provides several advantages over more traditional statistical techniques. Traditional regression techniques, for example, regress each criterion variable on a set of predictor variables. This type of statistical procedure is based on the assumption that the criterion variables are independent of each other. This assumption is generally unsubstantiated and often untenable, particularly in social research. In addition, traditional regression techniques do not allow the researcher to examine combinations of criterion variables simultaneously, and consequently it is impossible to analyze relationships between sets of variables. The ability of canonical correlation analyses to examine relationships between sets of variables is precisely the characteristic that makes it an important tool for visual research. From a more pragmatic perspective, canonical analysis provides information regarding:

1. The nature of the links or patterns of interdependency that join two sets of variables.
2. The number of (statistically significant) links between the two sets of variables.
3. The extent to which the variability of one set is dependent upon or redundant given the other set (Levine 1977).

POTENTIAL APPLICATION TO VISUAL RESEARCH

To illustrate this technique, we are using data gathered from backcountry users in Rocky Mountain National Park during the summer of 1980. A series of scales designed to measure the perceived psychological benefits (PPBs) of backcountry camping were included in a questionnaire administered to individuals as they left backcountry areas. In addition, a series of scales designed to measure the extent to which individuals perceived specific biophysical impacts (e.g. litter, vegetation trampling, etc.) were included in the questionnaire. Although the purpose of this study was to measure perceived biophysical impacts, visual impairment variables might just as easily might have been included.

The canonical analysis was performed using PPBs as the criterion set of variables and perceived biophysical impacts were used as the predictor set.[3] Results from the analysis are presented in Table 17.1.

The overall results from the canonical analysis indicated two significant canonical functions. Wilk's lambda indicates the strength of those functions; it is inversely related to the strength of the canonical function, so the smaller lambda is, the stronger the relationship between the two sets of variables. The canonical correlation squared (R_c^2) is defined as the variance shared by <u>linear composites</u> of two sets of variables. R_c^2 cannot, however, be interpreted as the variance shared by two sets of variables. The canonical correlation is actually a Pearson correlation coefficient be-

TABLE 17.1
Canonical Analysis Results

	First Canonical Variate	Second Canonical Variate	Third Canonical Variate
Canonical Correlation	.502	.478	.451
Chi-Square	277.141	218.011	165.102
Degree of Freedom	198.	168.	140.
P <	.001	.005	.071
Wilk's Lambda	.256	.342	.444

tween two linear composites and thus represents the amount of overlap between composites (Levine 1977).

Table 17.2 presents structure matrix coefficients which are used to identify those PPBs and biophysical impacts that contribute most substantially to each of the canonical functions. This procedure is generally necessary because the standardized coefficients produced by canonical analysis are often difficult to interpret if multicollinearity exists between variables in each set. This situation is analagous to the problem of interpreting beta

TABLE 17.2
Structure Matrix Coefficients

PPBs		R_c^2	Biophysical Impacts	
Nature experience	+.408		Litter	-.640
Social contact	-.214	→ .502 ←	Horse waste	-.424
Solitude/Privacy*	+.198	p< .001	Soil erosion/ Compaction	-.685
Slow down mentally	-.278			
Escape daily routine	+.285			
Solitude/Privacy	+.292	.478	Horse waste	-.769
Security	-.315	→ ←	Trail cutting	-.411
Family togetherness	-.448	p< .005		
Leading others	-.369			
			Litter	-.279
			Human waste	+.226
Nature experience	-.487		Horse waste	-.241
Escape daily routine	-.260	→ .451 ←	Soil erosion/	
Enjoy open space	-.295	p< .070	Compaction	+.285
			Veg. trampling	+.421
			Campfire rings	+.248

* Did not quite load on the canonical function using Tatsuoka's (1970) cut-off rule but was extremely close.

weights when predictor variables are intercorrelated in multiple regression. For this reason, a structure matrix was formed and each variable was correlated with the vector of the matrix that represented the canonical function (Cooley and Lohnes 1971).

DISCUSSION

It would be naive to assume, based upon existing research, that any single instrument will adequately predict the social costs of visual impairment in different situations at different times and under different environmental contexts. It does seem possible, however, that multivariate procedures such as canonical correlation analysis can be useful in identifying patterns or tendencies that will help us anticipate the effect of visual impairment and more clearly understand the relationship between various component parts.

Some specific instances in which canonical correlation analysis might be helpful include:

1. Examining relationships between PPBs and alternative mitigation techniques. In situations where several different mitigation strategies are feasible, this type of analysis might prove useful in selecting that procedure which will have the least detrimental effect on the quality of the recreation experience.
2. Examining relationships between impairment variables such as atmospheric transport, dilution, and deposition losses and specific environmental attributes such as degree of surface variation, atmospheric variations, and types of ground cover. The analysis might prove useful in developing some type of continuum capable of identifying resource types susceptible to varying degrees of impairment.
3. Examining relationships between physical resource characteristics and PPBs. While we know some work has been done in this area, the importance of the topic merits continued research.
4. Examining relationships between human perception and background characteristics such as length of stay, amount of prior use of a site, age, sex, education, and other factors which might influence perceptual measures of scenic beauty of visual impairment

These are only a few of the possible analyses which might prove interesting. Certainly, it is important to continue to improve our measurement instruments and to continue to isolate specific variables relevant to understanding visual values. Results will have to be checked and rechecked using different procedures at different times and in different settings. If, however, we can identify and develop models that will help us understand the social impacts of visual impairment, then we feel confident that strategies can be developed to minimize the social costs of unavoidable visual impairments.

NOTES

1. Clean Air Act Amendments of 1977, Public Law 95-95, Section 160, 169a.

2. International parks, wilderness areas, and memorial parks with more than 5000 acres, national parks with more than 6000 acres.
3. Although distinguishing between predictor and criterion sets may be useful for the researcher, this distinction is not statistically necessary.

BIBLIOGRAPHY

Buhyoff, G.J., and J.D. Wellman. 1979. "Seasonality Bias in Landscape Preference Research." Leisure Sciences 2(2):181-189.

Carls, E.G. 1974. "The Effects of People and Man-Induced Conditions on Preferences for Outdoor Recreation Landscapes." Journal of Leisure Research 6(2):113-124.

Cliff, N., and D.J. Kurs. 1976. "Interpretation of Canonical Analysis: Rotated vs. Unrotated Solutions." Psychometrika 41(1):35-42.

Cooley, W.W., and P.R. Lohnes. 1971. Multivariate Data-Analysis. Wiley Publishers, Inc., New York, NY.

Daniel, T.C. 1979. "Psychological Perspectives on Air Quality and Visibility in Parks and Wilderness Areas." In D. Fox, R.J. Loomis, and T.C. Green (Eds.), Proceedings of the Workshop in Visibility Values. U.S. Forest Service General Technical Report WO-18. Fort Collins, CO.

Driver, B.L. and R. Tocher. 1970. "Toward a Behavioral Interpretation of Recreation Engagements with Implications for Planning." In B. Driver (Ed.), Elements of Outdoor Recreation Planning. University of Michigan Press, Ann Arbor, MI.

Driver, B.L., D. Rosenthal, L. Johnson. 1979. "A Suggested Research Approach for Quantifying the Psychological Benefits of Air Quality." In D. Fox, R.J. Loomis, and T.C. Green (Eds.), Proceedings of the Workshop in Visibility Values. U.S. Forest Service General Technical Report WO-18. Fort Collins, CO.

Flachsbart, P.G. 1979. "A Framework for Assessing Human Responses to Proposed Impairments of Atmospheric Visibility in Class I Areas." In D. Fox, R.J. Loomis, and T.C. Green (Eds.), Proceedings of the Workshop in Visibility Values. U.S. Forest Service General Technical Report WO-18. Fort Collins, CO.

Fishbein, M., and I. Ajzen. 1975. Belief, Attitude, Intention, and Behavior: An Introduction to Theory and Research. Addison-Wesley Publishing Company, Reading, MA.

Horst, P. 1961. "Relations Among M Sets of Measures." Psychometrika 26 (June):129-149.

Levine, M.S. 1977. "Canonical Analysis and Factor Comparisons." In E. M. Uslaner (Ed.), Quantitative Applications in the Social Sciences. Sage Publications, Inc., Beverly Hills, CA.

Lowenthal, D. 1972. "Research in Environmental Perception and Behavior." Environment and Behavior 4(3):333-342.

Shafer, E.L. 1969. "Perception of Natural Environments." Environment and Behavior 1(1):71-82.

Tatsuoka, M. 1970. Advanced Topics in Statistical Analysis. Educational Testing Services, Inc. Champaign, IL.

Tinsley, H., and R. Kass. 1978. "Leisure Activities and Need Satisfaction: A Replication and Extension." Journal of Leisure Research 10(3):191-202.

18. Altering the Visual Quality of a Recreation Resource and Activity Displacement

Rabel J. Burdge
Leo McAvoy
James Absher
James H. Gramann

This research fits into the discussion of visual values by providing a methodology and data on how perceptions of changes in the physical recreation resource will, in turn, produce changes in recreation behavior. The recreation resource can change through site deterioration, over crowding, increased competition from non recreational uses, and, of course, visible deterioration in air and water quality. On the other hand, a recreation resource may be upgraded by the elimination or rehabilitation of the deteriorating effects.

The management question as always is which activities and recreation experiences will be altered or displaced as the nature of the recreation resource changes. If the water in a reservoir becomes visibly dirtier, swimmers may go elsewhere or choose other activities. On the other hand, neither the experience nor the location of fishing from a boat may change. The dirty water on the St. Croix River displaced the canoeists to the upper reaches, but the power boaters soon moved in on the navigable stretches and they now seem happy. This research paper deals with two somewhat related questions. First, what visual site characteristics seem important for selected recreation activities, and secondly, can alterations in the recreation resource lead to activity displacement? The methodology is immediately applicable to issues of scenic quality at national parks and wilderness areas.

Rabel Burdge is Professor of Environmental Sociology, Rural Sociology and Leisure Studies, University of Illinois, Champaign Urbana, IL. Leo McAvoy is Associate Professor of Parks and Recreation, University of Minnesota, Minneapolis, MN. James Absher is Assistant Professor of Leisure Studies and Institute for Environmental Studies, University of Illinois, Champaign Urbana, IL. James H. Gramann is Visiting Assistant Professor of Forestry, University of Wisconsin, Madison, WI. Data for this paper came from Research Project No. 884-305 between the River Recreation Research Consortium and the Upper Mississippi River Basin Commission. Project title was: "Evaluation of Impacts of Navigation and Associated Operation and Maintenance on Recreation, Potential Wilderness and Cultural Resources of the Upper Mississippi River Basin." Other principal investigators and members of the research consortium were: Robert Becker, Clemson University; Thomas Bonnicksen, University of Wisconsin.

RESEARCH SETTING

Our research focuses on the wilderness, cultural, and recreational resources of the navigable portions of the Upper Mississippi River System (UMRS) which extends from Cairo, Illinois, on the south to slightly north of Minneapolis, Minnesota. The Illinois River, which is also part of that navigational system, extends from Alton, Illinois, to Chicago. Other portions include about fifty miles of the Kaskaskia River in southern Illinois, and the Black, Minnesota, and St. Croix Rivers.

The Army Corps of Engineers is in the process of widening and lengthening the existing Lock and Dam No. 26 at Alton, Illinois, near the confluence of the Illinois and Mississippi Rivers. In addition, a second lock and dam will be built in the middle of the river at that same location. When both locks are complete and operational, it is estimated that the number of barge tows will increase 50 to 100 percent over the present levels. That increased traffic, along with associated operation and maintenance activity will alter every aspect of the Upper Mississippi recreation resource. In addition to more and larger barge tows, we will see increased numbers of fleeting areas, more dredging activity, lockage delays upstream, dirtier water, and more water fluctuation as the required channel depth is maintained.

This paper does not deal with the visual aspect of air quality as the independent variable in producing change in recreation activity. We assume that the changes in the recreation resource, brought about from increased commercial use, would require the same type of analysis concerning consequences for activities and experiences whether or not the alteration is air or water quality; although, the nature of change in the resource would affect the response. Consequently, the methods used in this research are of interest to those analyzing air quality impacts.

METHODS

Data for this paper came from two sources: persons that visit the UMRS for recreation, called here either recreationists or users; and the managers or owner/operators of public recreation facilities and recreation support services along the UMRS. The latter group is simply called manager/providers. During the spring and summer of 1981, survey crews contacted recreationists at presampled locations and asked them to complete a self-administered questionnaire. A questionnaire including the same items was dropped off by a separate crew to each manager/provider along the 1200 plus miles of the UMRS. On-site contacts were made with 1914 recreationists. Of those, 335 refused to complete the questionnaire or did not complete the instrument satisfactory for analysis. The resulting data set of 1579 cases yielded an 82.5 percent response rate. For the manager/provider survey, 366 were dropped off. With one telephone followup, 258 of those were returned, producing a response rate of about 70 percent.

SITE CHARACTERISTICS AND RECREATION ACTIVITY

The first phase of the analysis attempts to establish a linkage between visual characteristics of recreation sites and outdoor recreation activities common to the UMRS. Each of the recreationists contacted was given a list of fifteen characteristics important in choosing a recreation site along the river. Six of the site characteristics had visual attributes--nice vegetation, sandy beach, view or scenic quality of the site, absence of "no trespassing" signs, and not crowded. The overall ranking for all respondents for all fifteen site characteristics is shown in Table 18.1. This paper focuses on selected characteristics which deal with the visual aspects of the site selection.

Table 18.2 shows the percent of the recreationists that reported doing each of fifteen outdoor recreation activities during the sample year and a ranking for the river related activities they did the day they were interviewed. The most popular activities for the recreationists were river watching (sightseeing), swimming, picnicking on shore, power boating, and fishing from the shore. This observation reflects the reality that the UMRS is a water-based facility centering around boating and swimming. Table 18.3 shows the percentage of each of the recreation facilities inventoried along the UMRS that either offer a specific recreation opportunity or have that activity on site. In the second column of Table 18.3, the manager/providers are ranked according to their personal river related recreation activities. The manager/providers have use patterns very similar to the recreationists with fishing from the boat ranked number one, followed by power boating, river watching, fishing from the shore and water skiing. Other water-based and water-enhanced activities were included in the questionnaire, but not reported here.

Table 18.4 shows the relationship between each of the fifteen river-related recreation activities and how the recreationists ranked five characteristics used in site selection. The numbers refer to the percentage of the respondents that ranked that site characteristic either first, second, or third. The percentage is compared using a chi-square test between recreationists that did and did not do the activity. For example, nice vegetation was ranked as one of the top three characteristics by 8.6 percent of the respondents that reported river watching as an activity. This is slightly higher than those who did not report river watching as an activity. Reading across Table 18.4 we see that sandy beaches are less important, while the view of scenic quality of the area is significantly more important to river watchers, compared to recreationists who did not report that activity. Whether or not the area lacked barge fleeting areas was not an important site characteristic; while not being crowded was more often selected by river watchers. The strategy here is to establish a relationship between activities and various scenic qualities of the recreation site. It may be that certain activities or clusters of activities are done despite the visual aspects of the recreation resource.

View or Scenic Quality

The "view or scenic quality" of the area was ranked as important by persons reporting river watching, picnicking on shore, hiking, bicycling, bird watching, sailing, and to some extent, canoeing. These activities form a different cluster than the fishing, swimming, water skiing, boating syndrome reported for "sandy beaches." While the view or the scenic quality of the area was ranked as important to many respondents, those who ranked it higher were more likely to report involvement in water-enhanced rather than water-based activity.

TABLE 18.1
Ranking of Site Characteristics Used in Selecting Place for Recreation Activity (N=1545)

Site Characteristics	Overall Rank
Sandy Beach*	1
View or Scenic Quality*	2
Good Fishing*	3.5
Not Crowded*	3.5
Adequate Water Depth for Boat	5
Safe Swimming Location	6
Marinas/Docks/Slips	7
Boat Ramps	8
Nice Vegetation*	9.5
Bait Shops or Grocery Nearby	9.5
Developed Camping (full hook-up)	11.5
Simple Tent/Primitive Camping	11.5
Lack of Fleeting Areas/Barge Traffic	13.5
Absence of "No Trespassing" Signs	13.5
Proximity to Lock	15

* Items analyzed in this paper

TABLE 18.2
Percentage of Respondents Having Done Fifteen River Related Recreation Activities During Sample Year, 1981

Recreation Activity	Percent Having Done Activity During Sample Year (N=1,555)	Ranking of Main Activities on Day Interviewed
River watching (sightseeing)	66.7	2
Swimming	62.6	3
Picnicking on shore	61.2	4
Power boating	57.1	1
Fishing from boat	53.6	5
Fishing from shore	49.1	7
Water skiing	41.3	6
Picnicking on river island	40.6	9
Camping on shore	40.4	8
Camping on river islands	33.7	10
Hiking	28.5	11
Bicycling	26.8	14.5
Bird watching	21.0	14.5
Canoeing	20.5	13
Sailing	13.9	12

TABLE 18.3
Percentage of Recreation Facilities on the UMRS Providing Each of Fourteen River Related Activities (N=255) and Ranking of Manager-Provider Outdoor Activities (N=226)

Recreation Activity	Percentage of Sites with Activity	Ranking of Manager/Provider Most Frequent River Related Recreation Activities
Fishing from boat	94.9	1
Power boating	90.6	2
Fishing from shore	89.0	4
River watching (sightseeing)	82.0	3
Water skiing	81.2	5
Picnicking	76.9	6
Camping on river islands	73.7	7
Canoeing	69.0	11.5
Camping on shore	68.6	9.5
Swimming	65.5	9.5
Bird watching	64.7	8
Sailing	47.5	13
Hiking	46.3	11.5
Bicycling	42.7	14

Site Vegetation

"Nice vegetation" as a site characteristic was found to be important for picnicking on shore, hiking, bicycling, and bird watching and found not important for swimming, boating, fishing, water skiing, and camping. Nice vegetation seems important for activities which are water or river-enhanced rather than those that are water-based.

Sandy Beaches

"Sandy beaches" was seen as an important site characteristic for swimming, picnicking on shore, power boating, fishing from a boat, but not on shore, picnicking on river islands, water skiing, camping on river islands, and to some extent, bicycling. Sandy beaches are formed by channel dredging and by silt deposits in the upper stretches of the river system. They tend to be the focus for river and island water-based recreation.

Not Crowded

Although crowding may not be a visual characteristic, it appears that crowding is an important factor leading to the selection of alternate sites (see Table 18.1) These data show that it was an important characteristic for water-enhanced activities of river watching, hiking, bicycling, and bird watching. For the water based and the river island located activities, crowding was ranked but not as highly as for water-enhanced recreation.

TABLE 18.4
Relationship Between Participation in Selected River Related Recreation Activities and Ranking of Recreation Site Characteristics (Recreationists Only)

Activity (N=1555)		Nice Vegetation	Sandy Beaches	View or Scenic Quality	Not Crowded
River Watching (Sightseeing) N=1019	NO	6.6	46.9**	29.7**	29.0*
	YES	8.6	42.9	39.9	35.1
Swimming N=960	NO	11.0**	27.5**	48.3**	36.9**
	YES	6.2	54.0	30.7	30.8
Picnicking on Shore/N=938	NO	5.4**	40.7*	35.0**	32.2
	YES	9.5	46.4	38.6	33.6
Power Boating N=868	NO	11.4**	32.7**	46.2**	40.3**
	YES	6.4	52.9	30.4	27.6
Fishing from Boat/N=819	NO	11.2**	41.2**	44.0**	37.5**
	YES	5.2	46.8	31.4	29.3
Fishing from Shore/N=750	NO	8.9	47.2	41.0**	33.1
	YES	7.0	41.2	33.3	33.0
Picnicking on River Islands N=622	NO	10.4**	33.0**	41.0**	35.4**
	YES	4.5	59.0	31.7	29.7
Water Skiing N=631	NO	10.5**	32.9**	46.9**	37.2**
	YES	4.3	66.2	23.5	27.3
Camping on Shore/N=611	NO	8.7*	42.5	40.0*	34.4
	YES	6.9	46.7	33.0	31.1
Camping on River Islands N=514	NO	9.8**	37.8**	40.6**	33.3*
	YES	4.3	56.8	30.6	32.7
Hiking N=436	NO	6.1**	46.3**	31.3**	31.8**
	YES	12.6	38.9	52.1	36.2
Bicycling N=404	NO	7.1**	43.6**	34.7**	32.0**
	YES	10.4	46.0	44.0	36.1
Bird Watching N=318	NO	6.3**	45.7*	34.0**	32.3**
	YES	14.2	38.5	49.3	36.1
Canoeing N=312	NO	8.4	43.8**	36.4	30.9
	YES	6.2	45.7	40.4	41.5
Sailing N=211	NO	8.1	44.3	36.0**	34.5**
	YES	7.1	43.8	44.9	23.9

* 5 percent level of significance
** 1 percent level of significance

The analysis shown in Table 18.1 and Table 18.4 indicates that for all users of river recreation activities, "sandy beaches," the "view or scenic quality," as well as "not crowded" are important factors in site selection. Furthermore, dramatic differences are present between rankings and river recreation activities. In general, water-based activities such as swimming, power boating, fishing, picnicking on river islands and water skiing rank facilities higher than the visual characteristics in site selection. The reverse is true for water-enhanced activities such as river watching (sightseeing), hiking, bicycling, and bird watching.

RESOURCE ALTERATION AND ACTIVITY DISPLACEMENT

This section illustrates the methodology to explore the relationship between changes in the recreation resource and the resulting changes in the location of recreation activity using the UMRS example. We attempt to establish a link between recreation resource alteration and those persons that would seek the same recreation resource elsewhere. For the purposes of future visual impact assessments by the National Park Service, it would be ideal to undertake an analysis with data of perception of changes in air quality and visual aspects at national park areas as related to actual changes in the resource and to displacement of activities to other locations based upon this methodology.

The Analytic Approach

Table 18.5 lists the statements given both the manager/providers and the recreationists to measure the impact of resource alteration on activity displacement. First we obtained their evaluation of present navigation and associated operation and maintenance activity on the Mississippi (i.e., there is too much channel dredging done on this section of the river). Secondly, they were asked to indicate whether or not they would go elsewhere for recreation, if that activity increased substantially. Each respondent was asked to reply to the Likert format, with undecided or uncertain responses eliminated in the present analysis. Unfortunately, two of the items were not included on the recreationists' questionnaire. Both the manager/providers and the recreationists were asked to respond based on their own river related recreation activity. Remember the manager/providers are tied to a river location and as such displacement may not occur as readily as for the more mobile recreationists.

In our research, we explored the relationships between the statements in Table 18.5 and four activities--river watching, hiking, swimming and fishing from a boat. Based on the site quality information in Table 18.4, we found that activities seemed to divide nicely along the lines of water-based (swimming and fishing from a boat) and water-enhanced (river watching and hiking). The other activities in each of these two categories show similar results. In order to reduce presentation time with redundant findings, we have eliminated the rest of the activities shown in Tables 18.2 and 18.3. The only two activities from Table 18.4 which do not follow the water-based and water-enhanced pattern were canoeing and sailing. While water-based, their site selection ratings followed that of the water enhanced activities. We illustrate our technique and findings only for changes in water quality and water levels.[1] The emphasis for this paper is to illustrate that changes in the recreation resource will cause displacements that

TABLE 18.5
List of Perceptions of Management Issues Related to Expanded Navigation and Their Relationship to Recreation Activity on the UMRS

Item	Questionnaire Where Item Was Included
There is too much channel dredging done on this section of the river	Both
If channel dredging increased a great deal on this part of the river, I would go elsewhere for recreation	Both
The water in this part of the river is clean enough for swimming and other recreation	Both
If the water in this part of the river became any dirtier, I would go someplace else for recreation	Manager/Provider Only
Water levels in this area are occasionally too high for some recreational uses	Both
Water levels are sometimes too low in this area for some recreational uses	Both
I would go elsewhere for recreation if water levels in this area fluctuated more than they now do	Both
There is too much barge traffic on this part of the river	Both
If the amount of barge traffic increased a great deal on this part of the river, I would go elsewhere for recreation	Both
There are too many areas where barges are moored along the shores on this section of the river	Manager/Provider Only
If many more barges are tied up along the shore in this vicinity, I would go elsewhere for recreation	Both

* All questions are in the Likert response (Strongly Agree to Strongly Disagree). Two items were not included on the recreationist questionnaire.

will be a function of the change and the nature of the activities that are affected, all of which can be effectively measured by this approach.

Water Quality

The left-hand portion of Table 18.6 shows the relationship between perceptions of cleanliness of the water and activity displacement for the two water-based and the two water-enhanced recreation activities. The right hand portion shows the percentage of each category of respondents who reported doing the four activities and whether or not they agreed or disagreed with the statement that if water quality decreased they would go elsewhere for recreation. Data are shown for both the manager/providers and the recreationists.

A substantial increase in barge traffic will keep the water "stirred up" and increase the amount of bank erosion due to wave action. Unfortunately, the "what if" question was not included on the recreationists' questionnaire. From the response of the recreationists (bottom of Table 18.6), it appears that persons engaging in water-based recreation are more likely to perceive the water as being dirtier, than those who do not. That relationship is also true for the manager/providers. Apparently, if the river gets any dirtier, the displacement potential is quite high for manager/providers, even though most are tied to a particular location on the river. On the average, about two-thirds either agree or strongly agree with the statement that they would go elsewhere for their own recreation. However, water quality, or the appearance of the river appears to be an important factor in determining whether or not people will participate in recreation activities. At a minimum it represents a complaint about the visual aspects of the river.

Water Levels Too High or Too Low

In managing the Upper Mississippi River System for both navigation and flood control, the water levels are subject to considerable fluctuation. We asked the respondents whether or not the river level in their area was either too high or too low for some recreational uses. Table 18.7 reports the responses to the statement that the water is sometimes too high for certain recreation activities. The manager/providers, whether or not they have the recreation activity at their site, are likely to agree with the statement that water levels are too high. Futhermore, approximately 45 percent of those who do the activity, report that they would go elsewhere if the water levels fluctuate more than they do at the present. The activity displacement response is about the same for all recreation activities.

In contrast, the recreationists are somewhat less inclined to agree with the statement that water levels in that area of the river are too high for recreation. However, the recreationists are more likely to agree with the displacement statement that they would go elsewhere for recreation. The average sum of the agree statements in the right hand column is about 55 percent, as compared with about 42 percent for the manager/providers.

Low water levels also are seen as a problem for some recreation activities and that both manager/providers and recreationists view the problem in the same way. As expected, low water levels would lead to displacement about as much as high water. Displacement in the case of low water marks tends to be slightly less for the water-based as compared with the water-enhanced activities. These data show that water levels are

TABLE 18.6
Relationship Between Perceptions of the Cleanliness of the Water and Activity Displacement, if the Water Became Any Dirtier, for Selected Water Based and Water Enhanced Recreation

Manager-Providers Activity	Percent With Activity At Site (N=255)	THE WATER IN THIS PART OF THE RIVER IS CLEAN ENOUGH FOR SWIMMING AND OTHER RECREATION (N=158)						YES ACTIVITY					
		NO ACTIVITY		YES ACTIVITY				Strongly Agree and Agree		Disagree and Strongly Disagree			
								IF WATER IN THIS PART OF THE RIVER BECAME ANY DIRTIER, I WOULD GO SOME PLACE ELSE FOR RECREATION					
		SA/A	D/SD	SA/A	D/SD			SA/A	D/SD	SA/A	D/SD		
Water Enhanced													
River Watching	82.0	71	29	60	40			32	29	33	6 (N=124)		
Hiking	46.3	64	36	59	41			37	22	33	8 (N=73)		
Water Based													
Swimming	65.5	47	53	70	30			35	33	26	6 (N=96)		
Fishing (Boat)	94.9	42	58	63	37			34	29	31	6 (N=132)		

Recreationists' Activity	Percent Doing Activity (N=1555)	SA/A	D/SD	SA/A	D/SD	SA/A*	D/SD*	SA/A*	D/SD*
Water Enhanced									
River Watching	66.7	37	63	39	61				
Hiking	28.5	40	60	35	65				
Water Based									
Swimming	62.6	28	72	44	56				
Fishing (Boat)	53.6	30	70	45	55				

* Unfortunately, the displacement item on the perceptions of water quality was not included in the questionnaire that was distributed to the recreationists. (SA/A means the respondent checked strongly agree or agree and D/SD means the respondent checked whether disagree or strongly disagree.)

TABLE 18.7
Relationship Between Perceptions That Water Levels Are Too High and Activity Displacement, if Water Levels Fluctuate More, for Selected Water Based and Water Enhanced Recreation

Manager-Providers Activity	Percent With Activity At Site (N=255)	WATER LEVELS IN THIS AREA ARE OCCASIONALLY TOO HIGH FOR SOME RECREATION USES (N=231)				YES ACTIVITY				
		NO ACTIVITY		YES ACTIVITY		Strongly Agree and Agree		Disagree and Strongly Disagree		
						I WOULD GO ELSEWHERE FOR RECREATION IF WATER LEVELS IN THIS AREA FLUCTUATED MORE THAN THEY DO NOW				
		SA/A	D/SD	SA/A	D/SD	SA/A	D/SD	SA/A	D/SD	
Water Enhanced										
River Watching	82.0	87	13	72	28	29	40	15	16 (N=115)
Hiking	46.3	75	25	75	25	26	42	16	16 (N=55)
Water Based										
Swimming	65.5	70	30	77	23	30	42	12	16 (N=84)
Fishing (Boat)	94.9	92	8	74	26	30	40	14	16 (N=125)

Recreationists' Activity	Percent Doing Activity (N=1555)	SA/A	D/SD	SA/A	D/SD	SA/A*	D/SD*	SA/A*	D/SD*
Water Enhanced									
River Watching	66.7	55	45	54	46	31	23	21	25 (N=523)
Hiking	28.5	56	44	50	50	35	21	18	26 (N=212)
Water Based									
Swimming	62.6	56	44	54	46	34	22	21	23 (N=498)
Fishing (Boat)	53.6	56	44	53	47	36	21	21	22 (N=433)

seen as an important factor, which impacts directly on recreation activity. It represents a factor which is largely manipulated by management to the benefit of flood control.

CONCLUSIONS

This paper illustrates a particular methodology that may be applied in national park and wilderness settings to provide insight as to what might happen if the visual aspects of a resource are altered. Of course, air quality is one such alteration for which other papers in this volume have established its importance in terms of the recreation experiences. The methodology in this paper could be used to show if and how alterations in the park resource will lead to displacement for specific park-related activities.

We established a linkage between visual characteristics of the UMRS recreation sites and types of water-based and water-enhanced river recreation activities. It was found that visual characteristics of a recreation site are not as important in site selection for persons doing water-based activities as compared with those reporting water-enhanced activities. In fact, the differences were quite dramatic. People that come to the Upper Mississippi River resource for water-based activities do not list visual characteristics as important in site selection. The recreation activities that this study found to be related to the visual characteristics of the site were recreation activities such as hiking and bird watching, activities that are more likely to take place at National Park Service areas.

The second part of the paper illustrated the methodology used to attempt to establish a linkage between alteration in the recreation resource and the likelihood of recreationists leaving the site. In other words, taking their participation in a recreation activity to a different place. It was found that alterations did, in fact, lead to potential displacement, and that some variation was present, depending upon the type of recreation activity. In this study, activities that were water-based tended to lead to displacement if resource alteration occurred.

NOTES

1. Results for impacts from increased channel dredging low water, barge traffic, and fleeting areas are available from Rabel Burdge.

REFERENCES

Jubenville, A. 1978 "Visual Resource Management," in Outdoor Recreation Management edited by A. Jubenville. Saunders Press, New York.
Hammitt, W.E. 1981. "The Familiarity - Preference Component of On-Site Recreational Experiences." Leisure Sciences 4:177-193.
U.S. Forest Service. 1974. National Forest Landscape Management, Volumes I and II. U.S Government Printing Office, Washington, D.C.
Shafer, E.L., J.L. Hamilton, and E.A. Schmidt. 1969. "Natural Landscape Preferences: A Predictive Model." Journal of Leisure Research 1:1-19.

PART V
Economic Approaches to Value Assessment

Economic value assessment techniques attempt to put dollar quantifications upon attitudes and behaviors associated with changes in visual resources by measuring the trade-offs people are willing to make in order to enjoy a certain level of visual quality. This then provides one measure of how much people value protection of visual resources and how much protection is desirable. As with the social and psychological assessment tools, these techniques address the question of the importance (or adverseness) of perceived changes in visual resources. Economic measures can also be used in cost-benefit analysis of proposed changes.

The first two papers present applications of two specific techniques. Rae presents the results of two recent visibility valuation experiments conducted at national parks using the contingent ranked attributes approach. Johnson and Haspel present the results of an application that is a variation upon the travel cost approach and was used to value visual impacts from potential strip mining activities near Bryce Canyon National Park.

Rowe and Chestnut address issues and uncertainties in economic analyses of visual values and suggests priorities for future research. Among the most important issues are concerns about nonuser or preservation values, which preliminary evidence suggests in some cases may be more important than on site user values.

The concept of preservation values--values that exist in addition to the values of a good in its actual use--has become widely accepted and frequently discussed; however, much debate remains about their importance, their theoretical basis, and methods for measuring them. If nonuser values are to be used in policy making, they must be based upon a sound theoretical basis and accurate measurement. The final three papers in this section present perspectives, definitions, and evidence concerning nonuser values, particularly existence values. McConnell, Randall and Stoll, and Talhelm each present theoretical definitions and implications concerning the measurement and use of existence values. McConnell suggests that all existence values are derived from expected future use by others and suggests a two step contingent market estimation approach. Randall and Stoll suggest that existence values are a function of information and supply, not irreversibility. They also present evidence and concerns about empirical estimates of these values. Talhelm classifies existence values as one of six types of values unrevealed in markets and suggests it will be a function of the good's uniqueness, which can be measured by its substitutibility with similarly defined goods.

19. The Value to Visitors of Improving Visibility at Mesa Verde and Great Smoky National Parks

Douglas A. Rae

INTRODUCTION

Three basic approaches have been developed and applied to measure benefits of non-market goods: (1) use of data on actual market behavior such as travel costs and property values, (2) use of data from hypothetical or contingent market surveys, and (3) use of data from simulated market games using real money. Quantifying visibility benefits in Class I areas has been limited to contingent market survey approaches due to lack of data on actual market behavior and difficulties in setting up a simulated market to buy or sell a good or service with different levels of visibility.
The most widely used of the hypothetical or contingent market techniques is contingent bidding. Bidding methods are, by nature, quite simple and are open to various forms of bias by individual respondents and survey researchers. Careful design and implementation of the survey instrument can eliminate much of this bias, but there is always the opportunity for individuals in a hypothetical framework to influence the results by overbidding or underbidding. Consequently, economists have expressed considerable caution, even skepticism, in using the results from contingent bidding experiments.
An alternative to contingent bidding is the contingent ranking methodology described herein. This methodology employs a hypothetical choice situation in which individuals rank hypothetical alternatives contingent upon how they value each combination of characteristics. A number of characteristics or attributes may be evaluated by using a statistical technique to quantify how much each characteristic contributes to the overall satisfaction, as revealed in the ordinal ranking of each alternative. The benefit trade-off between attributes can be calculated from the relative contribution of each alternative. The choice situation can be made quite realistic by mirroring actual market choices that may depend on a number of attributes. The more the choice approximates an actual market choice, the more the revealed behavior is likely to reflect actual market be-

---

Douglas Rae is a Senior Research Associate at Charles River Associates, Boston, MA. This paper is the result of research funded by the Electric Power Research Institute under the technical supervision of Dr. Ronald Wyzga. Edison Electric Institute provided additional funding to facilitate presentation of findings at the Visual Values Workshop.

havior. Each alternative may include several characteristics or attributes that add to the reality of the choice while reducing the likelihood of strategic behavior by individuals to bias results through the increased complexity. This paper reports the results of two contingent ranking studies of visibility benefits that were conducted at Mesa Verde and Great Smoky Mountains National Parks during the summer, 1981. All benefits are reported in (quarter III) 1981 dollars.

CONTINGENT RANKING METHODOLOGY

The contingent ranking methodology is founded in the Lancastrian theory of consumer behavior and consumer preferences. This theory, which is based on a substantial body of literature in economics, psychometrics, and consumer research, attempts to describe how consumers choose from among similar products. To differentiate among products, such as cars, consumer choice theory focuses on attributes, which are basic characteristics of a product. The product or good then can be decomposed into a number of attributes, and most products can be described as being multiattribute.

The relative value or utility of any product or service depends on the level of each attribute and the importance of each attribute to the consumer. The level of an attribute is best described by considering the example of a car.[1] Cars vary in price, mileage, performance, seating capacity, maintenance costs, and many other characteristics. Thus, the attribute of "price" may vary over a continuum from $5,000 to $15,000 for most standard models. Similarly, "mileage," "performance," and other attributes may vary over a wide range. Some discrete attributes may be available or not available depending on the model. The relative importance of each attribute will be reflected in the final ranking of choices. In this sense a random consumer's preference for a given type of car can be predicted from the weighting of each auto characteristic relative to other consumers. By evaluating a number of choices by individuals, these weights, called parameters, can be estimated for an average consumer by various quantitative techniques.

It is the quantification of the trade-off between characteristics or attributes that provides a measure of the benefits of environmental goods such as visibility. If improving environmental quality, say visibility at a national park from 20 kilometers to 50 kilometers, is worth twice as much as a $5 increase in entry fee, then that improvement in visibility is worth $10 (assuming linearity). In this manner, a monetized estimate of a nonpriced attribute, such as environmental improvement or degradation, can be derived from a comparison of the utility trade-off between an environmental quality attribute and a price or entry fee attribute.

The quantification of the trade-off between attributes is accomplished by estimating a set of weights for each attribute that maximizes the likelihood that a random individual will rank the alternatives in the order they were actually chosen. These weights are known as parameter estimates, and are calculated by a logit-based discrete choice model that has been modified to use ranked or ordered data.[2] The ordered logit specification yields a set of parameter weights that maximizes the likelihood of the complete ordering rather than the likelihood of choosing only the most preferred alternative.[3] The model can be applied in an aggregate fashion utilizing the assumption that all respondents have identical tastes or individually to each respondent's rankings of alternatives. The individual ordered logit model allows estimation of individual parameters, and permits more precision in calculating benefit trade-offs on an individual basis where there is considerable variation in tastes and preferences across individuals.

The interpretation of these weights is fairly straightforward. Each parameter weight represents the marginal contribution of an attribute to the overall satisfaction or utility, as reflected in the ordinal ranking of the alternatives. In this case utility is measured in terms of an arbitrary utility base. Thus, utility is specified as the change in utility from the base level to the changed level. Mathematically, let ΔU_{ij} represent the change in utility derived by individual i in choosing alternative site j;, ΔE_j the change in environmental quality at site j; ΔP_j the change in price or entry fee to consume (visit) j; and β_1 and β_2 are parameter weights. Then

$$\Delta U_{ij} = \beta_1 E_j + \beta_2 P_j$$

Holding $\Delta U_{ij} = 0$ and realizing that $\Delta E_j = 1$, since it represents a discrete change from a base level of environmental quality to another level, then

$$0 = \beta_1(1) + \beta_2 \Delta P_j.$$

The change in price, ΔP_j, is equal in its marginal contribution to overall utility to the change in environmental quality, ΔE_j, is reflected in the ratio of the parameter weights, $-\beta_1/\beta_2$. For an improvement in environmental quality, $\beta_1 > 0$, this trade-off will be positive because the sign on β_2, the price parameter, is expected to be negative. It is important to note that this trade-off is computed at the existing level of utility, $\Delta U_{ij} = 0$, and therefore represents a compensating surplus measure of benefits.

As noted above, the weights that consumers attach to various attributes can only be estimated if a sufficient number of choices have been made to permit use of the statistical technique. Thus, the procedure requires data on consumer preferences for actual or hypothetical products or services. For hypothetical products or services, the ranking of choices is contingent upon the combination and level of attributers, and the data on the ranking of consumer choices are obtained by designing a survey in which consumers compare and rank a number of choices. In quantifying benefits of environmental changes, the advantage of a survey that is based on hypothetical products or options and contingent choices is the same as for new product design: it is possible to assess the value of a change before actually incurring the costs or risks of implementing any action. Of course, there are risks that consumers may not respond to a survey in the same manner that they respond in a market, but these risks can be minimized by survey designs that closely approximate market choices.

Survey Design

The contingent ranking methodology requires data in the form of choices of products or alternatives, specified in terms of attributes, that are ordered according to individual preferences. Since under normal circumstances market data reveal only first choices, a survey procedure is required to elicit second, third, fourth, and subsequent choices. A major advantage of this contingent ranking methodology is that an individual's ranking of many choices provides a large number of observations so that many fewer respondents are required to yield results within acceptable confidence limits.[4]

To test how important visibility is to visitors to Mesa Verde and Great Smoky National Parks, Charles River Associates (CRA) formulated a survey choice situation where survey respondents compared and ranked combinations of attribute levels in terms of overall desirability as a vacation destination. Different levels of visibility, entry fee, and other conditions at a given park were combined in hypothetical choice alternatives and presented to respondents. Sites other than Mesa Verde and Great Smoky were included in the survey ranking to reflect the fact that alternative sites are available and to cause respondents to focus broadly on all the characteristics of a site that contribute to overall enjoyment of national parks and other outdoor recreation areas.

Attribute levels were chosen so as to represent distinct increments. Entry fee ranges were chosen to force a trade-off between dollar costs and other attributes. Too small a range may result in large numbers of respondents willing to pay the maximum and more, a result that could underestimate benefits if widespread across the sample. Similarly, too wide a price range forces a trade-off, but reduces the precision of the estimate within the range. A visibility survey pretest was conducted at Mesa Verde in October 1980 to test the survey design, and adjustments in attribute levels and entry fee ranges were made based on the results of that pretest.

Choice of Visibility Conditions

CRA selected four slides at each park to represent different visibility levels. Care was taken to select slides that were significantly different from each other so as to represent discrete visibility levels. The four slides of Mesa Verde were chosen to represent an intense plume, an intense haze, a moderate haze and a clear condition with visual ranges varying from about 120 to 260 kilometers.[5] At Great Smoky, where discrete plumes are not normally visible, we selected four slides of generalized haze conditions that ranged from about 10 km to 100 km of visual range.[6] Care was taken to include only slides from the same processing batch.[7] Details of these slides are provided in Table 19.1.

Data on the occurrence of haze and clear conditions are available from the telephotometric measurements of inherent contrast that are recorded daily at each site. These data were used to estimate the percentage of time that the condition portrayed in each slide was likely to occur. Since individuals will differ on whether a 135 km visual range condition should be represented by a slide showing a 156 km or 119 km visual range, considerable judgment was used in expressing the frequency of occurrence data as a percentage time of occurrence that respondents could understand.

The existing distribution and several other improved visibility scenarios are shown for Mesa Verde and Great Smoky in Table 19.2. For Mesa Verde, plume and haze models were used to develop a control scenario that approximated Best Available Retrofit Technology (BART) at the 2000 MW Four Corners Plant and a no-plant scenario was also included. At Great Smoky hypothetical improved scenarios were developed, since visibility modeling revealed that BART controls on two nearby powerplants would have little effect on the existing frequency of occurrence of the four visibility conditions.[8] Precipitation occurred a significant percentage of the time at Great Smoky, and the percentage of significant precipitation days was incorporated in the four Great Smoky scenarios.

TABLE 19.1
Information on Slides Used in Visibility Survey

Number	Date	Time (a.m.)	Green Contrast	Plume Contrast	Visual Range (km.)	Description
MESA VERDE NATIONAL PARK						
MV 1633	7/17/80	09:10	-.04	.11	--	Intense Plume
MV 1771	7/24/80	09:30	-.07	--	119	Intense Haze
MV 91	9/16/79	09:15	-.12	--	156	Moderate Haze
MV 1414	7/05/80	09:05	-.24	--	256	Clear-Minimal
MV 22	9/04/79	09:00	-.24	--	256	Pollution
GREAT SMOKY MOUNTAIN NATIONAL PARK						
--	8/23/80	12:00	-.00	--	10	Intense Haze
--	7/19/80	12:00	-.04	--	20	Moderate Haze
--	10/6/80	12:00	-.27	--	50	Slight Haze
--	10/7/80	12:00	-.46	--	100	Clear

Note: Mesa Verde slides were taken from Far View Visitor Center toward Hogback Ridge, a target distance of 56 km. Visual range calculations were based on a $C_o = -0.8$. All Great Smoky slides were taken at noon from TVA's Look Rock monitoring station toward Cerulean Knob. Visual range calculations are based on a target distance of 14.2 km. and an inherent target contrast, $C_o = -0.8$.

Source: Charles River Associates, compiled from information supplied by the Visibility Research Center, Las Vegas, and the Tennessee Valley Authority.

ANALYSIS OF VISIBILITY BENEFITS: MESA VERDE

Independent variables used in both the aggregate and individual models for the Mesa Verde analysis included visibility, waiting time, and entry fee (price). A number of socioeconomic and trip characteristics, such as income, education, age, sex, ethnic origin, previous visits, number of adults and children in party, and number of children in family are used interactively in the aggregate model. The four visibility conditions: intense plume, intense haze, moderate haze, and clear are differentiated to enable valuation of the visibility change. Similarily, waiting time, as a proxy for congestion, is identified as either no wait or one hour wait at Balcony House, one of the popular ruins requiring a ranger-guided tour, and the parmeter weight reflects the value of an hour saved. Entry fee was left continuous over the $2.00 to $20.00 per vehicle range. All model spec-

TABLE 19.2
Frequency of Occurrence of Visibility Conditions

(Percent of Summer Daylight Hours)*

	Visibility Condition					
Scenario	Intense Plume	Intense Haze	Moderate Haze	Slight Haze	Clear Conditions	Rain
MESA VERDE NATIONAL PARK						
Existing	11	19	56	--	14	--
Agreed Controls**	1	20	60	--	19	--
No Four Corners Plant+	0	2	60	--	38	--
GREAT SMOKY NATIONAL PARK						
Existing	--	15	30	35	10	10
Improved 1	--	10	25	40	15	10
Improved 2	--	5	20	45	20	10
Improved 3	--	1	10	50	29	10

NOTES:
* At Mesa Verde, summer daylight hours are based on total daylight hours between May and September. At Great Smoky non-winter days were used based on total days between April and November. In fact, however, modeling results are not very different on an annual basis.
** Arizona Public Service Company has agreed to control seventy-two percent of the sulfur oxide emissions and to limit particulates to .05 lb/ million BTUs.
+ This scenario assumes no emissions from the Four Corners plant but assumes continued emissions from the San Juan plant at existing levels.
Source: Charles River Associates, based on monitoring data and modeling results.

ifications normalized the visiblity changes on the clear condition so that all other visibility parameters were expected to have negative signs; that is, other visibility conditions were expected to be less satisfying than the clear condition.

Deterministic Case

The initial survey task required respondents to rank order eight Mesa Verde alternatives on the assumption that the visibility, congestion, and price conditions specified on each card were known in advance of a visit. In analyzing these deterministic rankings over the aggregated sample of individuals CRA tested a variety of utility specifications. It was necessary to delete the moderate haze alternative to achieve the correct negative signs on degraded visibility conditions.[9] Specification A in Table 19.3 indicates that respondents attached a negative value to the intense plume and the intense haze, valued "not having to wait" positively and disliked paying higher entry fees. Parameter weights on the congestion and price variables were significant at the 95 percent confidence level, but the visibility parameters did not achieve that level of significance.

Incorporating socioeconomic variables as interaction terms yields only slight improvement in the quality of the relationship. In specification B income is interacted with entry fee in a specification where the negative utility of higher entry fees is lessened for individuals with higher incomes. The entry fee/income term in B has the correct sign and is significant at the 90 percent confidence level using a one tail test but the effect of income on respondents' willingness to pay for improved visibility is small, as shown in the last columns in Table 19.1.

Assuming tastes are constant over the entire sample, the value to the average respondent of improving visibility from the intense plume and intense haze condition to the clear condition is small, about $1.10 per vehicle for the former and $0.75 per vehicle for the latter. The wide standard errors reflect the low precision of the estimates, but the order of magnitude is consistent with values obtained from contingent bidding methods. The value of time saved is about $1.05 per vehicle for the average income.

Where individual's valuations of visibility and other attributes vary widely, more precise estimates can be obtained by dropping the assumption of identical tastes and estimating attribute parameters for each individual. It is then possible to compute benefit trade-off values for each individual and then to sum the over all respondents to find an average. For brevity of presentation only, these individual results are not presented in tabular form. The parameter weights for each individual vary substantially, and about 43 percent of the individuals ranked the choices in such a way as to yield a negative value to visibility improvement.[10] In computing the average benefit trade-off those individuals with negative values on visibility improvement were assigned a value of zero.[11] The average visitor (based upon 190 individuals) to Mesa Verde is willing to pay $5.10 (1.15) per vehicle to obtain clear visibility conditions in place of an intense plume, $4.57 (0.95) per vehicle in place of an intense haze, and $2.84 (0.60) per vehicle in place of a moderate haze. The value of an hour waiting time is $2.43 (0.35). The standard errors of the estimates (reported in parentheses above) indicate a fairly small variation around the mean, although this is somewhat misleading because the negative values were rounded up to zero.

Probabilistic Case

In another ranking respondents ranked seven Mesa Verde alternatives, including three cards that described overall visibility in terms of the percentage of summer daylight hours for which each of the four conditions was

TABLE 19.3
Aggregate Model Parameter Estimates: Mesa Verde Deterministic Case

Attribute	Specification		Benefits ($/vehicle) Income ($000)		
	A	B	20	35	50
Clear = 0	Base	Base			
Moderate Haze = 1	Omitted	Omitted	--	--	--
Intense Haze = 1	-0.1575	-0.1633	0.73	0.77	0.79
	(-0.8767)	(-0.9081)	(0.75)	(0.79)	(0.88)
Intense Plume = 1	-0.2306	-0.2340	1.05	1.11	1.13
	(-1.2845)	(-1.3035)	(0.73)	(0.76)	(0.78)
No Wait = 1	0.2188	0.2229	1.00	1.05	1.08
	(2.0700)	(2.1088)	(0.44)	(0.46)	(0.47)
Entry fee ($)	-0.2166	-0.1963			
	(-10.7881)	(-8.2700)			
Entry fee/Income	-	-0.5285			
		(-1.5123)			
Number of Observations	196	196			
Maximum Likelihood Estimate	-1055.5522	-1053.9932			

Note: 1) Numbers in parentheses in columns 1 and 2 are asymptotic t-statistics. These are calculated on a formula that is correct for infinite samples. For small samples, the measure is consistent, but not unbiased. At 190 degrees of freedom and a two-tail test.
$t_{0.995}$ (99 percent confidence interval) = 2.59
$t_{0.975}$ (95 percent confidence interval) = 1.97
$t_{0.950}$ (90 percent confidence interval) = 1.65
$t_{0.900}$ (80 percent confidence interval) = 1.29
2) Benefits reflect a compensating surplus estimate of willingness to pay per vehicle with numbers in parentheses representing standard errors of estimate.

Source: Charles River Associates, based on computer analysis using an ordered logit model.

likely to occur. The three probabilistic conditions were chosen to represent the currently existing visibility condition (Existing), an improved condition to approximate better sulfur oxide and particulate controls (Agreed Controls) at the Four Corners power plant,[12] and a case where the Four Corners power plant is shut down (No Plant). In addition, we included the moderate haze and clear condition from the first set of cards and explained to respondents that the one visibility condition specified, the clear condition for example, was guaranteed to occur with 100 percent certainty.

As was the case with the deterministic ranking, it was necessary to delete the moderate haze condition in the aggregate model. We postulate that the difference in the frequencies of occurrence between the "existing" and "controls" scenarios was insufficient to cause respondents to rank the two scenarios consistently, and insignificant parameter estimates were the result.[13] Deleting the "controls" scenario visibility condition yielded correct signs on the remaining visibility scenarios at modestly significant (80 percent) confidence levels, as shown in specification X in Table 19.4. However, the sign on the congestion attribute, "no wait at Balcony House," is incorrect.

The sign on not waiting should be positive, since saving time is usually associated with positive benefits. The incorrect sign is most likely due to having deleted the two moderate haze choices and no longer having sufficient "no wait" choices to balance "one hour wait" in all the probablistic alternatives. To rectify this problem we constrained the parameter on "no wait" to equal about 0.22, as estimated in specification B from the ranking under certainty (Table 19.3), and allowed all other parameters to find their own levels. Specification Y in Table 19.4 then yields estimates of the benefits of visiblity improvement that are consistent internally and with estimates in the deterministic specification. Evaluating the benefit tradeoffs, as shown in Table 19.4, we find that for a mean income $34,100 the benefit of improving visibility from the "existing" probabilistic condition to a guaranteed "clear" condition is $3.09 per vehicle and from "existing" to "no plant" is $0.93 per vehicle.

These benefit estimates from the probabilistic ranking are quite similar to a composite benefit estimate derived from the deterministic ranking. The individual model applied to the (deterministic) ranking under certainty yielded benefit estimates of $5.10 per vehicle to ensure a clear condition over an intense plume, $4.57 per vehicle to ensure clear over an intense haze, and $2.84 per vehicle to ensure clear over a moderate haze. Data on the existing frequency of occurrence of visibility conditons indicate that intense plumes occur about 11 percent of daylight hours, intense hazes (119 km) about 19 percent, moderate hazes (156 km) about 56 percent, and clear conditions (256 km) about 14 percent. Thus, about 14 percent of the visitors can be expected to encounter a clear condition, for which they would be willing to pay zero while the other 86 percent would be willing to pay between $2.84 and $5.10 per vehicle. The composite value of moving from the existing distribution of visibility events to a guaranteed clear condition is $3.02 per vehicle ($3.02 = .14(0) + .11(5.10) + .19(4.57) + .56(2.84)) compared to $3.09 per vehicle in the probabilistic ranking. Similarly, the composite value of moving from a "no plant" condition to a guaranteed clear condition is $1.79 per vehicle, ($1.79 = .38(0) + .00(5.10) + .02(4.57) + .60(2.84)) compared to the $0.93 per vehicle estimate obtained in the aggregate probabilistic ranking. This difference is about 1.5 standard errors and is not significant at a 95 percent confidence level.

Overall, the consistency of these benefit estimates is encouraging and reinforces the credibility of the contingent ranking technique. To our knowledge this is the first time visibility or any other environmental good that is subject to random variation has been presented in a complex probabilistic form to survey respondents.

ANALYSIS OF VISIBILITY BENEFITS: GREAT SMOKY

Independent variables used in both the aggregate and individual models include visibility, visitor center activities, and entry fee (price). Socioeconomic and trip characteristics, such as income, education, age, sex, ethnic origin, previous visits, number of adults and children in party,

TABLE 19.4
Aggregate Model Parameter Estimates and Benefits: Mesa Verde
Probabilistic Case

Attribute	Specification		Compensating Surplus Benefits ($)*
	X	Y	
Clear = 0	Base	Base	Base
Existing = 1	-0.4021	-0.6417	$3.09
	(-1.3893)	(-2.5116)	(0.95)
Agreed Controls = 1	Omitted	Omitted	--
No Plant = 1	-0.1916	-0.1935	0.93
	(-0.3776)	(-1.3868)	(0.61)
No Wait** = 1	-0.3459	0.2200	1.06
	(-1.7550)	(NA)	(NA)
Price	-0.1462	-0.1851	--
	(-3.9833)	(-6.2253)	
Price/Income	-0.7515	-0.7709	--
	(-1.5381)	(-1.5483)	
Number of Observations	195	195	
Maximum Likelihood Estimate	-506.593	-508.1489	

Notes: * Benefits are computed for the average income level of $34,100.
** There is no t-statistic on the "No Wait" parameter since the parameter value was constrained to be 0.22, as described in the text.
Source: Charles River Associates, based on estimation using an aggregate ordered logit model.

and number of children in family, were tested interactively in the aggregate model. Four different visiblity conditions: intense haze (10 km), moderate haze (20 km), slight haze (50 km), and clear (100 km) were selected to estimate the value of changes in visibility from one condition to another. Similarly, visitor center activities were specified as either "full program" or "limited program," and the parameter reflects the marginal value of the more diverse program of activities available to park visitors during the peak summer season. Entry fee was left continuous over the $0 to $30 per vehicle range.[14] As in the Mesa Verde case, all initial model specifications normalized the visibility changes on the clear condition so that all other visiblity parameters were expected to have negative signs; that is, other visibility conditions were expected to be less satisfying than the clear condition. The visitor center activity variable was normalized on the "limited program" so that the "full program" was expected to have a positive sign that indicated it is more valued than the "limited program".

Deterministic Case

The initial survey task required respondents to rank order eight Great Smoky alternatives on the assumption that the visibility, visitor center activities, and entry fee conditions specified on each card were known in advance. In analyzing respondent rankings over the aggregated sample of individuals, CRA tested a number of specifications, but none estimated correct signs for either the moderate haze (20 km visual range) or visitor program variables, and these were subsequently deleted. Again, we speculate that the 20 km visual range may not have been distinct from the 10 km visual range and caused enough inconsistencies in respondents' rankings to cause incorrect signs in the aggregate specification.

The resulting aggregate specifications, D and E in Table 19.5, indicate that the 100 km condition is preferred to the 10 km condition. However, the 50 km slight haze condition is preferred over the clearer 100 km condition at a 90 percent level of confidence (two tailed test). The benefit trade-off computed from Specifications D and E indicates that on average respondents were willing to pay about $9.50 per vehicle to obtain a clear condition over an intense haze (10 km) condition. Similarly, improving visibility to a slight haze condition from an intense haze condition is worth about $13.10 per vehicle-trip. Counterintuitively, a slight haze (50 km) condition appears to be worth about $3.65 per vehicle more than a clear (100 km) condition.[15]

The effect of income on willingness to pay for improved visibility is greater than at Mesa Verde. An individual with an average annual family income of $20,000 is willing to pay about $7.40 per vehicle-trip to obtain a 100 km visibility condition over a 10 km visibility condition whereas an individual with twice as much annual income is willing to pay $11.20 per vehicle-trip. For the average visitor to Great Smoky Mountain National Park, an extra thousand dollars of income results in an additional $0.19 value in improving visiblity from about 10 to 100 kilometers.

The individual ordered logit model was then used over 186 individuals to test the results obtained in the aggregate analysis. Individuals who ranked the choices solely according to the entry fee variable indicated no willingness to trade-off improved visibility for a higher entry fee and were therefore assigned a zero value to visibility benefits. There were thirty three of these individuals (about 18 percent of the sample).

An additional twenty six individuals ranked their alternatives solely in terms of visibility, clearest to haziest. For such a ranking we can only identify the minimum value respondents are willing to pay by evaluating the maximum trade-off in the various card decks. Assigning these individuals the minimum value yields average benefits of about $14.80 per vehicle-trip intense haze to clear, $10.00 per vehicle-trip moderate haze to clear, and $5.00 per vehicle-trip slight haze to clear.[16] In comparison, if those minimum benefit values were doubled for the subset of individuals who ranked according to visibility the resulting average benefit values would increase to about $18.00 intense haze to clear, $12.30 moderate haze to clear, and $6.70 slight haze to clear. It is important to note that by allowing tastes to vary in the individual model the expected orders of magnitude are obtained for the slight haze and moderate haze conditions, whereas these were insignificant in the aggregate specification. Similarly, the individual model permits us to identify those individuals that derive positive benefit from the "full" visitor program. The marginal benefit of a "full program" of visitor center activities is estimated to be in the range of $2.80 to $3.00.

Table 19.5
Aggregate Model Parameter and Benefit Estimates: Great Smoky Deterministic Case

Attribute	Specification D	Specification E	Benefits (E) Income ($000) 20	30.9	40
Clear = 0	Base	Base			
Slight Haze = 1	0.2447	0.2416		-3.64	
	(1.8679)	(1.8439)		(2.29)	
Moderate Haze = 1	Omitted	Omitted			
Intense Haze = 1	-0.6287	-0.2362	7.39	9.48	11.22
	(-2.7840)	(-0.8744)	(2.44)	(2.06)	(2.01)
Full Program = 1	Omitted	Omitted			
Price	-0.0666	-0.0663			
	(-5.7116)	(-5.6825)			
Intense Haze * Income	--	-0.0127			
		(-2.5898)			
Number of Observations	206	206			
Maximum Likelihood Estimate	-1293.8811	-1290.4617			

Note: 1) Numbers in parentheses under parameter estimates are asymptotic t-statistics. These are calculated on a formula that is correct for infinite samples. For small samples, the measure in consistent, but not unbiased. At 190 degrees of freedom, a one-tail test requires a t-statistic equal to or greater than
  $t_{0.995} = 2.59$ $t_{0.950} = 1.65$
  $t_{0.975} = 1.97$ $t_{0.900} = 1.29$

2) Benefits estimates represent a compensating surplus estimate, and numbers in parentheses are standard errors of estimates for the visibility/price trade-off.

Source: Charles River Associates, based on computer analysis using an ordered logit model.

Probabilistic Case

In another ranking, respondents compared seven Great Smoky alternatives, including four cards that described overall visibility in terms of the percentage of non-winter days for which each of the four conditions was likely to occur. The four probabilistic conditions were chosen to approximate the currently exisiting visibility condition (Existing) and three levels of improved conditions (Improved) that were defined in terms of increased percentages of time that the clearer visiblity conditions occur (see Table 19.2). As in the Mesa Verde survey, the slight haze and clear condition were included from the deterministic cards, and respondents were instructed to treat those cards as if the one visibility condition specified occurred 100 percent of the time.

The basic specification, with all the changes in visibility based on the clear condition, was consistent with the aggregate deterministic specification in that respondents preferred the slight haze condition to the clear condition with the parameter significant at the 99 percent confidence level. Consequently, in subsequent runs we rebased the visibility conditions on slight haze (50 km) and deleted the clear condition. Further testing necessitated omitting the Improved 1 and 2 conditions due to lack of significance.

Table 19.6 presents parameter and benefit estimates for specifications J and K in terms of dollars per vehicle-trip. Specifications J and K indicate that the benefits of improving visibility from an existing distribution of events to a guaranteed 50 km visual range are on the order of $7.80-$9.30 per vehicle for the average individual. Similarly, from an Improved 3 condition with 10 and 20 km visual ranges occurring only 11 percent of the time to a guaranteed 50 km visual range is worth about $3.60-$3.95 per vehicle-trip. Improving visibility from the Existing to the Improved 3 scenario is valued at about $4.20-$4.40 per vehicle-trip. The quality of these visiblity benefit estimates is fairly good, as reflected by the relatively small standard errors of the estimates.

The marginal value of the full program of visitor activities was valued by respondents at about $2.40-$3.40 per vehicle-trip, which is consistent with the $2.80-$3.00 estimates derived from the individual model using data from the ranking under certainty. The consistency of these results between the deterministic and probabilistic rankings is again encouraging.

The visibility benefit estimates at Great Smoky from the probabilistic ranking also agree quite well with results from the deterministic ranking. These cannot be compared easily because the slight haze condition was valued more highly than the clear conditions in the aggregate specifications. Nevertheless the orders of magnitude are comparable. For example, the benefit of moving from the existing to the most improved condition, which amounted to $4.20-$5.40 in the aggregate probabilistic specification, as reported above, is about the same as the benefit of moving from the moderate haze to the slight haze condition, which amounted to about $5.00-$5.60 in the individual deterministic specification. The comparability of these results suggests respondents are reasonably consistent in ranking alternatives both in a deterministic and probabilistic setting.

BENEFITS AND COSTS

The trade-off between improved visibility and increased entry fee represents a measure of willingness to pay for an environmental improvement; thus it is a measure of benefits. Benefit measures were obtained by analyzing the results of both an aggregate model specification and an individual model specification. The results from the individual models were more precise, and these results are used below to compare visibility benefits at Mesa Verde and Great Smoky National Parks to each other and to costs of control.

On a vehicle-trip basis visitors to the Smokies were willing to pay about two to three times as much as visitors to Mesa Verde to insure clear (90 percentile) visibility compared to more degraded conditions that occur with similar frequencies at the respective parks. However, given that the average vehicle-trip to Mesa Verde is a single day visit compared to an

TABLE 19.6
Aggregate Model Parameter and Benefit Estimates: Great Smoky
Probabilistic Case

Attribute	Specification			
	J	Benefits ($ per vehicle)	K	Benefits ($ per vehicle)
Slight Haze = 0	Base	--	Base	
Improved 3 = 1	-0.2879	3.60	-0.2858	3.94
	(-1.7211)	(3.74)	(-1.2875)	(3.90)
Improved 2 = 1	0.0541	--	Omitted	--
	(0.4245)			
Improved 1 = 1	-0.0434	--	Omitted	--
	(-0.2345)			
Existing = 1	-0.6268	7.84	-0.6739	9.31
	(-1.8525)	(3.20)	(-1.5896)	(3.44)
Full Program = 1	0.1943	2.43	0.2485	3.43
	(1.2503)	(1.77)	(0.9527)	(3.36)
Price	-0.0800		-0.0724	
	(-3.4030)		(-2.5057)	
Number of Observations	202		202	
Maximum Likelihood Estimate	-928.1841		-343.9580	

Note: 1) Numbers in parentheses under parameter estimates are asymptotic t-statistics. These are calculated on a formula that is correct for infinite samples. For small samples, the measure in consistent, but not unbiased. At 190 degrees of freedom, a one-tail test requires a t-statistic equal to or greater than
$t_{0.995} = 2.59$ $t_{0.950} = 1.65$
$t_{0.975} = 1.97$ $t_{0.900} = 1.29$
2) Benefits reflect a compensating surplus estimate holding utility constant. Numbers in parentheses are standard errors of estimate for the benefit trade-off.
Source: Charles River Associates, based on computer analysis using an ordered logit model.

average length of stay of three days at Great Smoky, willingness-to-pay on a daily basis is about equal.

The variable nature of weather patterns and other atmospheric conditions precludes the possibility of clear visibility all the time even if there were no emissions from man-made sources. A marginal reduction in emissions at a single powerplant will change the probability of occurrence of various visibility conditions by small amounts. The benefits of changing the distribution of visibility conditions at the two parks are compared in Table

19.7 on both a trip and an annual basis. Total visibility benefits to users depend, of course, on total numbers of visitors. Great Smoky has more visitors annually than any other national park in the United States; about 8.5 million visitor days are recorded at Great Smoky each year. Mesa Verde receives about 0.5 million visitors each year. This translates into about 1.0 and 0.17 million vehicle-trips per year, respectively, and provides the basis for the estimate of annual visibility benefits in Table 19.7.

The greater number of visitors to the Smokies results in a significantly larger visibility benefit. Under the more realistic scenarios that describe visibility as a change in the frequency of occurrence of various conditions, an increase in the percentage of clear days at Mesa Verde from about 14 percent (typical of existing conditions) to 38 percent (hypothetical) no plant scenario would be worth under $500,000. At Great Smoky, increasing the percentage of clear days from about 10 percent under existing conditions to 29 percent (hypothetical Improved 3 scenario) yields annual user benefits in the range of $3.35-$5.00 million. In contrast, BART controls on one 2,000 MW power plant are estimated to cost nearly $80 million annually.

POLICY CONCLUSIONS

CRA has demonstrated that there are real benefits to visitors at national parks from improving visibility. However, these _user_ benefits do not appear to be of the same order of magnitude as the costs of controlling point source emissions necessary to reduce particulate, sulfate, and nitrate emissions to achieve significant visibility improvement. Only if non-user benefits are demonstrated to be substantial will it be possible to justify substantial pollution controls on existing plants on the basis of visibility alone.[17]

In the absence of such data on option and existence values for most Class I areas it seems sensible to move cautiously in the area of requiring new marginal BART controls for existing plants unless they are justified in terms of total benefits, including health or other non-visibility benefits. Instead, efforts should be concentrated on minimizing the visibility impacts of new plants where significant increases in emissions could markedly reduce the frequency of clearer visibility conditions. This paper dealt only with the benefits of improving visibility from the perspective of a willingness-to-pay for improvements. Assuming changes should be measured from the existing utility level, any reduction in visibility would have to be measured on a willingness-to-be-compenstated basis; and this benefit measure could be substantially larger than willingness to pay.

TABLE 19.7
Benefits of Improving Visibility at Mesa Verde and Great Smoky Based on Changes in the Frequency of Occurrence of Four Visibility Conditions

Visibility Improvement Scenario		Per-Vehicle Trip ($)		Total Annual Benefits
From	To	Deterministic	Probabilistic	($ millions)
MESA VERDE				
Existing	Clear	$3.02	$3.09	.504–.516
Agreed Controls	Clear	2.68	--	.448
Existing	Agreed Controls	0.34	--	.057
No Plant	Clear	1.80	0.93	.301–.155
Existing	No Plant	1.22	2.16	.204–.361
GREAT SMOKY				
Existing	Clear	$8.50–10.50	--	8.50–10.50
	Slight Haze	--	7.80–9.30	
Improved 3	Clear	$5.15–6.50	--	5.15–6.50
	Slight Haze	--	3.60–3.90	
Existing	Improved 3	$3.35–4.00	4.20–5.40	3.35–5.00

Notes: Scenarios are described in Table 2.
Source: Charles River Associates.

NOTES

1. The contingent ranking methodology was first applied to the problem of estimating the demand for electric cars by evaluating the value of different characteristics. See Beggs et al. (1981).
2. This modification was first accomplished by Dr. Jerry Hausman, Associate Professor of Economics, Massachusetts Institute of Technology.
3. A formal mathematical derivation of the ordered logit model is presented in Beggs et al. (1981).
4. Standard errors decline as a function of

$$\sqrt{(\# \text{ observations}) (^*\text{alternatives} - 1) - (\# \text{ parameters})}.$$

In the case of 200 observations, 8 alternatives and 5 parameters, doubling the number of observations from 200 to 400 will reduce standard errors by a factor of about 30 percent. Similarly, doubling the number of alternatives from 5 to 10 will also reduce the standard errors by about 33 percent. This approximates

$$1 \sqrt{\# \text{ alternatives} - 1}.$$

5. Slides of Mesa Verde were supplied and analyzed by the Visibility Research Laboratory, Las Vegas.

6. Slides of Great Smoky were supplied by the Tennessee Valley Authority, Muscle Shoals, Alabama, and analyzed by the Visibility Research Lab, Las Vegas.

7. A problem did arise early in the Mesa Verde survey, and we were forced to substitute a moderate haze slide with slightly different coloration from a different processing batch.

8. It was estimated that eliminating both plants would reduce sulfate loading by less than 5 percent, and the number of clear days would increase by only about 1 percent.

9. Given the transitivity assumption, deleting an alternative has no effect on the ordering of the remaining alternatives. If A is preferred to B, and B to C, then with B deleted A will still be preferred to C. The rationale for deleting the moderate haze condition is that, after switching clear condition slides for reasons described above, the moderate haze slide came from a different slide batch and its coloring may have biased respondent rankings.

10. Of the total sample, about 25 percent of the respondents ranked the cards wholly on the basis of price, and therefore indicated no willingness (zero value) to trade-off higher entry fees for improved visibility. For many others the ranking closely followed the ordering by price and caused the parameters on the poor visibility conditions to be positive.

11. If respondents are equally likely to err in both directions this assumption biases the benefit estimates upward. If one assumes that respondents who attached no value to visibility in the ordering were confused and ranked the cards incorrectly, then the values reported would increase by about 75 percent. On the basis of recent work in urban visibility where few respondents ranked on the basis of price, we do not feel it is justified to delete those who attached zero value to visibility improvement at Mesa Verde.

12. The consent agreement between Arizona Public Service Company, owners of the Four Corners power plant, and the State of New Mexico calls for the company to achieve 72 percent control of sulfur oxide emissions by December 31, 1984. Baghouses are already under construction that will control the particulate emission to 0.05 lbs. per million BTUs.

13. In addition to the coloration problem with the moderate haze slide in the deterministic analysis, it may be that respondents with no previous visits to Mesa Verde were unable to appreciate the differences in visual ranges--156 km for moderate haze and 256 km for clear--which may have been much greater than their everyday experience.

14. A wider price range was used at Great Smoky, since it was anticipated that the comparison of clear with more degraded haze conditions typical of the area would result in a higher willingness to pay and necessitate higher fees to force a trade-off. We started at zero, since there is currently no fee to enter the park.

15. Before concluding that the Smokies are well-named, we direct the readers' attention to the individual analysis.

16. The only evidence available to indicate the value of visibility to individuals that ranked their alternatives by visibility from best to worst is in the amount individuals were willing to pay to insure a clear condition on a future visit (bidding question). This insurance bid evidence suggests that the value assigned to individuals, for whom a visibility parameter cannot be estimated because of perfect ordering, is probably not far above the minimum necessary to justify the ordering. An option value analysis was undertaken using insurance payments (bids) but is not reported here.

17. Schulze et al. (1981) reported large non-user benefits in a University of Wyoming study of the Grand Canyon and other southwest parks, but results obtained by Randall et al. (forthcoming) suggest that these magnitudes may depend greatly on the order in which the questions are asked. More work is needed to clarify this area and to identify the magnitudes of the potential bias.

BIBLIOGRAPHY

Beggs, S. S. Cardell, and J. Hausman. 1981. "Assessing the Potential Demand for Electric Cars." Journal of Econometrics 16: 1-19.

Bishop, R. C., and T. A. Heberlein. 1980. "Simulated Markets, Hypothetical Markets, and Travel Cost Analysis: Alternative Methods of Estimating Outdoor Recreation Demand." Agricultural Economics Staff Paper, University of Wisconsin, Madison, WI.

Randall, Alan, John Hoehn, and George Tolley. Forthcoming. "The Structure of Contingent Markets: Some Results of a Recent Experiment." American Economic Review.

Schulze, W. D., D. S. Brookshire, E. Walther, and K. Kelley. 1981. Methods Development for Environmental Control Benefits Assessment. Vol. X: The Benefits of Preserving Visibility in the National Parklands of the Southwest. U.S. Environmental Protection Agency. Office of Research and Development, Washington, D.C.

20. Economic Valuation of Potential Scenic Degradation at Bryce Canyon National Park

F. Reed Johnson
Abraham E. Haspel

Energy independence through the development of domestic energy resources has been a goal of the federal government since 1974. Nevertheless increased oil and gas drilling, the tapping of vast coal deposits through surface mining, the development of synthetic fuels from oil shale and tar sands, and the expansion of nuclear facilities inevitably involve substantial environmental costs. While energy development is often accompanied by air and water pollution, development may also damage rare and delicate geological and ecological characteristics. Among the values that may be lost as a result of development activities are the visual and scenic benefits of certain landscapes. Given the peculiar difficulties in quantifying aesthetic and consumptive uses of natural resources in commensurate terms, trade-offs between such values pose serious dilemmas for policy makers legally responsible for managing public lands under multiple use criteria.

A recent conflict over proposed surface mining near Bryce Canyon National Park provided the impetus for studying these kinds of trade-offs. The Bryce Canyon case is precedent setting in that it marks the first unsuitability petition filed under the provisions of the Surface Mining Control and Reclamation Act of 1977. The way in which issues are ultimately resolved in this case bears on future implementation of the act and the current review of associatied regulations ordered by Interior Secretary James Watt.

This paper reviews the background to the Bryce controversy, describes data obtained by the Department of the Interior, and reports some recent analytical results derived from the data.

BACKGROUND TO THE CONTROVERSY

Even before the oil embargo of 1974 and the subsequent energy shortages, the need for additional sources of electric power was evident for the southwestern portion of the country, especially Southern California. In

---

F. Reed Johnson is an Associate Professor of Economics, U.S. Naval Academy, and Abraham E. Haspel is an economist with the Office of Policy Analysis, U.S. Department of the Interior. The views represented in this paper are those of the authors and not necessarily those of the U.S. Department of Defense or the U.S. Department of the Interior.

1973 the Nevada Power Company, the City of St. George, Utah, the Washington County Water Conservancy District, Utah, and the Department of Water and Power of the City of Los Angeles (later replaced by Southern California Edison Company and Pacific Gas and Electric Company) notified the Bureau of Land Management (BLM) of their intention to develop the Allen Warner Valley Energy System (AWVES). The multicomponent AWVES, as envisioned by its designers, included the 2000 megawatt (MW) Harry Allen powerplant (near Las Vegas), the 500 MW Warner Valley powerplant and a 55,000 acre-foot reservoir near St. George, Utah, and two coal slurry lines from a new coal mine on coal leases issued in 1963 and 1968 in the Alton coal field near Bryce Canyon National Park (BCNP). AWVES's primary purpose was to generate power for transmission to Southern California.

As BLM and the participants in the system began the process of preparing the necessary environmental information required by the National Environment Protection Act and Department of the Interior regulations, Congress passed the Surface Mining Control and Reclamation Act (SMCRA), requiring lands that were surface mined to be reclaimed. Additionally, SMCRA provided a mechanism in Section 522 whereby lands could be declared unsuitable for mining on the basis of certain criteria. This mechanism provides an opportunity for any person adversely affected by mining to petition the appropriate regulatory authority. The petition must allege unsuitability on the basis of at least one of the following criteria:

1. Reclamation pursuant to the requirements of SMCRA is not technologically and economically feasible.
2. Mining operations are incompatible with existing land use plans.
3. Mining operations will "affect fragile or historic lands in which such operation could result in significant damage to important historic, cultural, scientific, and aesthetic values and natural systems."
4. Mining operations will "affect renewable resource lands in which such operations could result in a substantial loss or reduction of long range productivity of water supply, or of food..."
5. Mining operations have an effect on natural hazard lands.

If proven, the first criterion requires mandatory designation of unsuitability; otherwise designation is discretionary and can be based on the other four criteria.

On November 28, 1979, the Office of Surface Mining (OSM) received a petition, submitted by the Sierra Club, Friends of the Earth, Environmental Defense Fund, and seven landholders, asking that certain federal lands abutting BCNP be designated unsuitable for all types surface coal mining operations. Arguments in support of their request alleged that following mining it would be impossible to reestablish a suitable vegetation cover because of unfavorable soil and climatic conditions and lack of an adequate supply of irrigation water of suitable quality. The petitioners claimed that mining could damage the aesthetic values and natural system of the park and Dixie National Forest by scarring scenic vistas, reducing visibility, degrading air quality, threatening the geologic formations in BCNP, damaging the area's ecological system, diminishing the area's recreational value, and disrupting essential wildlife habitat. They also argued that mining in the petition area would damage the hydrology of arid lands and injure crop and pastureland by excessive withdrawal of groundwater.

The petition set in motion a series of studies to investigate the merits of the petition allegations and to provide information to then Interior Secretary Cecil Andrus upon which to base a response. Study results failed to substantiate claims regarding the infeasibility of reclamation. A finding of unsuitability was therefore discretionary and the decision ultimately turned on the question of whether surface mining should be permitted on nearby federal land when mine operations would be visible and audible within BCNP. Thus only the allegation concerning aesthetic values emerged as the critical factor in the secretary's decision.

In December 1980, Andrus announced his decision to declare the portion of the disputed lands visible from the park unsuitable for mining. The secretary observed in his decision that the area contained less than 10 percent of the coal proposed for development under federal leases. Both the developers and the environmental groups have subsequently disputed the decision in the courts. Within the last few months a federal judge ruled not to remand the decision to Secretary Watt for review. The outcome of this pending litigation will ultimately decide the disposition of the contested resources, although interest in AWVES appears to have diminished on the part of the developers.

GEOGRAPHY OF THE AREA

A short geographical description of the park and surrounding area will assist in understanding the issues involved in this study. The park is located in southern Utah, about twenty miles east of Panguitch. It is a narrow strip of land about twenty miles long, situated on the edge of a plateau, and oriented in a north-south direction. Bryce Canyon is not a canyon at all, but rather the highly eroded pink sandstone face of the plateau. It is the last and highest step in the "Grand Staircase," a geologic structure that rises between the Grand Canyon in Arizona and Bryce Canyon in Utah. The unusual topography and geology of the park attract visitors from all over the world.

All but one of the designated viewpoints in the park look down on the eroded cliffs and distant scenery to the east. Visual range of over ninety miles is not uncommon. The remaining viewpoint, Yovimpa Point, faces southwesterly toward the lower portions of the Grand Staircase. The eastern section of the currently undeveloped Alton coal field is visible from Yovimpa Point, a portion of the field coming within five to seven miles of and 2000 feet below the viewpoint.

THE 1980 BCNP VISITOR SURVEY

In order to assess the impact of proposed surface mining on park visitors, a survey was conducted by the National Park Service and the Office of Policy Analysis between June 15 and September 12, 1980. The survey was designed to elicit information in three areas: (1) the travel and recreation behavior of visitors to BCNP; (2) their socioeconomic characteristics; and (3) their views about potential changes in the view from Yovimpa Point if surface mining were to take place.

The survey consisted of three parts administered to different samples of the visitor population. Bus visitors were not sampled. Survey A was administered to exiting visitors completing their visit to the park. Respondents provided detailed information on their trip itinerary, participation in park activities, and socioeconomic characteristics. A comparison of value estimation techniques was facilitated by obtaining information on respondents' incremental willingness to drive to the park as well as actual distance traveled. The part A sample provided 722 observations.

Survey B also sampled exiting visitors. This part elicited information about travel within the park as well as activity involvement. Specific attention was paid to whether or not the party visited Yovimpa Point, and incremental willingness to drive data were obtained to provide a basis for estimating perceived benefits of that particular viewpoint. The part B sample provided 663 observations, of which ninety-two, or 29 percent, had visited Yovimpa Point.

Survey C was administered to 1311 visitors at Yovimpa Point itself. (Buses do not stop at this viewpoint.) Visitors responded to questions on a standard set of socioeconomic characteristics and were then shown three photographs of the view from Yovimpa Point depicting (1) the status quo, (2) the view with surface mining simulated, and (3) the view simulated after the mine site was reclaimed upon completion of mining. The photographs were used to ask visitors how the "value" (undefined) of both the specific vista and the park overall would be affected and how the length of time they stayed at the point and at the park would be affected if the view were altered as depicted in the photographs.

Table 20.1 shows the comparison among several common variables from the three survey parts. The Yovimpa Point subsample of the B survey (B(Y)) and the C survey sample both involved Yovimpa Point (YP) visitors. At the 10 percent level of significance or better, the sample means are the same for number of passengers in the vehicle, age, and income across all three samples. Since the driver was the respondent in the exit surveys, proportionately more males were sampled at the gate. Oddly, the proportion of respondents who had previously visited the park is greater in the B sample, and the average length of stay is much greater in the C sample.

ANALYSIS OF THE DATA

Travel Cost and Direct Willingness to Drive Value Estimates

We have recently reported estimates of the value of BCNP derived from Survey A data (Haspel and Johnson, 1982). These estimates were calculated from trip itinerary and origin data and from responses to a question on how much further people would have been willing to drive to get to the park. The latter data can be converted directly into a demand curve and willingness to pay values by assigning an appropriate per mile cost of travel.

Actual travel behavior provides a basis for applying the well-known Clawson-Knetsch travel cost method (TCM) for deriving demand for a site. (See, for example, Clawson and Knetsch 1966 and Water Resources Council 1979.) Unfortunately, one of the basic assumptions of this procedure is violated in the case of BCNP. Visits to the park are typically part of a trip that involves stops at several destinations. The joint costs of travel cannot therefore be attributed to BCNP alone. It is possible to modify the basic procedure to adjust for this problem. Details of this modification are discussed in Haspel and Johnson (1982).

Under the assumption that visitors from given states sampled in surveys B(Y) and C traveled similar distances and had similar itineraries as those in Survey A, we have applied the modified TCM procedure to Yovimpa Point visitors. Results of these calculations are shown in Table

TABLE 20.1
Comparison of Samples from the Three Surveys

	Means (Standard Deviation)			T-Test on Difference of Means		
	A	B(Y)	C	A:B(Y)	A:C	B(Y):C
PASS	3.02 (1.51)	2.93 (1.59)	2.97 (1.50)	0.63	0.57	-0.29
PREV	1.63 (0.48)	1.71 (0.45)	1.23 (0.42)	-1.92	14.99	12.53
PDVAC	1.45 (0.53)	1.45 (0.50)	----	0.00		
AGE	4.41 (1.24)	4.35 (1.23)	4.32 (1.25)	0.53	1.24	0.28
INCOME	6.34 (1.86)	6.74 (1.81)	6 53 (2.37)	-2.41	-1.58	1.27
SEX	1.16 (0.38)	1.22 (0.42)	1.31 (0.46)	-1.61	-6.26	-2.45
STAYT	20.37 (21.89)	25.67 (22.05)	41.00 (31.58)	-2.65	-13.65	-7.44
TRIPT	17.66 (20.55)	---- ----	47 (155.89)		-5.35	
#OBS.=	472	184	787			

PASS: number of passengers in vehicle
PREV: previous visit (1=yes, 2=no)
PDVAC: on paid vacation (1=yes, 2=no)
AGE: age category (5= 30-39)
INCOME: income category (6= $20-25,000; 7= $25-35,000)
SEX: sex (1=male, 2=female)
STAYT: length of stay at park in hours
TRIPT: length of entire trip in days

20.2. While the per vehicle willingness to pay of the B(Y) sample is about $20 less than that of the C sample, the two estimates bracket the values of $62 and $71 per vehicle derived from Survey A. These results consequently do not suggest that the value of the park to Yovimpa Pont visitors is different from BCNP visitors in general.

Survey B asked a direct willingness to drive to Yovimpa Point question analagous to the question in Survey A for the whole park. Respondents provided answers in minutes, which are converted into costs by assuming a given average speed and per mile costs. The speed limit in the park is 35 MPH. Table 20.3 shows the sensitivity of the estimates over a range of 30 to 40 MPH. The average willingness to pay of $3.40 over and above actual access cost is about 5 percent of the value per vehicle for the entire park. Yovimpa Point thus represents about 1.5 percent of the total value of the park based on 1980 visitation.

It should be emphasized that these "lower-bound" estimates are derived from data on non bus park visitors and obviously do not reflect the existence, option, and bequest value of the park to nonvisitors (see Greenly et al. 1981). These values are also willingness to pay estimates rather than willingness to be compensated estimates. The latter values can be expected to be larger and are based on different assumptions about the legitimate distribution of property rights in our society (Gordon and Knetsch 1979). Nevertheless, these estimates are based on empirical evidence and provide value measures in terms comparable to commercial values that may be involved in many multiple use allocation problems.

Unfortunately, the data collected on the expected value losses associated with visual degradation of the Yovimpa Point vista were not in a form that allowed further application of the foregoing estimation techniques. We have attempted to analyze the qualitative data from Survey C given its limitations, however, and report some of these results in the following section.

Analysis of Qualitative Response Data

Survey B obtained data on additional willingness to drive to Yovimpa Point (DDRIVY) and length of stay at the vista (STAYY), under existing conditions. The means for these variables are twenty-nine and forty-four minutes, respectively. Survey C obtained data on changes in value and length of stay in the presence of visual quality changes expected with surface mining (DVALY and DSTAYY)[1]. Respondents chose among three alternative responses concerning the impact of mining upon the park value: decreased, stay the same, or increased. They were offered six choices for the effect of mining on the length of visit at the point ranging from "no visit" to "significantly more." We expected these values and consumption variables to be functions of prior information, time in the park, and various socioeconomic variables reflecting tastes and resource constraints.

Ordinary least squares regression and probit analysis were used to investigate the relationship between these value and length of stay measures and various hypothesized determinants[2]. Only four independent variables were statistically significant in explaining the dependent variables: previous visit (PREV), length of stay in the park (STAYT), age category (AGE), and income category (INC). The results are shown in Table 20.4.

We mention here only those those coefficients that were significant at the 10 percent level or better. The PREV coefficient is negative for STAYY. Those who had never been to BCNP before tended to spend more time at Yovimpa Point, presumably reflecting the novelty value of seeing the vista for the first time. The coefficient for STAYT is positive for STAYY, as one would expect. Those who stay in the park longer also tend to experience greater loss in value and greater reductions in time at YP in response to the worst case photograph. STAYT may then reflect a taste factor favoring existing characteristics of the park. The AGE coefficient is positive for both the DVALY and DSTAYY, indicating stronger prefer-

Table 20.2
Comparison of Per Vehicle and Total Willingness
to Pay for BCNP by Yovimpa Point Visitors, 1980

Sample	Ave. WTP All Visitors ($)	Ave. WTP YP Visitors ($)	Total WTP YP Visitors ($ Mill.)
A Survey			
Travel Cost Est.	62		2.23
Will. to Drive	71		2.56
B(Y) Survey		59	2.12
C Survey		81	2.92

Estimated 1980 visitation to BCNP = 125,000 vehicles (excl. bus visitors).
Estimated 1980 visitation to YP = 36,000 vehicles.
Assumes $.20/mile travel cost; no imputation for time costs.

Table 20.3
Per Vehicle Willingness to Pay for Yovimpa Point, B(Y) Sample

Assumed Ave. Speed	Ave. WTP ($)	Total WTP ($1000)
30 MPH	2.93	105
35 MPH	3.40	122
40 MPH	3.93	141

Mean additional willingness to drive to YP = 29.29 minutes.
Estimated 1980 visitation to YP = 36,000 vehicles.
Assumes no time cost.

ence on the part of younger people for environmental preservation. This result is also consistent with intuition. Holding age constant, however, the coefficient on INC for DVALY and DSTAYY is positive. Much of the literature on income elasticity of environmental quality suggests that the benefits of conservation accrue largely to wealthier groups. Such groups may also bear a disproportionate share of the costs of improving environmental quality, however, and the net effect on welfare can possibly be negative (Johnson 1980).

Note that none of the independent variables are useful in explaining DDRIVY, and only PREV and STAYT are significant for STAYY. The demand determinants do tend to contribute significantly to explaining changes in value and consumption because of surface mining, however.

The qualitative responses of DVALY and DSTAYY are crosstabulated in Table 20.5. The largest cell in the table is for respondents who indicated no change in value and no change in length of stay if surface mining occurred: 29 percent indicated no change for either question. Of those who indicated a loss in value, only 8 percent would stay less or not visit Yovimpa Point at all, and 5.7 percent would actually stay longer. About the same percentage of those who indicated an increase in value would stay less time.

These results suggest that the impact of surface mining does not cause a simple parallel shift in demand curves, but both the slope and intercept change. The apparent perverse tendency for those who experience no change or an increase in value to decrease their consumption is consistent with responses of about 6 percent of the B(Y) sample that YP was worth less than the access cost (at an average net loss of $2.17). In the absence of complete prior information about the viewpoint, some visitors simply make mistakes in calculating the net benefits of going there. The costs of getting to the site may therefore exceed benefits with or without mining for some respondents. For those who experience a loss in value, but would not decrease their consumption, the reduced benefits are still greater than the access cost.

The Economic Cost of Scenic Degradation

Given the absence of any cardinal measure of the value loss to Yovimpa Point visitors, is it possible to justify any economic estimate of this loss? Presumably the total estimated value of Yovimpa Point to visitors places an upper bound of about $120,000 annually on the possible cost, adjusted appropriately for growth in demand and discounting over the life of the mine (Haspel and Johnson 1982). Since about 70 percent of the visitors expected no loss in value, the upper bound may arguably be reduced to $36,000. In order to go beyond this, it would be necessary to calculate actual changes in the slopes and intercepts of the demand curve for Yovimpa Point. Considerable experience has been accumulated in collecting and analyzing data to obtain such estimates. Unfortunately the particular design of Survey C does not allow us to go much further[3].

It should be noted in conclusion that the amenity loss to the park discussed here does not include losses from noise or dust at either Yovimpa Point or other locations within the park. Added to nonvisitor and bus visitor losses, the value losses could be considerably larger than those we have identified here.

Table 20.4
Regression and Probit Results

Dependent Variables	Independent Variables			
	PREV	STAYT	AGE	INC
OLS Regression Coefficients and t-ratios				
1. STAYY	-23.1259 (-2.51)	0.6214 (4.22)	-0.9914 (-0.29)	-1.8435 (-0.78)
2. DDRIVY	-0.1325 (-0.36)	-.0003 (-0.51)	-0.1774 (1.32)	0.1351 (1.45)
Probit Beta Coefficients and t-ratios				
3. DSTAYY	3.7384 (0.40)	-0.5658 (-4.69)	9.0852 (2.73)	2.4942 (1.76)
4. DVALY	-13.6419 (-1.24)	-0.5795 (-3.51)	13.0777 (3.42)	3.7157 (2.36)

1. R bar squared = .14
2. R bar squared - .01
3. -2 times log likelihood ratio = 28.79 (4 D.F.)
4. -2 times log likelihood ratio = 30.56 (4 D.F.)
   (Chi-squared tests significant at 1% level)

STAYY: Length of stay at YP (minutes)
DDRIVY: Additional willingness to drive to YP (minutes)
DVALY: Change in value of YP for worst case photograph (decrease, same, increase)
DSTAYY: Change in length of stay at YP for worst case photograph (no visit, significantly less, slightly less, same, slightly more, significantly more)
PREV: Previous visit (1=yes)
STAYT: Length of stay at park (hours)
AGE: Age category
INC: Income category

Table 20.5
Crosstabulation: Change in Stay at YP and Change in Value of YP for Worst Case Photograph

	Change in Value			
Change in Stay	Decrease	Same	Increase	All
No Visit	2	56	0	58
	3.45	96.55	--	100.00
	0.76	9.00	--	6.42
Significantly Less	6	144	0	150
	4.00	96.00	--	100.00
	2.27	23.15	--	16.59
Slightly Less	13	148	1	162
	8.02	91.36	0.62	100.00
	4.92	23.79	5.56	17.92
Same	228	259	7	494
	46.15	52.43	1.42	100.00
	86.36	41.64	38.89	54.65
Slightly More	13	14	9	39
	36.11	38.89	25.00	100.00
	4.92	2.25	50.00	3.98
Significantly More	2	1	1	4
	50.00	25.00	25.00	100.00
	0.76	0.16	5.56	0.44
All	264	622	18	904
	29.20	68.81	1.99	100.00
	100.00	100.00	100.00	100.00

Cell Contents-- Count
% of Row
% of Column

NOTES

1. Parallel questions were asked about the effect of mining on value and length of stay of the whole park. Since it was unclear how respondents included the change in YP value in their calculation of change in park value, we have not included these responses in our analysis.

2. The assumptions of OLS regression are violated when the dependent variable is discrete. Probit derives maximum likelihood coefficients by constructing a continuous probability index for the dependent variable. The coefficients can be transformed into probabilities that the dependent variable will have a given value for specified values of the independent variables. (See Pindyck and Rubinfeld 1976.)

3. We have attempted to weight responses to produce a kind of "fully-impacted equivalent" number of YP visitors, but the weights used are necessarily arbitrary.

BIBLIOGRAPHY

Clawson, M., and J. L. Knetsch. 1966. Economics of Outdoor Recreation. The Johns Hopkins Press, Baltimore, MD.

Gordon, I., and J. Knetsch. 1979. "Consumer Surplus Measures and the Evaluation of Resources." Land Economics 55 (February).

Greenly, D. A., R. G. Walsh, and R. A. Young. 1981. "Option Value: Empirical Evidence from a Case Study of Recreation and Water Quality." Quarterly Journal of Economics 96 (November): 657-673.

Haspel, A. E., and F. R. Johnson. 1982. "Multiple Destination Trip Bias in Recreation Benefit Estimation. Land Economics 58 (August).

Johnson, F. Reed. 1980. "Income Distributional Effects of Environmental Policy." Atlantic Economic Journal 8 (December).

Pindyck, R. S., and D. L. Rubinfeld. 1976. Econometric models and Economic Forecasts. McGraw Hill, New York, NY.

U.S. Water Resources Council. 1979. "Procedures for Evaluation of National Economic Development (NED) Benefits and Costs in Water Resources Planning (Level C): Final Rule." Federal Register 44 (December 14): 242.

21. Priorities for Economic Analysis of Visibility Values

Robert D. Rowe
Lauraine G. Chestnut

INTRODUCTION

This paper takes a first cut at identifying and categorizing the importance of problematic issues currently faced in economic analysis of visibility values, including many that have been brought up in this volume and many that have not. The paper is to serve as a starting point in the economics area to identify issues needing future analysis and discussion which are most important. We do not expect this to be the final word on the current state of affairs, but rather look forward to the readers' suggestions and proposed revisions. The comments here are directed specifically at visibility benefit analysis, but we believe they are equally applicable to visual impact analyses.

To rank issues and directions for future research we first suggest the following goals for benefit analyses.

1. Establish defensible benefit estimates for specific case studies.
2. Establish defensible benefit estimates for answering broad national policy questions.
3. Establish a set of defensible benefit estimates with sufficient accuracy, reliablity, and variety to be transferred for new uses while meeting the first two goals.

To date, much progress has been made. Results of visibility benefit analyses have shown that visibility benefits can be quantified, that the amounts are substantial enough to influence benefit-cost analyses of air pollution controls, and that they improve the information available for resource allocation decisions. Complete confidence cannot as yet be placed in the exact values so far estimated, but ongoing progress warrants continued efforts. Much work is especially needed to meet the third goal. To date, there are insufficient studies, with insufficiently precise informa-

---

Robert D. Rowe and Lauraine G. Chestnut are economists with Energy and Resource Consultants, Inc., Boulder, Colorado. Support for presentation of this paper was provided by the American Petroleum Institue and the Electric Power Research Institute. This paper draws upon a more general treatment of economic approaches to visibility analyses in Rowe and Chestnut (1982b).

tion, to be able to transfer results to new situations with a great deal of confidence. Transferability of results is an important goal because it will facilitate analysis of new issues at considerable savings of time and money.

The economic analysis issues addressed can be divided into two general categories: (1) general economic and policy issues and (2) technique specific technical issues. Previous work has made great progress in the development of estimation techniques leaving the issues in the first category as the most crucial for consideration in future research to meet the three identified goals. Within these two catagories, we further rank the needs for future work as those that are "most important," "important," and "useful."

Our rankings are illustrated by the following example. We recently participated in a regulatory impact analysis concerning integral vistas for the National Park Service (Chestnut and Rowe 1982). This analysis examined the benefits of preventing the potential occurrences of several plumes in integral vistas that extend outside, but are visible from viewpoints within, national parks. Given limited resources and time for the first round of analysis, we hoped to establish a feel for what benefits of protection might be by transferring results from related research efforts. We were quickly able to assess that estimates of on-site user benefits at Southwestern national parks are reasonably consistent--say ± fifty percent for the most similar changes (Rowe and Chestnut 1982b). It would be "useful" to have more studies using alternative techniques and analyses of more carefully developed scenarios, but this is not a critical need. Any plume blight visible to park visitors would also affect local residents, quite possibly in greater numbers and frequencies. More benefit analyses for these populations would be very beneficial or "important" to meeting our goals. A third concern was the potential existence of preservation values (existence values, option values, bequest values, etc.) Even if these values for protecting national parks from such impacts were extremely small for the typical U.S. household, but existed nationwide, with eighty million such households the aggregate preservation values would overwhelm user values by a factor of 100 or more. More work on preservation values is therefore "most important."

GENERAL ECONOMIC AND POLICY ISSUES

Eleven issues are identified in this area.

Most Important Issues

1. <u>Visibility aesthetics are one of many types of benefits to compare to emission control costs.</u> It is most important to keep this in mind. One should not necessarily expect, and certainly not require, visibility aesthetic benefits alone to meet or exceed costs of controls, although the focus of the regulations requiring controls may not include all of the benefits that may result. This issue is far too often overlooked in pollution control evaluation studies. It also suggests that benefit analyses for all types of impacts need to move forward simultaneously.

2. <u>All preservation value issues need more examination.</u> Because preservation values (option, existence, bequest, etc.) to prevent visibility impacts, if they exist, may greatly overwhelm user values, much more work is needed. Previous studies indicate that preservation values may exist in some circumstances, but many questions have yet to be addressed. We

need to better understand the philosophical/psychological basis of what these values are and why they exist. We also need more work to understand the effects on preservation values of differences in local/non-local use patterns, site uniqueness and the availability of substitute sites, reversibility of impacts, and characteristics of the impacts being evaluated. Finally, much more work is needed to develop and refine techniques to estimate preservation values.

3. Move toward marginal analysis. While physical scientists have made considerable strides in modeling marginal changes in air quality and perception, economic analyses have not. For example, economic analyses have only looked at values for the existence or non-existence of plumes, all the while knowing that changes in the size, color, duration, and location may affect the estimated values. Another example is illustrated in Figure 21.1 which shows potential relationships between preservation values and the number of days protected from adverse impacts at a site. There have been estimates for the preservation value of protecting a site from visual aesthetic impacts on most or all days of say $X (Schulze et al. 1981) and we know preservation values will be $0 if zero days are protected from adverse impacts. Yet to date there is no information concerning the shape of the function between these two endpoints and it is therefore difficult to answer such questions as, "What are preservation values if we protect a site from a specified impact that could occur twenty times per year?" Similiar issues must be addressed in nearly all aspects of economic benefit analyses for visiblity protection.

4. Consider benefit-cost analyses of alternative locations versus alternative equipment levels. Analyses to date have focused upon the sensitivity of benefits and costs for controlling alternative levels of emissions from a source at one site under one set of control strategies. For example, current Clean Air Act regulations often require a specific reduction in emissions or a specific type of control equipment. Benefits and costs of controls are usually estimated on the assumption that these types of control strategies will be used. Source siting locations are also often assumed to be given. It would be extremely beneficial to perform similiar analysis, but to also examine the sensitivity of benefit and cost estimates for the same source at several alternative sites and the sensitivity of benefits and costs to changes in control strategies. Perl and Dunbar (1982) show that changes in control strategies, such as instituting regional versus source specific emissions limits, may significantly alter benefit-cost ratios without substantive changes in the intent of pollution control legislation.

5. Continue validation of estimates across studies. Because economic benefit analyses are still rather new, especially in terms of marginal impact analyses and examining impacts across a variety of different sites, it would be most beneficial to continue verification of results through comparison and validation using alternative techniques and through repetition with the same technique.

6. Specific case studies. As specific issues or resource utilization conflicts arise, economic benefit methodologies must be employed to the best of their abilities to address immediate resource allocation issues.

Important Issues

1. Estimate user values in more diverse situations. User value estimates have shown a broad consistency for visibility impacts in Southwestern national parks (Rowe and Chestnut 1982b). However, little is

FIGURE 21.1
Total Preservation Values Related to Number of Days Of Adverse Impact Prevented

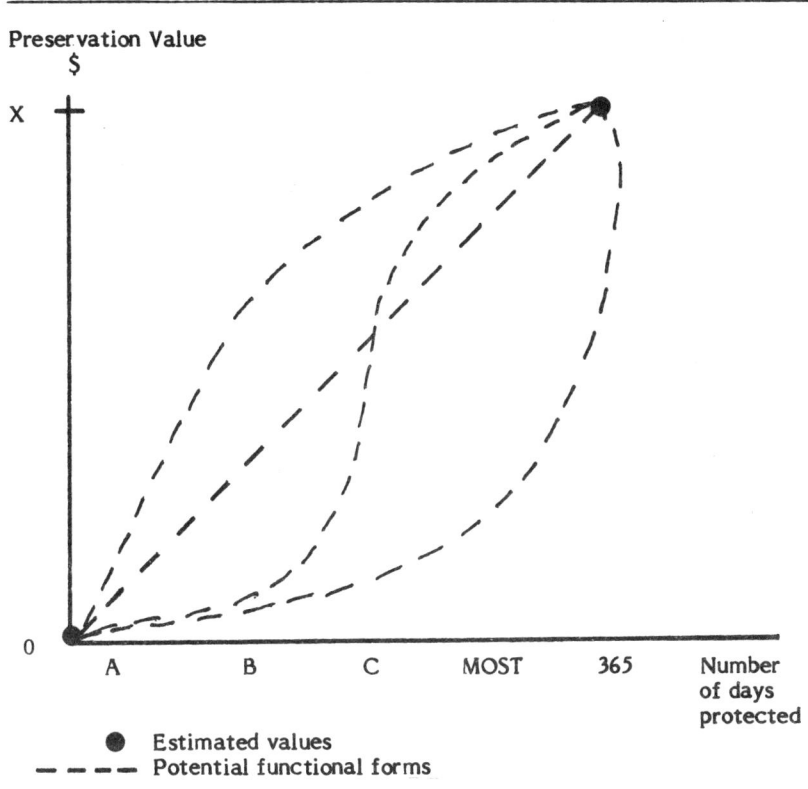

known about user values for similiar impacts in other parts of the country, at wilderness areas, or at recreation areas where there is a substantial share of use by local rather than non-local residents. More work is needed to estimate user values for impacts at urban and non-urban residential areas. For example, user values associated with protecting visibility at a national park may be far outweighed by values local residents may have for preventing the same impact from occurring simultaneously at their places of residence. Local residents may simply value the same level of protection more than park visitors or there may be more local residents affected and effects may occur more frequently than for than park visitors.

2. Better definitions of behavior and attitude impacts and group impacts. Economic benefit analyses tend to focus upon the current effects to the individual and thus may miss group or societal effects, or impacts that occur at a later time, such as personal development and future sharing of the experience, as discussed by Opie, Craik and others in this volume. We also know that due to externalities and spin-off effects, the sum of benefits

to those individuals directly affected may not accurately represent total benefits. Economists have focused upon readily identifiable behavior expected to be affected by changes in visibility often without a thorough understanding of all the behaviors and attitudes that may actually be affected. Finally, it has been assumed, probably incorrectly, that if an individual does not perceive or hypothesize that there will be an impact, then no benefits from protection will occur. While the economist's focus probably captures the majority of benefits, more work on these topics is still very important to ensure that significant benefits are not being overlooked.

Useful Issues

1. Recreation user value studies in Southwest Class I areas. Because current legislation specifically addresses impacts to visibility in Class I areas and because of the projected rapid growth in energy development in the Southwest, continuation of such efforts is still useful.
2. Examine preference trends. Current economic analysis focuses upon present day values and forecasts them into the future. With increasing trends toward recreation as an important social activity and with increasing demand upon scarce resources, one might expect future benefits from protecting recreation resources to exceed current benefits at an increasing rate. Costs of protecting recreation resources are also likely to change over time, although technological improvements and increasing resource scarcity will have opposite impacts on the direction of change. In either case, benefit-cost ratios are likely to change over time. There is little concrete evidence in these areas and evaluation of historical trends could prove insightful.
3. Quasi-option values. Quasi-option values refer to the value of deferring a control decision or reserving the right to evaluate any potential impacts at a later date, to more accurately evaluate them at that time. These benefits occur because information concerning and techniques for evaluating future impacts may improve over time, thus yielding more accurate resource allocation decisions if deferred as long as possible. In some sense this is the essence of the National Park Service integral vista program, which requires that potential impacts to integral vistas be considered case by case in permitting decisions. To justify such programs it needs to be demonstrated that the quasi-option values exceed administrative costs. The problem is that little has been done to estimate these values or separate them from total preservation value estimates.

TECHNIQUE SPECIFIC TECHNICAL ISSUES

Eight "most important" and "important" issues are identified and discussed below. Several "useful" issues are mentioned briefly.

Most Important Issues

1. Validation across techniques. One of the most important tests for the validity of these techniques in application is whether they yield similiar results for similar impact analyses. This point was mentioned above under General Issues, but applies equally to the development and examination of specific techniques.
2. Validation of intended versus actual behavior. All contingent valuation techniques rely upon stated or intended behavior rather than actual

behavior. If this intended behavior is not an accurate reflection of what actual behavior would be in the hypothesized situation, the results of these techniques will be flawed, although one may still be able to determine if the estimates are upper or lower bounds upon "true values." There is some evidence that such hypothetical biases or inaccuracies may exist (Bishop and Heberlein 1979). Because contingent market techniques seem to be the only techniques available to address certain issues, more work is certainly merited in this area.

3. Technique development to capture preservation values. Variations on bidding methods are the only approaches thus far used to estimate preservation values for environmental goods (Schulze et al. 1981, Walsh et al. 1978 and 1982). Due to limitations in these approaches and for the sake of validating results, creative thinking will be required to develop alternative approaches to address these issues.

Important Issues

1. Consideration of adjustment processes and averting behavior. Depending upon the specific application of a benefit estimation technique, these techniques tend to estimate per period benefits (damages) resulting from either short term or long term adjustments by the respondent. In most cases a respondent's per period benefits will differ in the long run from in the short run because individuals will learn to minimize adverse impacts or maximize beneficial impacts as they adjust to the changed circumstances. Benefit estimation techniques have tended to ignore these adjustment costs and estimate benefits for one period in time. Extrapolating these one period estimates through time misstates total benefits. While this has at times been considered for air pollution health impacts, we are unaware of thorough consideration of this issue for visibility aesthetics. (See Rowe and Chestnut 1981 and 1982a for more on this issue.)

2. Bidding method application issues. Several specific issues of concern, beyond those identified above, include considerations of how the implicit assignment of air quality property rights in bidding method questions affects the bids received, how incorporation of budget constraints will affect bids, and how the order of presentation of environmental goods to be examined affects the values estimated for each good. These issues apply to all bidding method applications including user and preservation value analyses.

3. Contingent ranked attributes application issues. Initial applications of this approach have proved very promising (Rae, Chapter 19 in this volume), but much development and refinement is needed. A concern with this approach is the need for respondents to simultaneously consider changes in several different attributes of the experience being considered, such as price, congestion and air quality for a recreation experience at a certain site. If respondents focus upon changes in only one attribute, i.e., use a lexicographical ordering, accurate valuation of the attribute is impossible. This has been a problem in the first applications of this technique for air quality benefit analyses. Further, if alternatives are not clearly understood, as may be the case for individuals who have never experienced the alternatives under consideration, the ability to rank alternatives accurately is seriously impaired. This problem is similar to that of hypothetical bias.

Initial applications of this approach have estimated visibility aesthetic values roughly consistent with similar studies using alternative approaches; however, these results, which include numerous statistical anomalies and contradictory results, such as some negative values for improved visibility, indicate that much refinement in the technique is needed.

4. Travel cost application issues. The travel cost approach has been very defensible in many applications because it uses actual observed or typically undertaken behavior. The most important problem for measuring visibility aesthetic values is that the approach has been designed to measure value for the existence versus nonexistence of a site, rather than for changes in characteristics of the site. Two general adaptations of the approach could be used to measure the effects of changing air quality upon demand--surveys obtaining actual visitation data and surveys obtaining hypothetical visitation data, both related to changes in air quality. How these types of approaches could be developed is examined in Rowe and Chestnut (1982b). To date there have been no applications of travel cost approaches to visiblity aesthetics, although pioneering attempts to use it to value other recreation attributes are described in Haspel and Johnson (Chapter 20 in this volume) and Desvousges and Smith (1982).

5. Property value and other hedonic approach application issues. Property value and other hedonic approaches are usually of limited usefulness for the examination of air quality issues for Class I areas due to limited market data. Nevertheless, in instances where they are applicable, more work is needed to examine the separation of benefit measures into health, aesthetic, and other benefits; to consider the sensitivity of value estimates using alternative functional forms; and to consider time series approaches, rather than point in time approaches. Other issues are identified in Rowe and Chestnut (1982b).

Useful Issues

There are many useful issues to be addressed in all of the techniques. These include more work on eliminating biases (vehicle, information, starting point, etc.) in bidding method user value questions, the choice of appropriate consumer surplus measurers to be used in all analyses, and data selection issues in market approaches such as actual travel cost and property value approaches. We consider these concerns worth pursuing, but not as the sole focus of future work.

CONCLUSIONS

We have identified many, but surely not all, important issues for economic method development related to quantifying benefits of protecting visibility aesthetics, particularly at Class I areas. Given limited resources to develop these research tools, and given a pressing need to determine defensible, though not necessarily perfect, benefit estimates we suggest that the primary focus of future efforts should first be to those issues identified as "most important" and to those identified as "important" general and economic policy issues. The other issues should not be ignored, but should be pursued, for now, as the opportunity arises while attention is focused on more pressing concerns.

BIBLIOGRAPHY

Bishop, R.C., and T.A. Heberlein. 1979. "Measuring Values of Extra Market Goods: Are Indirect Measures Biased?" American Journal of Agricultural Economics 61 (December): 926-930.

Chestnut, L.G., and R.D. Rowe. 1982. Integral Vista Benefits Analysis. Draft. Abt Associates report to the National Park Service, Air Quality Division Washington, D.C.

Desvousges W.H. and V. K. Smith. 1982. "A Comparative Analysis of Travel Cost and Contingent Valuation Methods for User and Non-User Benefit Estimation." Paper presented at the Visual Values Workshop, Keystone, CO May 10-12.

Perl, L. J. and F. C. Dunbar. 1982. "Cost Effectiveness and Cost-Benefit Analysis of Air Quality Regulations." American Economic Review 72(2): 208-213.

Rowe, R.D., and L.G. Chestnut. 1981. Issues in Visibility Benefit-Cost Analysis. Abt Associates report prepared for the U.S. Environmental Protection Agency, Office of Air Quality Planning and Standards, Durham, NC.

_____. 1982a. Health Benefits Analysis for Carbon Monoxide Control. Abt Associates report for the U.S. Environmental Protection Agency, Office of Air Quality Planning and Standards Durham, NC.

_____. 1982b. The Value of Visibility: Economic Theory and Applications for Air Pollution Control. Abt/Books. Cambridge, MA.

_____. 1982c. "Economic Measurement of Impacts from Visibility Impairment: Where is the State of the Art?" Proceedings of the 75th meeting of the Air Pollution Control Association. Pittsburgh, PA.

Schulze, W.D., D.S. Brookshire, E.G. Walther, and K. Kelley. 1981. Methods Development for Environmental Control Benefits Assessment. Volume VIII. The Benefits of Preserving Visibility in Natural Parklands of the Southwest. Prepared for the U.S. Environmental Protection Agency, Office of Research and Development. Washington, D.C.

Walsh, R.G., R.A. Gilliman, and J.B. Loomis. 1982. "Wilderness Resource Economies: Recreation Use and Preservation Values." Prepared for the American Wilderness Alliance, Denver, CO.

Walsh, R.G., D.A. Greenley, R.A. Young, J.R. McKean, and A.A. Prato. 1978. Option Values, Preservation Values and Recreation Benefits of Improving Water Quality: A Case Study of the South Platte River Basin, Colorado. EPA-600/5-78-001. Prepared for the U.S. Environmental Protection Agency, Office of Research and Development, Research Triangle Park, NC.

22. Existence and Bequest Value

Kenneth E. McConnell

Economists are occasionally accused of being ritualistic in their devotion to economic concepts. A newly developing ritual concerns the use of existence and bequest values. It occurs in applied benefit-cost analysis and it is usually evoked by the finding that the measurement of costs and benefits shows that costs exceed benefits. The ritual is most often found in the conclusion and it involves saying: "It appears that costs exceed measured benefits but we know that there are substantial unmeasured benefits in the form of existence and bequest values and hence the project should/should not be undertaken." The practicing economist is then free to choose the outcome consistent with his preferences or those of his sponsors.[1]

While this ritual may provide some short run comfort to economists, in the long run economics as a partial guide to policy will thrive or perish according to the soundness and testability of the concepts it uses. This paper discusses some of the theoretical underpinnings of existence and bequest demand and values. The immediate purpose is to sharpen the concepts of demand for and value of resources that may not be directly used. The ultimate purpose is to develop concepts that can be used in benefit-cost analysis.

Non-use benefits as they are discussed in the current literature come from basically two sources. First is the desire to preserve resources simply for their existence. This is the essence of existence value. Second is the desire to preserve resources because of the uncertainty of demand. Option value (Weisbrod 1964; Long 1967; Lindsay 1969; Schmalansee 1972) is the value of an option to consume the good when preferences are uncertain. In addition, quasi-option value (Arrow and Fisher 1974; Conrad 1980; Miller and Lad 1981) is the expected value of information gained from postponing an irreversible development.

Almost all of the theoretical literature in the area of non-user benefits has been devoted to the issues of uncertainty and irreversibility, which are central to the decisions about whether to preserve natural environments. Less effort has been devoted to the task of defining existence de-

---

Kenneth (Ted) McConnell is a Professor of Agricultural and Resource Economics at the University of Maryland, College Park, MD. This research stems from a project supported by the Great Lakes Fishery Commission. Support of the American Petroleum Institute is also gratefully acknowledged.

mand and value, though casual observation might suggest that existence value may be relatively more significant than option value. A brief discussion of the nature of existence value is offered by Walsh et al. (1978) and the results of attempts to measure existence values are reported by Greenley et al. (1981). Schulze et al. (1981) develop and estimate a model of existence demand for air quality in national parks of the Southwest. Thus it seems relevant to ask: are there benefits that accrue to some individuals from the continued existence of unique natural resources such as Sequoia National Park, Old Faithful, the ecological health of the Great Lakes, and other natural environments? That is, are people willing to give up claims on income to preserve a resource, though the prospect of experiencing the resource at first hand will never come? What kind of preferences lead to existence demand? What are the estimation issues raised by preference structures that give rise to existence values? This paper attempts to define existence demand and raise a number of questions about its estimation and use.

It seems clear that we as a society value many natural resources for their own sake. The spirit of the frontier and the challenge of wilderness are part of America's historical myths. The preservation of some monuments to these myths, in the form of natural environments, seems attractive to many people. When these natural environments are not preserved, one can expect a decline in the well being of some individuals, even though they do not change their behavior in any way. In an era of growing resource scarcity, the preservation of natural environments imposes opportunity costs on society. The optimal allocation of resources under certainty poses no serious conceptual problems, even where the resources are valued only for their existence. However, conceptual resource allocation models assume that the values of alternative uses of resources are known. Systematic thinking about the nature of existence value may help economists to contribute to the decision-making process which determines the fate of natural environments.

THE DEMAND FOR GOODS THAT GIVE EXISTENCE VALUE

Although existence demand has been used occasionally in the applied economics literature, the structure of preferences that gives rise to existence value has not been explored. The demand for the existence of goods has been defined implicitly. Representative use of existence value would be the willingness to pay (or to accept compensation) for the existence of a resource without the prospect of using the resource. For example, Krutilla and Fisher (1975) define existence demand implicitly: "In the case of existence value, we conceived of individuals valuing an environment regardless of the fact that they feel certain they will never demand in situ the services it provides" (p. 124). The key to the definition is the in situ condition. People value services of the resource, but they need not go to the site to enjoy the services.

To be more precise about the nature of existence demand and existence value, let x be an n-dimensional bundle of goods purchased on the market at fixed market prices, denoted p. Let R be the resource for which there is existence demand. Individuals have preferences for different bundles x, R. These preferences are defined by the utility function $u(x,R)$, where u is quasi-concave and increasing in x and R.

The demand for the existence of R can be said to occur when $u(x,R)$ is weakly separable in x and R but that $\partial u/\partial R > 0$. That is, marginal rate of substitution between any two goods purchased on the market is independent of the level of R. Defining u_j to be $\partial u/\partial x_j$, existence demand occurs where

$$\partial u_i(x,R)/u_j(x,R)/\partial R = 0 \text{ for all } i,j \text{ and for any } x. \tag{1}$$

There are several consequences of this structure of preferences. First, Marshallian demand functions for x will be independent of R. Second, R will not be weakly complementary to any x.

To see the first characteristic, note that the system of n demand functions can be solved from the n-1 ratios of marginal utilities and the budget constraint and that each of these equations is independent of R. Weak complementarity, as defined by Mäler (1974) and Freeman (1979), does not hold by definition, $\partial u/\partial R > 0$ for all x. As noted by Smith (1981), weak complementarity is a characteristic of preferences that must be assumed, rather than tested for.

The definition of an existence good implies that changes in the availability of the resource do not influence behavior at all. Some simple examples reveal how extreme this condition is. The ecology of the Great Lakes is news. Reporting on it helps sell magazines, newspapers, and TV shows. Presumably the entrepreneurs running these media choose to report on the ecological state of the Great Lakes to the extent they do because it enhances the relative marginal value of their output to consumers, thus violating the definition. While this example suggests the extremely restrictive nature of the definition, it also shows the importance of pursuing existence demand as a separate topic. It would be clearly impossible to establish a scheme for collecting rents for the existence of the Great Lakes based on tariffs on media products. The problem of estimating the differential effect of a particular kind of news on an enterprise with a joint output such as a newspaper or magazine seems particularly overwhelming.

A careful inspection of the definition of existence demand suggests another reason to be pragmatic. Knowledge of the existence of goods does not fall like manna from heaven. It must be sought. Individuals must make some effort, absorb some costs to obtain information about the existence of goods. Hence the purchase of some commodities (or in the household production framework, inputs) may be necessary to obtain knowledge of any resource. This requirement violates the essence of the definition of an existence good. For most practical applications, it seems plausible to suppose that the information about the resource comes to individuals as a public input, and that existence demand means an individual values a resource even when there is no *in situ* use.

BEQUEST DEMAND

Part of the value of preserving natural environments is occasionally attributed to the bequest motive. It is the desire by individuals to leave natural resources for their heirs. Bequest value is implicitly defined by Krutilla (1967): "A bequest of maximum value would require an appropriate mix of public and private assets, and, equally, the appropriate mix of opportunities to enjoy amenities experienced directly from association with the natural environment along with readily producible goods" (p. 784).

Suppose we define an individual's bequest demand in some formal way. Let u_H be the utility of the individual's heirs. Then an individual's bequest motive might appear as utility to him from the enhanced utility of his heirs:

$$u = u^*(x, u_H).$$

Suppose that u_H depends (among other things) on the amount of resources the heirs have:

$$u_H = u_H(R)$$

Then the individual's preference function is

$$u^*(x, u_H(R)) = u(x, R).$$

Since we can recover only monotonic transformations of u at best, we have no way of distinguishing u from u^*. The motive is immaterial. We cannot tell the difference between existence demand and bequest demand except perhaps to note that discounting might make the bequest motive less strong. Consequently, the previous discussions, which dealt with existence demand, apply equally well to bequest demand. Economics is independent of motives.

WHAT IS EXISTENCE VALUE?

Suppose that a good is valued only for its existence. That is, there is no on site use. Then defining existence value is a straightforward exercise in deriving the cost or expenditure from the preference function and market prices for the commodities. The cost function is defined as:

$$C(p, R, _x u) = \min \{p \cdot x \mid u(x, R) = u.\} \tag{2}$$

Since we are dealing with quantity changes and not price changes we define _equivalent surplus_ as the amount of compensation paid or received that leaves the individual in his subsequent welfare position in the absence of the quantity change in the resource; and _compensating surplus_ as the amount of compensation paid or received that leaves the individual in his initial welfare position following the quantity change in the resource. (Apologies to Currie et al. 1971, p. 746.)

Let R be the change in the quantity of the resource. The equivalent surplus denoted, E, is defined by

$$E = C(p, R + \Delta R, u^*) - C(p, R, u^*) \tag{3}$$

where u^* is the level of utility afforded by the vector p, R+ Δ R, y; i.e.,

$$u^* = V(p, y, R + \Delta R). \tag{4}$$

Here, V(p,y,R) is the indirect utility function. Since we are not concerned with changes in income we can suppress y. Thus indirect utility and the reference level of utility will depend only on R and p: V(p,R). Compensating surplus, denoted C, is given by

$$C = C(p, R + \Delta R, u^o) - C(p, R, u^o) \tag{5}$$

where u^o is the initial level of well being, i.e.,

$$u^o = V(p,R). \qquad (6)$$

Compensating and equivalent surplus correspond to willingness to pay or to accept compensation depending on whether R is positive or negative. Table 22.1 shows the correspondence. Table 22.1 is quite general, while this paper is concerned with potentially irreversible reductions in natural environments. Our goal is to analyze policies that reduce the quantity of the resource. Hence we can interchange willingness to pay with equivalent surplus and willingness to accept compensation with compensating surplus.

Defining existence value in terms of the cost function is convenient analytically, but it provides no bridge to measurement due to the absence of weak complementarity. In the presence of weak complementarity, it can be shown that changes in the surplus of a public good, as measured by the cost function, can be captured as the change in the area under an appropriate Hicksian demand curve. Given the approximations, such as those of Willig (1976) or Randall and Stoll (1980), one can measure these surplus changes under the Marshallian curve. Thus with existence goods, we have no set of data to reveal the preferences of individuals, and we must rely on some type of hypothetical or contingent valuation scheme.

TABLE 22.1
Surplus Measure Assuming that R is a Good Resource

		Equivalent Surplus	Compensating Surplus
Changes in R	Decrease	Willingness to accept compensation	Willingness to pay
	Increase	Willingness to pay	Willingness to accept compensation

The notion that a good is valued only for its existence, that it provides no in situ services, is farfetched. In most cases, resources are valued for their use. Existence value occurs only insofar as bequest or altruistic notions prevail. We want resources there because they are valued by others of our own generation or by our heirs. Thus use value is the ultimate goal of preferences that yield existence demand, though the existence and use may be experienced by different individuals.

DISTINGUISHING BETWEEN USE AND EXISTENCE VALUE

Many resources, such as national parks, give use values as well as existence value. Use value accrues to those who visit the site. In fact, it is difficult to conceive of such a resource that gives existence value but is not worth visiting. This section explores the difficulties of dealing simultaneously with existence and use values.

We will distinguish between existence and use value with the cost function. Value will be assumed to mean the compensating surplus measure of value. Analogous results can be derived for equivalent surplus. We will use the cost function to define first, the total resource value, second, use value, and third, existence value. We will then show that existence value plus use value is equal to resource value.

Suppose that R is complementary to the first good (any good would work; the first keeps notation down). If there is existence value, then we would expect that

$$u(0, x_2, ..., x_n, R) > u(0, x_2, ..., x_n, 0) \quad \text{for } R > 0. \tag{7}$$

This condition clearly violates weak complementarity as defined by Mäler (1974) and Freeman (1979). If the resource is valued both for its <u>in situ</u> services and for its existence, then changes in the resource will cause changes in both values.

First, define the value of a change in the resource from R^o to R^*, when both existence and use demand are possible. Letting ΔRV stand for the change in the resource value, we have

$$\Delta RV(p^o) = C(p^o, R^*, u^o) - C(p^o, R^o, u^o) \tag{8}$$

Note that resource value depends on the initial price vector p^o because ΔRV is the compensation needed to return the individual to the $u^o = V(p^o, R^o)$ level of utility when the resource is changed to R^*. Both existence and use values are included in ΔRV as given in this expression.

The use value of the resource is the change in compensation required to make the individual indifferent between the compensation and access to the site, induced by the resource change. To make sense, use value should be zero if x_1 is zero. The compensation function depends on the prices of commodities, not the commodities. Thus use value should also be zero when the price of x_1 drives x_1 to zero. Denote p^* as the price that drives the Hicksian demand for x_1 to zero. But recall that R may change, and if R is low, p^* may be low. Thus we give p^* as a function of R:

$$p^* = p^*(R).$$

Use value can be defined as the change in the compensating variation associated with a change in R. The compensating variation of eliminating x_1 is the difference in the cost function evaluated at the current price and the price that drives x_1 to zero. This compensating variation is the use value associated with R^o

$$UV(p^o) = C(p^*(R), R^o, u^o) - C(p^o, R^o, u^o) \tag{9}$$

Note that this expression gives the area behind the Hicksian demand for x_1 and above the price line p^o. To determine the change in use value induced by a change in R, compute expression (9) at p* and R* and subtract the value at R^o:

$$\Delta UV(p^o) = C(p^*(R^*), R^*, u^o) - C(p^o, R^*, u^o)$$
$$- (C((p^*(R^o), R^o, u^o) - C(p^o, R^o, u^o)) \tag{10}$$

Expression (10) is the change in the area behind the appropriate Hicksian demand curve. It defines use value in such a way that $\Delta UV = 0$ when $p^o = p^*$ and x_1 is zero, i.e., by definition of p^*, $UV(p^*) = 0$ so that $\Delta UV(p^*) = 0$.

Now let us define existence value in the general case. The definition should have several characteristics. First, existence value plus use value should equal resource value. Second, resource value should equal existence value when there is no use value. Third, resource value should equal use value when there is no existence value and as a consequence, weak complementarity holds. Let EV be the existence value, and ΔEV the change in existence value induced by a change in the resource. Then ΔEV is defined by the relationship:

$$\Delta EV = \Delta RV - \Delta UV$$

and by subtracting the expression for use value in (10) from the expression for resource value in (8), we should obtain an expression for existence value:

$$\Delta \text{existence value} = \Delta \text{resource value} - \Delta \text{use value}$$
$$\Delta EV(p^o) = \Delta RV(p^o) - \Delta UV(p^o)$$
$$= C(p^o, R^*, u^o) - C(p^o, R^o, u^o) \qquad (11)$$
$$- C(p^*(R^o), R^o, u^o) - C(p^o, R^o, u^o)$$
$$- C(p^*(R^*), R^*, u^o) - C(p^o, R^*, u^o)$$

Note the first and last terms cancel out and the second and fourth terms cancel out. Thus we have

$$\Delta EV(p^o) = C(p^*(R^*), R^*, u^o) - C(p^*(R^o), R^o u^o) \qquad (12)$$

The definition given in (12) makes sense because it comes down to the compensation needed to return the individual to his original utility from the p^*, R^* state (no use, reduced resource) less the compensation needed to return the individual to his original utility level from the p^*, R^o state (no use, original resource).

The definition of existence value can be examined for the three criteria. It is clear that use value + existence value = resource value, because that equality is the source of the definition. What if use value is zero? Suppose that use at the going price is zero. When $p^o = p^*$, in situ use is driven to zero and hence use value becomes zero. Suppose an individual is in equilibrium in the state p^*, R^o and we change R from R^o to R^*. Then the resource value is the compensation needed to take the individual back to p^*, R^o or p^o, R^o, since these prices are the same. Making these changes in (11) gives

$$\Delta RV(p^o) = C(p^o, R^*, u^o) - C(p^o, R^o, u^o) = C(p^*, R^*, u^o) - C(p^*, R^o, u^o)$$
$$= EV(p^*) = EV(p^o). \qquad (13)$$

This expression is existence value. Thus, the definition fits the criterion that resource value equals existence value when use value is zero.

To determine whether resource value equals use value when there is no existence value, recall the definition of existence demand. There is existence demand when changes in the resource give more utility, regardless of the level of commodities consumed. Suppose the resource is complementary to the first good. Then there is no existence demand if

$$\partial u(0, x_2, \ldots, x_n, R)/\partial R = 0 \tag{14}$$

Suppose that p^* again is the price vector that sets $x_1 = 0$. Existence value as given will be zero because no compensation for changes in R is needed when x_1 is zero or $p = p^*$. Consequently, (12) implies that

$$\Delta RV(p^o) = \Delta UV(p^o) \tag{15}$$

when (14) holds.

The definition of existence demand given in (12) meets the basic criteria of the concept:

1. $\Delta EV(p) = \Delta RV(p) - \Delta UV(p)$
2. $\Delta EV(p^o) = \Delta RV(p^o)$ when p^o satisfies $\Delta UV(p^o) = 0$
3. when preferences imply that $\Delta EV = 0, \Delta RV = \Delta UV$.

There are two consequences of this result. First for individuals who have both use and existence value, one may design internal consistency tests for measurement of value. One may measure use value via the travel cost method, total resource value by contingent market mechanisms, and compute existence by subtracting use value from resource value. Second, suppose we have two individuals: A, who lives close to and fishes in Lake Michigan, and B, who lives far away from Lake Michigan and does not fish there because the implicit price (travel plus time) is too high. Other than the distance, the two individuals are identical in tastes, income, and to the implausible extent of having the same substitutes. Both are assumed to have existence demand for a clean Lake Michigan. Of course, A's compensating surplus for changes in the cleanliness of Lake Michigan is higher than B's because A also fishes there, and the cleaner the lake, the more fish, the better the fishing, etc. Now suppose that we have a contingent valuation mechanism that will induce A to tell us the true total willingness to accept compensation for reduced water quality in the lake. Also suppose that we perfect the travel cost method such that we discover what effect the inferior quality of the lake has on A's use value. We cannot subtract use value (computed from the travel cost method) from total value (via contingent valuation) to compute existence value even when each technique works perfectly. Knowing A's total value and use value will not let us compute B's existence value unless they face the same price and source vector.

The significance of this result is striking when one considers the logistics of measuring existence value. Such measurement must sample users and non-users. However, the conclusion that existence value differs for users and non-users implies that estimates of willingness to accept compensation for those who have no in situ use cannot be extrapolated from in situ users. In effect, two different types of surveys are needed: one for users and one for non-users. In addition, even for identical preferences and weak separability of R, changes in the price vector change the marginal value of R. Thus, in the case of resources that have national prominence, it is necessary to sample people from all over the country.

IMPLICATIONS

In many respects, existence goods do not differ from the general class of public resources: resources that provide existence value enter many individuals' preference functions simultaneously, without detracting from the availability or value of the resource to others. Two aspects of existence goods deserve special attention: demand estimation and policy.

Perhaps the most disturbing aspect of public goods is the apparent difficulty of demand estimation. Definitions of existence goods violate the concept of weak complementarity, which Mäler (1974) and Freeman (1979) have shown to be so crucial in the estimation of the demand for environmental services. Thus the definition of existence demand practically precludes demand estimation using observations on revealed preferences such as the travel cost method for recreation or the hedonic method for air pollution.

The notion of existence value developed in the context of large and perhaps irreversible changes in natural environments. Many researchers felt that existence value as a concept applied to potentially irreversible reductions in a resource. In essence, existence resources are pure public resources that are not complementary to any privately-purchased goods. Nothing about this definition connotes its applicability to situations involving irreversibility or uncertainty. The concept of existence value as discussed in this paper applies to many different resources and institutions. There is existence demand for large natural wonders, small natural wonders, historical buildings, historical districts, speech dialects, species of birds, and many of the other characteristics that make up the quality of life. Because nothing about the theory of value for existence resources distinguishes these resources from other public goods, the crux of the existence value issue lies in its measurement.

Granted that the estimation of the demand for and value of existence goods cannot be easily based on observed behavior, do we have any indirect evidence of existence value? Unlike the case for local public goods, there can be no voting with the feet for existence goods. It doesn't make economic sense to move to enjoy an existence good if all one wants is its existence anyway. Hence we cannot judge by visitation, residential location, or property values that individuals have exercised their existence demand. The best evidence of the demand for existence goods comes from the formation of voluntary associations such as the Sierra Club, the Audubon Society, the Save our Bay groups, the Environmental Defense Fund, etc. In addition, there are many historical preservation groups whose continued function is partly in response to the phenomenon of existence demand. The willingness on the part of individuals to contribute to these groups is in part an expression of demand for existence goods.

What should economists do about measuring existence value? It is apparent that only contingent market mechanisms will provide estimates. In designing these mechanisms, we must be careful not to exclude the possibility of choosing among different types of existence goods. The policy questions economists will most likely be faced with are of the type: which resources should be preserved simply for the sake of their existence? Both the problem of choosing among existence goods and the separable nature of existence goods make a case for a two-stage contingent market mechanism. In the first stage, the budget to be spent on existence goods is determined. In the second stage, a consumer chooses among various types of existence goods. Such a two-stage maximization process is often appealed to in modeling systems of consumer demand functions, and it is justified on the basis of weak separability. It has been argued above that the nature of existence demand implies separability between the resource and private goods. Thus, a two stage approach could be justified, but it would require that the number a resources being investigation be predetermined.

In closing, it is hard to resist speculating on the logical consequence of the pervasive use of existence values in benefit-cost analysis. One view might be quite pessimistic. Suppose that we settle on compensating surplus as the appropriate criterion for project evaluation. Any project which changes any of the characteristics that make up the quality of life will require measurement of compensation. Some kind of contingent market mechanism would have to be used. Because in the case of existence value, by definition we have no tests for bad answers, we in effect have no way of distinguishing between bribery and compensation. One might argue that any tool that can apparently be used to thwart measures of progress so easily is of questionable value. Another view of the concept might come from other disciplines in the social science: sociology and anthropology. These disciplines have documented the social and emotional disruptions that accompany economic change. They have argued that the distribution effects of change are often dramatic and escape the economist's measurement. Such a perspective would lead one to welcome a tool such as existence value.

NOTES

1. There are numerous examples of this ritual. I quote from a document prepared in part by Dan Talhelm who is aware of the ritual because he is leading a research effort to develop methods for measuring existence values. In speaking of the unmeasured benefits in the form of existence and option values of rehabilitating the Great Lakes, the report states, "Judging from public and private support for rehabilitation efforts, these values may be in the billions of dollars annually" (Francis et al. 1979, p. 61).

BIBLIOGRAPHY

Arrow, K. J., and A. C. Fisher. 1974. "Environmental Preservation, Uncertainty and Irreversibility." Quarterly Journal of Economics 88:313-319.
Conrad, J. 1980. "Quasi-Option Value and the Expected Value of Information." Quarterly Journal of Economics 96:813-819.
Currie, J. M., J. A. Murphy, and A. Schmitz. 1971. "The Concept of Economic Surplus." Economic Journal 81:741-799.
Francis, G. R., J. J. Magnusson, H. A. Regier, and D. R. Talhelm. 1979. "Rehabilitating Great Lakes Ecosystem." Technical Report No. 37, Great Lakes Fisheries Commission (December).
Freeman, A. M. III. 1979. The Economics of Environmental Improvement. John Hopkins Press, Baltimore, MD.
Greenley, D. A., R. G. Walsh, and R. A. Young. 1981. "Option Value: Empirical Evidence From a Case Study of Recreation and Water Quality." Quarterly Journal of Economics 96:657-673.
Krutilla, J. V. 1967. "Conservation Reconsidered." American Economic Review 57:777-786.

Krutilla, J. V., and A. C. Fisher. 1975. The Economics of Natural Environments. John Hopkins Press, Baltimore, MD.

Lindsay, C. M. 1969. "Option Demand and Consumer Surplus." Quarterly Journal of Economics 83:344-346.

Long, M. F. 1967. "Collective Consumption Services of Individual Consumption Goods: Comment." Quarterly Journal of Economics 81:351-352.

Mäler, K. G. 1974. Environmental Economics: A Theoretical Enquiry. John Hopkins Press, Baltimore.

Miller, J. R., and F. Lad. 1981. "Uncertainty, Irreversibility and Quasi-Option Value." Paper presented at the 1981 Western Economics Association Meetings, San Francisco, CA.

Randall, A., and J. Stoll. 1980. "Consumer's Surplus in Commodity Space." American Economic Review 70:449-455.

Smith, V. K. 1981. "Introduction to Advances in Applied Microeconomics and Some Perspectives on Volume I." In Advances in Applied Microeconomics. Volume I. V.K. Smith (ed.), JAI Press, Greenwich, CT.

Schulze, W. D., D. Brookshire, and M. Thayer. 1981. "National Parks and Beauty: A Test of Existence Values." Environmental Protection Agency project report, Washington, D.C.

Schmalensee, R. 1972. "Option Demand and Consumer's Surplus: Valuing Price Changes under Uncertainty." American Economic Review 62:813-824.

Walsh, R. G., D. A. Greenley, R. A. Young, J. R. McKean, and A. A. Prato. 1978. "Option Values, Preservation Values, and Recreational Benefits of Improved Water Quality: A Case Study of the South Platte River Basin, Colorado." U.S. Environmental Protection Agency, Research Triangle Park, NC.

Weisbrod, B. A. 1964. "Collective-Consumption Services of Individual-Consumption Goods." Quarterly Journal of Economics 78:471-477.

Willig, R. D. 1976. "Consumer's Surplus Without Apology." American Economic Review 66:587-597.

23. Existence Value in a Total Valuation Framework

Alan Randall
John R. Stoll

In the last two decades, the range of resource value concepts recognized by economists has been considerably expanded. Traditionally, it was those values recorded from exchange of resource commodities in the marketplace that were recognized as legitimate measures of economic value. Some time ago, direct demands for in situ resources (eg., recreation sites, and ambient air and water of acceptable quality) were recognized and estimated. Later, as awareness spread of the potential for wholesale modification of natural environments and the accelerated extinction of species, the adequacy of traditional concepts of economic value became a frequent topic of discussion (Weisbrod 1964, Krutilla 1967, Schmalensee 1972, Arrow and Fisher 1974). It was frequently argued that legitimate economic values may be derived from maintaining the option for future use, or from simply knowing that continued existence of the resource in its current state is assured.

While considerable ingenuity and analytical rigor have been devoted to this reconsideration of resource value concepts, the task is not yet complete and some of the inevitable false starts have left a legacy of confusion. Value concepts have proliferated--use value, option price, option value, expected consumer's surplus, quasi-option value, existence value, preservation value, bequest value, etc.--but some of these are overlapping in concept while many are empirically elusive so that validation of estimates is difficult and often incomplete. Thus, confusion in some quarters is matched by skepticism in others.

In this paper, we attempt clarification of the relevant resource valuation concepts. Starting with total value, a categorization of value concepts is developed. Existence value is emphasized and the conditions influencing existence demand, supply, and value, in total and at the margin, are considered. Some estimates of existence value and option price, gleaned from previous empirical research, are presented and briefly discussed. Finally, we draw attention to some aggregation problems which arise in the transition from partial to more encompassing analyses. Our focus is, for the most part, quite general. Nevertheless, some observations on application

---

Alan Randall is a Professor of agricultural economics at the University of Kentucky. John Stoll is an Assistant Professor of agricultural economics at Texas A&M University. Support by the Electric Power Research Institute for the presentation of this paper is acknowledged.

to visual values are provided, and some of the empirical results concern visual values.

A BASIC CONCEPTUAL FRAMEWORK

Individuals engage in activities because they derive satisfaction from them. In economics a state of satisfaction is described by a utility function in which the activities are arguments within the function. When the arguments are assigned specific values, a level of utility or satisfaction is defined by the function. That is,

$$U = f(Z) \tag{1}$$

where U represents utility or satisfaction and Z is a vector of activities consumed by the household.

The concept of activity is broadly defined, and includes work and other income generating activities, formal educational activities, reading and watching television and movies (which may have an educational or informational component), eating, household maintenance, hobbies, and recreational activities (including sports, physical exercise, hiking, nature study, sightseeing, etc.).

These activities are produced by the household, in a process which combines purchased goods and services, environmental amenities and other public goods, and the household's time and effort. This production process is governed by the household's activity production technology.

A household production function for activities using a natural resource amenity is represented as

$$Z = z(X, Q \mid T) \tag{2}$$

where

$Q =$ a specific natural resource amenity, eg., a scenic vista;
$X =$ a vector of goods and services other than the specific natural resource Q; and
$T =$ the household's production technology.

The household's production technology at any given time is a function of the activities produced during previous time periods. Thus, it may be represented as

$$T_t = h \left[z_{t_0}(X,Q \mid T_{t_0}), ..., z_{t-1}(X,Q \mid T_{t-1}) \right] \tag{3}$$

where

t_0 = the initial time period;
t = the current time period; and
$t-1$ = the time period preceding t.

This formulation explicitly recognizes the development of skill in activity production through conscious acquisition of information and instruction and through the less deliberate process of "learning by doing." Past activity production influences the capacity to achieve satisfaction from current activities. Given the nature of the processes by which activity production technology is acquired, it is immediately clear that T may differ substantially, across households, as well as across time periods in a given household.[1]

Finally, it must be noted that these processes of household activity production are constrained. Clearly, the household has a limited budget and must also allocate a fixed amount of time among alternative activities.

TOTAL VALUE AND ITS COMPONENTS

The total value of Q is determined by the quantity provided (i.e., supply) and the various demands for Q as an input in household production processes. For the individual household, total value is the consumer's surplus from all Q-using activities, where consumer's surplus is defined as the net benefits remaining after all household production costs have been incurred.

To consider the components of total value, we focus on the various kinds of demands for Q. Two major components of value are use value and existence value.

Use Value. In general, any activity produced in a process such as that defined in Equation 2 may generate use values. Activity production may take place on-site (which generates visitor values, on-site use values, recreation values, etc.), but that is not essential. Use value occurs anytime Q is combined with one or more elements of the X vector. Thus, we consider the values generated by reading about Q in a book or magazine, looking at it in photographic representations, for example, to be use values. Clearly, our definition of use includes vicarious consumption.

Several kinds of use values may be defined. The most frequently observed and measured kind is current use value, which is generated by use in the present time period. Use values may also be generated by anticipation of future use. In the absence of risk preferences (positive or negative), the value of future use is expected consumer's surplus (E). The household may be willing to pay for an option for future use, if that would assure availability in the event that future use was demanded. The value of such an option is called option price (OP). There is another concept, option value (OV), defined so that $OP = E + OV$, where OV is a risk premium which would be paid by risk averse households.[2] There has been considerable controversy as to whether OV may take positive or neative signs (Schmalensee 1972). The problem is that there are two kinds of risk: demand risk, in which a purchased option would prove useless if future demand did not eventuate; and supply risk, in which future availability is not assured unless the option is purchased. In buying an option, one may encounter demand risk as a result of an act taken to avoid supply risk. Only in the case where demand is certain and supply uncertain can we be assured that, for a risk averse household, $OV>0$ and $OP>E$ (Bishop 1982).

An additional type of use value is termed quasi-option value. It is defined as the value gained by delaying a decision until a future time period when more information may be available (Arrow and Fisher 1974, Conrad 1980). While not confined to such cases, quasi-option value is often invoked in the case of destructible resources: if such resources were destroyed now, the possibility is eliminated that new technologies developed in future times may render them increasingly valuable. Quasi-option value is the value attributable to avoiding such a risk.

Existence Value. It is possible that value may be generated by simply knowing Q exists. Instead of the activity production function (Equation 2), the production function for pure existence activities is

$$Z = g(Q \mid T). \tag{4}$$

That is, existence values for Q are generated by Q alone, subject to an activity production technology (T) which permits an understanding and appreciation of Q. No elements of the X vector are involved, in the current time period. However, activities combining Q and X in some previous time periods seem essential to the acquisition of the kinds of T which permit existence activities (Equation 3).

Pure existence value excludes any values which arise from current use or anticipated future use. We defined vicarious consumption, above, as a kind of use. Thus, it seems that all pure existence demands must be altruistically motivated. One can conceive of <u>interpersonal altruism</u> which would generate existence values from knowing Q was available for others to use; <u>intergenerational altruism</u> from knowing Q will be available for future generations; and <u>Q-altruism</u>, in which the household enjoys the feeling that Q itself is benefiting from being undisturbed. These latter two kinds of altruism generate <u>bequest values</u> and <u>intrinsic values</u>, respectively, which are (of course) categories of existence value.

UNIQUENESS AND IRREVERSIBILITY: ARE THEY ESSENTIAL TO THE GENERATION OF OPTION AND EXISTENCE VALUE?

The early literature conveyed the impression that option and existence values were rather specialized phenomena, confined to unique natural objects threatened with irreversible destruction at the hands of humankind. Weisbrod (1964) assigned a key role to irreversibility in his conceptualization of option demand, while Krutilla (1967) argued that irreversibility and uniqueness were both essential to existence value. While these claims seem to have gained considerable acceptance, we believe that option and existence values have much more general application. Our argument will be addressed to existence value, although much of it is obviously adaptable to option value as well.

We conceptualize demand and supply curves for existence as in Figure 23.1. For resource amenities with many close substitutes, existence demand will be weak; for amenities in great supply, marginal existence value will be low; and if the project under evaluation will cause only a small reduction in existence supply, the aggregate loss of existence value will be small. Thus, even commonplace artifacts of human civilization (e.g., drink cans) may have existence value, although the circumstances which would make it large are unlikely. Empirically significant existence values are not confined to natural objects; we believe they occur for human artifacts and cultural manifestations, from historic buildings to grand opera. Nor are existence values confined to "the last few ___s on earth." We would expect to find positive existence values for local amenities, local subpopulations of flora and fauna, and for local cultural amenities. Opera buffs may place genuine value on knowing that grand opera is performed live in medium-sized cities like Des Moines. Irreversibility is not essential. Although it is true that visibility at the Grand Canyon could be rapidly restored following a decision to eliminate emissions, we have no conceptual difficulties with the Schulze et al. (1981) finding of substantial existence value for good visibility at the canyon. Since it is hard to conceptualize visibility impairment as strictly irreversible, our argument that irreversibility is not a nec-

FIGURE 23.1
Existence Value and Relative Scarcity

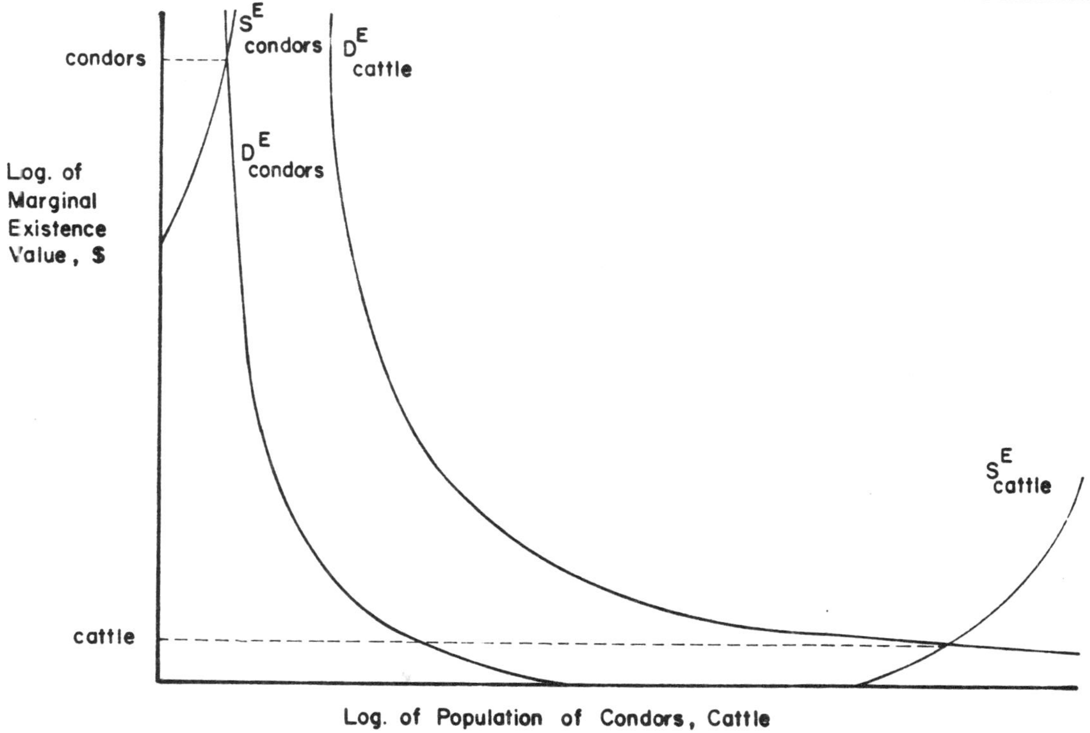

essary condition for existence value has important implications for the economics of visibility.

Consider the existence values of California condors and ordinary cattle (Figure 23.1). Given the world-wide range of cattle and their significance in many cultures, it may well be that the existence demand for cattle lies to the right of that for condors. However, the existence supply of cattle is enormous while that for condors is very limited. The existence value of the marginal condor surely exceeds that of the marginal cow. The loss of forty condors would eliminate their supply (irreversibly, in this example), while the loss of forty cattle would involve a trivial loss of existence value. However, the existence value of the whole cattle population may well exceed that for condors, and an event which destroyed all cattle would cause a greater existence loss than the destruction of the last remaining condors.

INFORMATION AND THE VOLATILITY OF EXISTENCE VALUE

Existence demands depend on T, which is itself influenced by Xs in previous time periods. Many relevant Xs are in some sense informational. The more one knows about the merits of some Q (and, perhaps, the precariousness of its circumstances), the larger existence value one may place on it.[3] Consider the snail darter. Until its discovery in 1973, its existence demand was zero. Its existence demand rose rapidly as it was accepted as a separate species and listed as endangered, and as knowledge of its existence and its plight spread rapidly among the public. Given its limited supply, it acquired a substantial marginal existence value. More recently, snail darter populations have been found in some streams where they were previously unknown, shifting the (known) supply to the right and, presumably, reducing the existence value of the marginal snail darter.

The general principle underlying this example is that, starting from an initial state of little or no information, small increments to the information base may produce large shifts in existence value, in total or at the margin. Existence value is, therefore, quite volatile in the face of new information. It is important to realize that this volatility has nothing to do with measurement error or bias. It is not that the "estimates" are volatile; the problem is that the perceived reality of existence value is volatile, especially when (relatively) large increments may be forthcoming anytime, to a small initial information base.

ESTIMATES OF OPTION AND EXISTENCE VALUES

There have been several reported attempts to empirically estimate option and existence values for resources. In each case, contingent valuation methods were used in survey situations. The results of four of these studies will be briefly reported here.

Brookshire et al. (1978) estimated option prices for grizzly bear and bighorn sheep hunting in Wyoming. These are the first researchers we know to have attempted to undertake this type of estimation. Their respective estimates were $22 and $30 per year.[4] They also estimated existence values for preserving three alternative local or regional habitats (estimates ranged from $77 to $89 per year) and prevention of 50 percent declines in several species type (estimates were in terms of annual utility bills and ranged from 25 to 47 percent). This study utilized a very small sample and was viewed as a pretest.

The South Platte River Basin in Colorado was used to study values associated with preserving water quality (Greenley, et al. 1981). Option prices associated with future recreational uses were estimated to be $23 per year, existence value to be $42 per year (interpersonal altruism and Q-altruism were $25 per year and intergenerational altruism was $17 per year). Aggregation of these values and other use values resulted in a total value of sixty-one million dollars for maintaining present water quality in a river basin. 57 percent of this value estimate was attributable to current recreational users.

Schulze et al. (1981) interviewed residents of four major metropolitan areas (Albuquerque, Chicago, Denver, and Los Angeles) to obtain information for estimating the existence value of clean air in the Grand Canyon. Their estimates ranged from $34.08 to $51.84 across the four cities (including current users of the Grand Canyon). When aggregated for the entire nation, the total value of clean air in the Grand Canyon was estimated to be about six billion dollars. The contribution of current user value to this total value was a very small proportion ("on the order of tens of millions of dollars," p. 12).

The fourth study to estimate option prices and existence values was again conducted by Brookshire et al. (forthcoming). Using a mail survey of Wyoming hunters they again estimated both option prices and existence values for grizzly bear and bighorn sheep hunting. The conditions under which these values were estimated considered variations in the probability of supply and probability of demand, and accounted for whether the individual was a hunter or wildlife observer of these species. Option price estimates range from $10 to $20 and existence values from $7 to $24 on an annual basis.

Clearly, all four studies represent preliminary estimates of option prices and existence values. Yet, they are important because they present evidence that these theoretical categories of value are measurable and generate empirically significant values. In some cases, the amounts involved are large relative to the more traditionally measured current use value.

COMPLEX PROGRAMS AND NON-MARKET VALUES

Customary benefit estimation procedures implicitly treat each project or program as the marginal increment to the existing package of goods, services, and amenities. If a non-marketed good is involved and a shadow-price is placed thereon, it is implicitly treated as the marginal addition to the vector of priced goods. But, in a world where many complex programs are under consideration at any time, each program cannot be the marginal program. In a world with many kinds of non-marketed and unpriced amenities, it makes little sense to value each, one-by-one, as though it were the only addition to the set of priced, or economically valued, goods.

Two recent studies have demonstrated the problems which arise when overly simple methods are used in truly complex situations. Schulze et al. (1981) estimated the annual value to Chicago residents of one particular increment in visibility at the Grand Canyon at $86 for a typical household. Randall et al. (1981) also working with Chicago residents, considered a sequence of air quality programs in which the affected region was incrementally expanded. Starting with the immediate Chicago region, the visibility increment was valued at about $325 per household annually. When the region was expanded to include all of the U.S. east of the Mississippi, the program was valued at about $355. When a visibility improvement program for the Grand Canyon was added, the whole package was valued at about $373. The incremental value of the Grand Canyon program was eighteen dollars, compared to $86 when that program was considered alone.

Majid et al. (forthcoming) considering proposals to add certain designated areas of land to a park system, found that the estimated benefits of proposed parks considered in isolation were much larger than the benefits of the same proposed parks explicitly treated as increments to an existing system of established parks.

These two empirical studies are merely examples of a general problem. Suppose we categorize all goods, services, and amenities into two groups, priced and unpriced. A little reflection is enough to establish that the unpriced category is surprisingly large: consider for example, how many kinds of services are produced within the household or provided by the open-access environment, and therefore unpriced. Now imagine we were to start pricing, or economically valuing, all of the unpriced goods, one-by-one. Before long, all prices (those for priced goods and those for previously unpriced goods now included in the priced category) would start changing. In principle, one could calculate the new vector of prices which would apply after all goods, services, and amenities had been shifted into the priced category. But we can be sure that no current price would be unchanged.

This problem has attracted little research, but will gain more attention as the complexities of programs to augment unpriced goods are recognized. If our argument that option and existence value concepts apply to quite a broad array of goods and amenities is correct, attempts to adequately represent these values in benefit cost analyses would rather quickly encounter this problem.

CONCLUSION

We have identified four categories of use value and three sources of demands for existence. While bouyant existence demand, uniqueness, and the threat of irreversible damage are together sufficient to generate substantial existence values, it is not true that strict uniqueness or irreversibility is necessary. Large existence values may be associated with, for example, visibility at certain scenic locations, even though strict irreversibility is not involved. In the case of scenic values, however, the possibility of irreversible scenic change does nothing to reduce existence value.

Several exploratory studies have estimated option price and existence value, in some cases for visual and scenic resources. In every case, contingent valuation methods were used. These methods are readily applied and are based on sound theory. However, they are sometimes susceptible to various data quality problems, and adequate validation is not always possible. Thus, the state of the empirical art indicates that estimation of option price and existence value is feasible, but validation problems have not yet been completely resolved. As observed in the preceding section, difficulties emerge as one moves from partial analysis of the benefits and costs of individual program components to a more general analysis of the multitude of complex programs being considered by various agencies.

The crucial role of information in generating existence values is recognized, along with its implication that existence values will often be quite volatile in the face of emerging information. It seems clear that the concepts of information and discovery, substitution possibilities, and relative scarcity (rather than strict irreversibility) will eventually make important contributions to a more complete theory of the values attributable to future use and existence.

NOTES

1. Stigler and Becker (1977) have argued that the concept of activity production technology is of great potential fruitfulness in explaining differences in activity choice among households. If, as they claim, T is analytically more tractable than the process of preference formation, an appropriate research strategy would focus on T while essentially ignoring tastes and preferences.
2. OP, E, and OV are all "sure payment" concepts. Building on the literature on the economics of insurance, Graham (1981) has shown that some schemes of contingent payments may generate values greater than the larger of OP and E. This recent, promising contribution may add new terms to the already crowded language of value categories.
3. The informational programs of many "nature lobby" organizations are clearly designed to increase existence value for particular natural amenities.
4. Unless otherwise indicated, dollar values reported here and below are annual values for a typical household.

BIBLIOGRAPHY

Arrow, K.J., and A.C. Fisher. 1974. "Environmental Preservation, Uncertainty, and Irreversibility." Quarterly Journal of Economics 55(2):313-319.

Bishop, R.C. 1982. "Option Value: An Exposition and Extension." Land Economics 58(1):1-15.

Brookshire, D.S., L.S. Eubanks and A. Randall. 1978. "Valuing Wildlife Resources: An Experiment." Transactions of the Forty-Third North American Wildlife and Natural Resources Conference

Brookshire, D.S., L.S. Eubanks and A. Randall. Forthcoming. "Estimating Option Prices and Existence Values for Wildlife Resources." Land Economics.

Conrad, J.M. 1980. "Quasi-Option Value and the Expected Value of Information." Quarterly Journal of Economics 95(2):812-820.

Graham, D.A. 1981. "Cost Benefit Analysis Under Uncertainty." American Economic Review 71(3):715-725.

Greenley, D.A., R.G. Walsh and R.A. Young. 1981. "Option Value: Empirical Evidence from a Case Study of Recreation and Water Quality." Quarterly Journal of Economics 97(4):657-673.

Krutilla, J.V. 1967. "Conservation Reconsidered." American Economic Review 57(4):777-786.

Majid, I., J. Sinden, and A. Randall. Forthcoming. "Benefit Evaluation of Increments to Existing Systems of Public Facilities." Land Economics.

Randall, A., J. Hoehn, and G.S. Tolley. 1981. "The Structure of Contingent Markets: Some Results of a Recent Experiment." Presented at the American Economic Association Annual Meeting, Washington, D.C.

Schmalensee, R. 1972. "Option Demand and Consumer's Surplus: Valuing Price Changes Under Uncertainty." American Economic Review 62(5):813-824.

Schulze, W.D., D.S. Brookshire, and M.A. Thayer. 1981. "National Parks and Beauty: A Test of Existence Values." Paper presented at the American Economic Association Annual Meeting, Washington, D.C.

Stigler, G.J., and S. Becker. 1977. "De gustibus Non Est Disputandum." American Economic Review 67(2):76-90.

Weisrod, B.A. 1964. "Collective-Consumption Goods." Quarterly Journal of Economics 78(3):471-477.

24. Unrevealed Extramarket Values: Values Outside the Normal Range of Consumer Choices

Daniel R. Talhelm

INTRODUCTION

It seems intuitively apparent that the benefits and costs of "drastic" choices of natural resources allocation--such as "irreversible" alteration of "unique" resources through species extinction, natural feature destruction, or "severe" ecosystem alteration--would not be limited to the present and future direct and indirect users of the resource. Nonusers also seem to experience direct gains or losses. For example, both "saving" Lake Erie from "dying" and protecting visibility at the Grand Canyon appear to be widely supported by people who have no intention of using either. There is some evidence that, at least in the Lake Erie example, nonuser values comprise a significant portion of the benefits. Despite widespread public support for rehabilitation, the benefits to users appear to be only about as great as rehabilitation costs (Francis et al. 1979). Unless such "unrevealed" values can be described and estimated, the various forms of benefit cost analysis will sometimes be inadequate.

Unrevealed extramarket values are those values that are not revealed by observing resource use occasions, because the values (1) are not explicitly established through market transactions, and (2) normally require no significant allocation of time, travel, or other resources economists might observe to estimate implicit values (Talhelm et al. 1980). However, the fact that there are these values implies that individuals would allocate corresponding resources if necessary.

Such activities are normally only revealed through actions such as voting, political activities, and verbal statements, although occasionally they are partially revealed through voluntary contributions or through compliance with "involuntary" assessments like taxes or ransom payments. The values of all indivisible goods (i.e., "public goods") and many goods other

---

Daniel R. Talhelm is an Assistant Professor of Fisheries and Wildlife at Michigan State University, East Lansing, MI. This paper is based on a project sponsored by the Great Lakes Fishery Commission involving D. R. Talhelm, R. C. Bishop, L. W. Libby, K. E. McConnell, A. J. Randall and A. A. Schmid. Contributors to the project included P. Bohm, K. Goran-Mäler, R. Johnson, C. Plott and V. K. Smith. Financial support for the author's participation in the Visual Values Workshop was in part provided by the American Petroleum Institute.

than those we normally think of as public goods also have values that are unrevealed in this respect.

One purpose of this paper is to examine the theoretical economics of various forms of unrevealed values. I propose six forms of these values, defined on the basis of the mechanisms causing the values to be unrevealed. The most important mechanisms are indivisibility and ambiguity.

The other purpose is to examine one of these, existence value, in detail. Here, I propose that the existence value of a good is attributable to the consumer's state of knowledge about the good; not necessarily to the state of the good itself. Awareness of a good's existence is information capital that may technically change the consumer's utility function. The existence of unicorns, for example, could not be valued unless we discover (or hypothesize) their existence. Upon gaining awareness, the consumer may care about the good's future existence. The magnitude of the existence value of a good depends in part on how unique consumers feel the good is. Consumer views of uniqueness may differ greatly from physical, or supply, perspectives of uniqueness. Therefore, uniqueness and substitutability are also examined in detail.

The next section examines existence value. The succeeding sections briefly examine other forms of unrevealed values: options for future consumption, public goods, unpredictable events, values obscured by ambiguous information and altruistic values.

EXISTENCE VALUE

Existence value may be simply defined as the value of knowledge of the existence of a good, apart from any use of the good. Existence is an indivisible, nonexclusive attribute of a good, conceptually separable from but jointly produced by some "uses" of the good. Consider wild polar bears, for example. Few people ever expect to see a wild polar bear or to otherwise use one, yet many apparently value the fact that they exist. They would be willing to pay to maintain the bears' existence apart from any potential use. Apparently, many also value maintaining the options of seeing or hunting the bears, or perhaps having a polar bear rug, but these values fall into the option value category.

Although existence is an indivisible, nonexclusive attribute of a good, knowledge about and awareness of the good may be both divisible and exlusive, and may have a positive marginal cost. One acquires knowledge of the good either by observing or otherwise using the good, or through someone else's experience with the good. Therefore, using a good jointly produces knowledge and any other products of that use.

An experience that produces an awareness of good P, say the experience of either reading about or viewing a particular set of scenic vistas, may be represented as the household production and consumption of commodity Zp (viewing or reading):

$$Z_p = f_p (X_i, X_p, T_p) \qquad (1)$$

where the production inputs are X_i, a vector of time and goods resources; X_p, the scenic vista(s); and T_p, time spent consuming Z_p produces new knowledge or awareness of P. If the new knowledge changes one or more of the ways a consumer produces utility from his environment, it in effect technically transforms the consumer's utility function from $U = U(Z_i)$ to U'

$= U'(Z_i)$, where Z_i represents the set of commodities consumed. The set Z_i may or may not change as U changes. The consumer may either maintain U' without further inputs of P, or acquire further information that transforms U' to U". At each stage, the consumer may value the existence of the good differently. Each input of P may produce (1) direct utility, (2) an intermediate product for later use, and/or (3) information capital that transforms U. If the input is simply the knowledge that P no longer exists, the only consumption activity might be the acquisition of that knowledge. The consumer's change in utility and/or utility function would indicate any change in value. The existence value of the good (if positive) would be measured as the equivalent surplus the consumer would pay to prevent the change, or as the compensating surplus the consumer would accept to compensate for the change.

Consider, as another example, the "honeymoon effect." Suppose an event would occur that would significantly reduce or eliminate good visibility in the Poconos. An economist estimating the loss in value attributable to the change in the character of air quality in the Poconos would not only have to estimate values for present and future use, but the value of the loss felt by the couple from Iowa who honeymooned there in 1923, but have never been back and have no intention of ever going back. They may feel a loss in utility just from knowing the scenic vistas are not like they were in 1923. The magnitude of their loss may depend on whether they enjoyed their honeymoon and whether they associate their happy or unhappy marriage with the visit. Perhaps they could even gain utility from the change!

If existence value is unrevealed by consumer resource allocations, $U(Z_i)$ and $U'(\ _i)$ could be the same except for the new information and the different willingness to pay to prevent extinction and willingness to accept compensation for extinction. However, $U(Z_i)$ could differ in other ways from $U'(Z_i)$ and still not reveal existence value. For example, knowledge of the scenic beauty at the Mt. McKinley National Park in Alaska might lead one to believe that other national parks in Alaska are also interesting and to acquire information about them. That second allocation of resources is attributable to the expected value of the information about other national parks in Alaska. Knowledge of these national parks may lead the consumer to read more about them or to view them again. The activities would reveal the total of the expected value of those activities and the expected value of additional knowledge.

Knowledge, like existence, may be a one-time event. Once knowledge is acquired, repetition is redundant. Values revealed by consumers in such one-time acquisitions could easily be misjudgments.

From the perspective of the household production function, consumption of a good (such as Z_p) enters one's cost constraints and one's utility function. This consumption choice might reveal the existence value of the good because part of the demand for the good is the demand for more knowledge about and awareness of the good. In other words, perhaps the demand for news, books, and other information sources could be separated into components, one of which could be attributed to existence value. Intuitively, it seems unlikely that the portion of knowledge and awareness attributable to a given resource is separable, particularly since (1) knowledge in an individual may be viewed as a technical change or a non-disposable capital acquisition (i.e., moving from U to U'), the value of which might be difficult for the consumer to estimate until after the fact, and (2) the knowledge/awareness content of news, books, etc., is hopelessly intertwined with other content related to other values. It would probably be as easy for consumers to estimate the value of the existence of the resource in question as for them to estimate the value of their knowledge and awareness of the resource.

In short, existence may be considered a pure public goods component of resources. Standard definitions of public goods are appropriate, provided that they admit the role of information in the household production function for these goods, which they typically do not. For example, Randall (1981) defines a public good as "a good that, once produced, is available without rivalry" (p. 179). He also points out that most definitions imply non-exclusiveness as well as indivisibility (p. 190).

Market goods may have existence value as well, and as with natural phenomena, I would expect that the more unique and admired a good is, the greater would be its existence value. Classic examples are the 1957 Chevrolet, various coins, the Mona Lisa, national defense, the Empire State Building, and Beethoven's Fifth Symphony. Commonplace examples include tomatoes and Boeing 747 airliners. All of these have well-known uses; viewing the Mona Lisa, for example. Yet, we can at least conceive of existence values for each. The good produces the usual products, Z_i, as well as knowledge. Once the consumer has seen the Mona Lisa, or an illustration or description of it, his utility function has been transformed from U to U'. Even if the consumer never intends to view the Mona Lisa again, he may feel a loss if he is informed that it has been slashed, destroyed, or lost. The same holds for scenic conditions at national parks.

Uniqueness and substitutability

Significant natural phenomena like polar bears, Niagara Falls, and the Grand Canyon are obviously unique. Yet uniqueness and similarity are merely products of the human tendency for generalization. Every object is unique in some respect from each other one. We notice some differences more than others because those differences seem important to us. We abstract to the level that suits our purpose. The point is that uniqueness is not so much a natural phenomenon as a human invention for generalizing. The well-known biological and physical classification systems were developed to generalize about reproduction, genetics, physical structure, and other attributes more related to supply than to demand. Consumers' classification systems, determining substitutability and uniqueness in demand, may be based on a different set of attributes. Classification systems for man-made products sometimes relate more to production than consumption, although most seem to compromise between consumer awareness of differences and producers' desires for product differentiation.

Public values are determined more by uniqueness from the consumer's perspective than by biophysical uniqueness, even though the former is sometimes overlooked. If, for the purposes of an economic analysis, all units of a proposed group substitute "perfectly" for each other, they are the same good. Partial substitutes and complements are related goods, and poor substitutes are unrelated goods. Classes in any classification system are presumably assigned on the basis of the system developer's purpose and perspective. For example, from some perspectives, all scenic vistas are the same. From other perspectives, all national park scenic vistas are the same, all western national park scenic vistas are the same, all Grand Canyon scenic vistas are the same, or no scenic vistas are the same (each is unique). Mathematically, economists measure substitution in demand as cross-price coefficients, cross-price derivatives, or cross-price elasticities

between goods, implicitly assuming perfect substitution within each given good. For example, let

$$Q_i = f(P_1, P_2, P_3, I) \qquad (2)$$

represent the demand for good X_i, where Q_i is the quantity of X_i, P_i are the prices of good X_i and two other relevant goods, and I is consumer income per capita. If $S_{12} = dQ_1/dP_2 > 0$, goods X_1 and X_2 are substitutes. If S_{12} is infinite or "sufficiently great," goods X_1 and X_2 are the same good, according to standard definitions. However, S_{12} can be infinite only at the point where P_1 and P_2 are equal. If the prices differ significantly, S_{12} will be zero because the consumer will never substitute. He/she would have no reason to pay a higher price for one unit of a good when another identical unit is available at a lower price (assuming, of course, perfect information and zero or identical transactions costs). Therefore, perfect substitutes have $S_{12} = 0$, except at the switching point at $P_1/P_2 = R_{12}^*$, where S_{12} is infinite. Indifference curves between perfect substitutes X_1 and X_2 would be a straight line, and if $R_{12} \neq R_{12}^*$ a corner solution results in specialization in either X_1 or X_2. If X_1 and X_2 have the same units of measure, $R_{12}^* = 1$.

It follows that for close substitutes, S_{ij} may also vary depending on R_{ij}. The indifference curves are relatively flat but eventually intersect or parallel the axes. Here, $\alpha > S_{12} > 0$ for a range of R_{12}, say $R_1 < R_{12} < R_h$. A most effective rate of substitution is possible at S_{12}^* where S_{12} is maximum at a corresponding R_{12}^*.

Uniqueness may be defined in terms of S_{ij}^*: good X_1 is unique if there is no good X_j for which S_{ij}^* is greater than some arbitrary level. However, since S_{ij} depends upon the definitions of the units of the goods, elasticity would be more useful for this purpose. Let ε_{ij}^* be the maximum point cross-price elasticity, where $\varepsilon_{ij} = (P_j/Q_i) S_{ij} = P_j dQ_i/Q_i dP_j$ for any corresponding R_{ij}. Then, a better definition of uniqueness is: good X_i is unique if there is no good X_j for which ε_{ij}^* is greater than some arbitrary level.

Given sufficient data, this definition could help us determine which resources are unique enough to consider special attention to preservation. Product definitions could be hypothesized and accepted or rejected using this criterion. For example, although fishing in Michigan's Great Lakes for salmon and trout is unique is some respect in each of the thirty-three counties in which such fishing is regularly available, salmon/trout angling in one county may be a close substitute for the same fishing in another county (see Jordan and Talhelm 1982). Perhaps some counties offer unique opportunities according to this definition; alternatively, perhaps salmon/ trout angling is only unique relative to other kinds of angling, perhaps any Great Lakes angling is only unique relative to non-Great Lakes angling, or perhaps angling is only unique relative to other forms of recreation. Such hypotheses could be tested by observing choices made by anglers from different locations, since the "prices" of these kinds of angling vary with travel distance.

With market goods, however, an arbitrary definition of uniqueness is unnecessary. Uniqueness appears as a economic rent that becomes capitalized into the market value of a product. Accurate estimates of willingness to pay also reflect uniqueness in nonmarket goods such as Great Lakes angling.

Furthermore, in analyzing existence value, the change in quantity (dQ_i) is essentially meaningless, since the question at hand is the existence or nonexistence of something. Existence value may be considered the same as the all-or-none value of existence of the good (apart from any use or option values). Therefore, perhaps the best definition of uniqueness considers the all-or-none value of existence of the good, measured as either equivalent surplus or compensating surplus, and abbreviated here as WTP. This may be visualized as the sum for all consumers of the entire area under each consumer's demand curve for existence, since the price is zero (existence is a public good, provided at zero marginal cost to the consumer). However, the quantity axis must be a dummy number indicating existence at zero price and nonexistence at some price. We shall return to the quantity issue later.

There is some indication that although consumers may understand species extinction and differentiate between canyons, their willingness to pay to preserve the existence of "unique" phenomena (or goods) is highly contingent on the number of phenomena they have recently been asked to contribute to, and perhaps even on the number of charitable causes they have contributed to. For example, if the total "Save the Unicorn" group asked me to help save unicorns, I might contribute $A if I had not made any other significant contributions recently. I would probably contribute much less, say ($A-B), where $A > B > 0$, if I had recently contributed to other similar groups. My short run discretionary income is more limited than long run, but I would make similar trade-offs in the long run. If unicorn preservation effort is good X_1, and other groups are X_2 and X_3, I suspect that my ε^*_{12} and ε^*_{13} would be relatively high. All three goods may appeal to my sense of altruism. In Lancaster's terms, all three goods may be independent inputs in my overall production of the same characteristics, altruism, or my sense of obligation to the "outer world."

An indication that this is a real phenomenon was given by Randall and Stoll (Chapter 23 in this volume). Whereas Schulze et al. (1981) found that the average U.S. household would pay up to $86 per year to assure clean air in the Grand Canyon, they reported a University of Chicago study showed that Chicago people would pay an average maximum of $325 per year to assure clean air in Chicago, $355 per year to assure clean air in Chicago and the rest of the Eastern U.S., and $371 per year to assure clean air in Chicago, the Eastern U.S., and the Grand Canyon. Apparently, the substitution effect is great, although there could be some income effect. While both clean air and the Grand Canyon are unarguably unique, valuable resources, consumers apparently find satisfactory substitutes.

Note that in the unicorn example, I was not contributing precisely for the continued existence of unicorns, but for an organization which I believed would represent me in various ways in its efforts to save unicorns. It might purchase unicorn habitat, lobby Congress, advertise and/or do other things. While my contribution may indicate my unicorn existence value, it does not measure precisely that value.

One of the key issues in understanding existence value is specifying the quantity measure of existence. I suggested a "dummy variable" approach above, but perhaps there could be other approaches. Do only rare and/or physically unique organisms or physical features have existence value? Can we attribute existence value to each individual scenic vista or only to all. Consider the existence of one particular scenic vista. People who have knowledge of a particular scenic vista may value its existence, and the magnitude of that value may depend on the extent of that knowledge. As with any good, the value any particular individual vista depends upon (1) the desirability of the attributes of that vista, (2) its uniqueness, which depends upon the extent which that vista is perceived as being different from other vistas, (3) the prices of the perfect and/or imperfect substitutes and of compliments, and (4) other factors such as the consumer's income. The existence value of a scenic vista depends upon the substitutability in demand of any other individual scenic vista for that scenic vista. In this case, many other sites may be practically perfect substitutes. Therefore the scenic vista's existence value would probably be low relative to that for all scenic vistas of a particular type. The existence values of (1) a particular set of scenic vistas, perhaps at one particular national park, (2) all similar scenic vistas, or (3) all scenic vistas depend on the same kinds of factors. The broader the definition of the good, the greater its existence value is likely to be. No matter how we define a good, its units of existence are the same; either that a specific good exists or it doesn't exist. Existence value is always an all-or-none question.

The existence value for a population may be divided by the number of units in the population to calculate average existence value. Marginal existence value, $d(WTP_i)/dQ_i$, would be zero because total existence value does not depend upon the number of units in the population (holding knowledge of the good constant).

Finally, existence values may conceivably be positive, zero, or negative. However, because it may be difficult to separate use-values from existence values, it may be difficult to detect negative or zero existence values. Goods that might have negative existence values are goods that apparently have negative "use" values, such as rampaging, life-threatening polar bears; smallpox; nuclear bombs; and mosquitoes. Even these might have positive existence values after we subtract their negative use values, attributable to bear attacks and damage, disease, destruction, and pesky bites. However, the empirical problem of restricting consumer's evaluations strictly to existence and not associated broader contexts may prove extremely difficult.

OPTION VALUE

Option value is the value of an option that keeps available the possible future use of a resource, apart from the value of using the resource. For example, I would be willing to pay up to $A if necessary to maintain the possibility (or option) of visiting a particular national park that might otherwise be closed, although my probability (π) of visiting the park may be almost nil even assuming the option is available. My net all-or-none value (i.e., Hicksian-compensated consumer surplus (CS) of the visit) would be $B. Therefore, my total willingness to pay for park visitation rights, my option price (OP), or $A, is my option value (OV) plus my expected net value of visiting, $\pi(CS)$, or $B, equation (3). If either my

$$OP = OV + \pi(CS) \qquad (3)$$

π or my CS is zero with certainty, then my OV would also be zero because I have no reason to maintain the option. However, OV could be positive even if I never exercise the option by visiting the park. The original concept of option value was that people who will never use a given resource may nevertheless be willing to pay to maintain the option to use the resource. This implies that the value of resource use may not be equivalent to observable consumer surplus, as we had previously assumed; that people would be willing to pay more than their expected future value. That may seem unreasonable, so let us examine it further.

Current economics literature suggests that option value (1) may be merely the value of reducing the uncertainty about future use, much like the value of insuring our home or car (i.e., an option makes the possibility of future use more certain, reducing the risk of lost value); and (2) may be positive, zero, or negative under given conditions, depending in part upon whether demand or supply is uncertain. The analysis is essentially the same for users as for nonusers: Both may have positive values that are not revealed by normal user activities. For a more thorough discussion of most of the points here, see Bishop (1982).

First, consider that people would be willing to pay up to their expected future consumer surplus, $\pi(CS)$, for an option for future use. (Assume the discount rate is zero so we can ignore discounting. It would add nothing and only complicate the argument. Similarly, let one possible future use occasion represent all possible future use occasions.) This includes eventual nonusers as well as users, by the simple nature of probability. Whether my π is .99 or .01, I could still eventually either be a polar bear hunter or a polar bear nonhunter. In either case, I would be willing to pay up to $\pi(CS)$ to maintain this option. If people are good judges of π, accurately projected consumer surplus based on presently observed use would be the same as aggregate $\pi(CS)$ estimated through a survey of the entire population of participants and nonparticipants. Participants receive $CS > \pi(CS)$ values, whereas nonparticipants receive $0 < \pi(CS)$ value. It is not clear whether this is the nonuser value Weisbrod (1964) had in mind when he proposed option value, or whether he was proposing OV as a risk premium.

Next, let us consider supply uncertainty and OV. In the example of my polar bear hunting option value, one of the reasons I was willing to pay to maintain my options was that I was uncertain whether sufficient numbers of bears would be available to permit hunting in the future. If my payment of OP now would help insure sufficient numbers in the future I may pay it. The reasoning here is precisely the same as the well known Friedman-Savage (1948) analysis of insuring. They showed that if the marginal utility of income is decreasing, the expected utility of a combination of uncertain outcomes is less than the utility of the equivalent certain outcome. For example, suppose my $\pi = .01$ because there is a $\pi' = .99$ that not enough bears will be available, but otherwise my $\pi = 1.00$. Then my expected consumer surplus is $.01(CS) + .99(0) = .01(CS)$. Friedman and Savage showed that my utility under those initial conditions would be the same as my utility would be if I paid OV + .99(CS) to receive B with 100% probability. In each case, my net utility is worth .01(CS).

Next, consider demand uncertainty. Most of the literature has dealt with this source of option value rather than supply uncertainty. Here we assume the consumer is uncertain about his/her future utility function. Under one set of future preferences, the consumer will use the resource, under another set the consumer will not, all other conditions being equal. Again, the consumer would be willing to pay OP = OV + $\pi(CS)$ to maintain the option in case he wishes to exercise it in the future. The conclusion so far appears to be that this kind of OV could be positive, negative, or zero. Basically, the analysis is that purchasing an option for $A could provide a future dollar gain of CS - A (good demanded) or a loss of A (good not demanded), and not purchasing an option could produce a "loss" of CS (good demanded but not available) or no effect (good not demanded). Without the option, expected utility, π(CS), is the weighted average of two points on the same utility curve, one point for income I and the other for (I - CS). With the option and certainty, expected utility is the average of utilities for two utility functions, but at the same income level (I - OP). Expected CS without the option could be greater than, equal to, or less than OP, so option value could be positive, negative, or zero.

The literature has not considered another likely source of "demand" uncertainty: the possibility that the consumer's constraints or household production technology will someday make a visit more attractive. If, for example, consumers have unforeseen reasons to be near the Grand Canyon, they may be able to take advantage of a low cost opportunity to visit the park. Another possibility is that future technical changes could reduce transportation costs. These possible cost reductions are actually supply changes from the household production function point of view (see Talhelm 1972, 1973, 1980), rather than demand changes, but they represent consumers' dreams about possible future use. Although the consumer remains on the same utility curve, a cost reduction increases his/her expected consumer surplus. The analysis here is the same as our supply uncertainty analysis above.

Finally, although we have recognized that a risk premium may be appropriately added to consumer surplus in certain cases to estimate project benefits or costs, we must also recognize that the risk premium may be negative or zero in some cases. Risk premiums should be routinely considered in benefit-cost analyses. There is also ample evidence of such premiums in the insurance market. Furthermore, options are commonly purchased for many assets.

PUBLIC GOODS

Public goods may be defined as indivisible goods. One person's consumption does not affect consumption by others. A classic example is national defense. The fact that I consume national defense does not add or detract from anyone else's consumption. Likewise, my viewing the flock of geese overhead does not detract from someone else's view.

Some authors assume or imply that public goods are also nonexclusive; that consumers cannot be excluded from consuming and cannot exclude themselves. We consume the level of national defense that is provided whether we want to or not. We cannot exclude ourselves without moving to another country, and even then we are affected by it. Existence is another indivisible nonexclusive good. Values of indivisible, nonexclusive goods are not revealed by normal consumer activities, although they may be at least partially revealed through voluntary contributions or political processes such as tax allocations, elections and assessment by elected representatives. These points were discussed earlier, and are further elaborated in most current microeconomics textbooks; so we will only go into further detail here on two more points. Existence was elaborated upon separately because it has not been widely recognized as a public good and because, like most other public goods, some analytical issues are peculiar to existence.

Many natural resources or attributes of natural resources are at least partly public goods. Services of the environment such as clean air and waste assimilation are examples. Environmental protection is generally also a public good, much like national defense or police protection. Scenic beauty is only partly public. It is indivisible up to the point where crowding becomes a problem, and often people can be excluded. Direct use of scenic beauty may sometimes be measured and consumer values estimated--consider Grand Canyon viewing, for example.

Finally, individuals generally "consume" public goods only indirectly. Individuals "consume" national defense in concept only, even though our concepts are based on actual weapons and human capabilities. Existence is similar; we consume the concept, not the resource. In other words, public goods are characterized by "nonconsumptive" use of some form, in that the consumer does not actually take possession of the good.

UNPREDICTED EVENTS

Have you ever been working in your yard or driving to work and are suddenly surprised by a flock of geese flying overhead? Suppose you allocated no time or money resources to enjoy the geese, but that you would have if necessary. The value of this unpredictable event was unrevealed by your overt actions.

This form of unrevealed value--consuming unpredictable goods without allocating any resources--is probably not an important source of consumer value. Practically all events may be predicted with some probability, and we allocate our resources based on expected results. However, there are events with positive and negative values that are unforeseen or for which the probabilities are so small that we ignore them. The values of such events are unrevealed by our overt actions.

AMBIGUOUS INFORMATION

Many things are improperly valued because we lack sufficient information. Ecological relationships and endangered species are among resources often cited as having unknown values. Bad experiences with introductions of exotic species, such as European carp throughout the U.S., sea lamprey in the Great Lakes, and starlings in North America may be readily cited as having had negative previously unrevealed values.

The value and role of information is well known. It's economic analysis is discussed in many textbooks. Our education system, our research investments and many other aspects of our lives involve information gathering for decision-making purposes.

Decisions may be deferred until further information becomes available (the value of this increased information is "quasi-option" value). Values of "irreversible" decisions are particularly sensitive to the lack of information.

ALTRUISM AND EQUITY

What is it worth to you to help the poor, starving people in Africa? To fight birth defects? To preserve your (or someone else's) religious ideals? Do we provide wilderness, scenic vistas and hospital care only for those who can pay their full cost? Such values are obviously important and are unrevealed or only partly revealed by normal market mechanisms. Economists tend to ignore them, particularly the values of fairness in resource allocation and equity in income distribution. They present special difficulties because although contingent valuation techniques by which such values might be accurately estimated are now becoming available, the values themselves depend on both the existing income distribution and the social conventions we have chosen to live by. This web of cause and effect for such values presents substantial analytical difficulties that may be with us for a long time--some say forever. However, we seem to be fast approaching the day when many such values may be estimated, and no longer ignored.

CONCLUSION

Unrevealed or partially revealed values may be the rule rather than the exception in our lives. At first it may seem bewildering and preposterous to imagine being faced with all of these values with our limited incomes. If we had to pay in all instances where we could, our incomes would seem Lilliputian until we consider that our incomes would also be correspondingly greater. Both sides of the ledger would expand. Existence value, for example, could be thought of as part of our "real" wealth. We are enriched by the existence of things and diminished when they cease to exist.

Explicitly including "unrevealed" values in decision making and in "real income" accounts is revolutionary. It is an important component of the information explosion. The critical component of this revolution is the development of satisfactory valuation techniques. I am awed to contemplate the impact of widespread successful use of contingent value techniques or other techniques that disclose many of the multitude of "unrevealed" values we hold and unthinkingly exercise daily.

BIBLIOGRAPHY

Bishop, R.C. 1982. "Option Value: An Exposition and Extension." Land Economics 58(February).
Francis, G.R., J.J. Magnuson, H.A. Regier, and D.R. Talhelm. 1979. Rehabilitating Great Lakes Ecosystems. Great Lakes Fishery Commission, Technical Report 37.
Friedman, M., and L.J. Savage. 1948. "The Utility Analysis of Choices Involving Risk." Journal of Political Economy 56(August): 279-304.
Jordan, S.W., and D.R. Talhelm. Forthcoming. "The Economic Feasibility of Fisheries Management Options for Michigan's Great Lakes." Journal of Great Lakes Research.
Randall, A.J. 1981. Resource Economics: An Economic Approach to Natural Resource and Environmental Policy. Grid Publishing, Inc., Columbus, OH.
Schulze, W.D., D.S. Brockshire, E.G. Walther, and K. Kelley. 1981. Methods Development for Environmental Control Benefits Assessment. Volume VIII. The Benefits of Preserving Visibility in the National Parklands of the Southwest. Prepared for the U.S. Environmental Protection Agency, Washington, D.C.
Talhelm, D.R. 1972. Analytical Economics of Outdoor Recreation: A Case Study of the Southern Appalachian Trout Fishery. Unpublished Ph.D. dissertation. N.C. State University, Raleigh, NC.
_____. 1973. "Defining and Evaluating Recreation Quality." Transactions of the 38th North American Wildlife and Natural Resources Conference 38:183-191.

_____. 1980. "Estimating Tourist Demand for Prospective Developments of Natural Resource for Recreational Use." in D. Hawkins, E. Shafer, and J. Rovelstad, ed. <u>Tourism Planning and Development Issues.</u> George Washington University Press, Washington, DC.

Talheim, D.R., R.C. Bishop, L.W. Libby, K.E. McConnell, and A.A. Schmid. 1980. <u>Understanding Unrevealed Extramarket Values of Natural Resources.</u> Mimeo research proposal, Michigan State University, E. Lansing, MI.

Weisbrod, B.A. "Collective-Consumption Services of Individual-Consumption Goods." <u>Quarterly Journal of Economics</u> 78(3):471-477.

PART VI
Management Perspectives and Case Studies

The resource management issues addressed in this volume have been approached primarily from the point of view of the land manager who must implement the policy. Industry and environmental groups have their own perspectives as to the underlying goals and proper implementation of the legislative mandates concerning resource management within the broader context of the choices concerning environmental protection and resource development. Defining and implementing these goals will require the cooperation of federal and state agencies, industry, and environmental concerns and must take into account the perspectives of each.

The first two papers in this section highlight industry and environmental perspectives concerning the overall resource management process to protect air quality and visual resources in national parks and wilderness areas, and the state of current research to implement these legislative requirements. Taylor, presenting the view of the Edison Electric Institute, concentrates upon the goals and implementation of the Clean Air Act. He finds many of the goals for Class I area protection to be unclear, several current regulatory programs to be unnecessary given current air quality conditions and available technology, and current research to be inadequate to meet the needs of industry in evaluating potential impacts. Yuhnke, presenting the views of the Environmental Defense Fund, finds the goals of the Clean Air Act and regulations for Class I areas to be well established in legislative history and through citizen concerns. He describes these goals as protecting a fundamental part of the American heritage.

The real test of the research results and theories concerning visual and visibility impacts is in their application to resource management decisions. In many cases there is still a gap that must be bridged between research results and their application within the existing legal and decision making environment. The two case studies described here supplement the preceding chapters by providing illustrations of the kinds of decisions and problems faced by the resource manager. They discuss the problems encountered in both visual value analysis and in the procedural process of implementing visual and visibility regulations. Understanding problems encountered in these cases should help focus future regulations and research efforts. Brady describes the development of environmental management process that focuses upon visual impacts in the highly developed Lake Tahoe recreation area. Haddow and Blankenship discuss the U.S. Forest Service program for monitoring and mitigating visibility impacts in the Colorado Flat Tops wilderness area.

PART VI
Management Integration
and Case Studies

25. Managing Our Visual Resources

John A. Taylor

This chapter discusses the perspective of the Edison Electric Institute (EEI) concerning visibility protection in Class I federal areas (principally, our large national parks). I will focus on EEI's general approach to environmental planning, the provisions of the Clean Air Act that protect visibility in mandatory Class I federal areas largely from existing sources, the Environmental Protection Agency (EPA) regulations that provide that protection, the prevention of significant deterioration (PSD) program that is designed to protect visibility in the Class I federal areas from the adverse impacts of new sources, and EPA's regulations to implement that statutory program.

A GENERAL APPROACH TO ENVIRONMENTAL PLANNING

There are three fundamental components of a rational approach to planning for environmental protection. The first is an identification of the goals that have to be achieved. For the purposes of today's discussion, this component requires the identification of the air quality related values that are to be preserved in Class I federal areas. Congress has focused attention on only one air quality related value: visibility. We must also decide how much visibility is enough.

The second component of a rational environmental planning program is the development of a valid and accurate methodology for relating emissions from sources to our air quality related goals. To do this we need:

1. An objective and well-defined set of criteria that can be used to quantitatively characterize the important air quality related values. For visibility, visual range and discoloration are important.
2. An adequate characterization of the "baseline" conditions for these criteria in terms of frequency distributions for various levels of visibility, the duration and extent of visibility impairments and the probability that impairments will coincide with the times during which there will be visitors to the national parks

---

John (Jack) Taylor is the Manager of the Air Quality Section at Virginia Electric and Power Company, Richmond, VA. He spoke on behalf of the Edison Electric Institute.

seeking to appreciate their scenic beauty and other attributes. Baseline conditions should account for the inherent variation in visibility due to natural and man-made phenomena such as meteorological factors, forest fires, dust storms, isolated point sources, and urbanized areas.[1]

3. A technique to predict, with reasonable confidence, what a particular action (or set of actions) would do to alter baseline conditions in Class I federal areas.[2]

4. A technique for determining how the public would react to these alterations in baseline conditions.

The third component of EEI's general approach to environmental planning concerns the political process. That process must be free to consider all relevant factors, such as each aspect of environmental quality, energy, economics, employment, and other social objectives. Only with the consideration of all relevant factors can we have a process designed to advance the public interest.

An important issue that the political process must take into account is the distribution of costs and benefits of our environmentally related decisions among different segments of the population. In the visibility context, a population segment living near Class I federal areas may strike one balance between visibility levels and energy costs while the nation as a whole may strike another. As examples, local population may want to deprive the nation of energy resources in an effort to preserve local scenic beauty or the same population may not want to bear high electrical costs necessary to preserve scenic beauty for the nation. As to the distributional question, the identity of the decision maker may dictate what decision is made.

In any event, the decision should be based on a consideration of all relevant factors, otherwise the public interest may not be served. The public interest can be prejudiced by environmental decisions that misdirect our finite resources by calling for the expenditure of funds that may not produce sufficient social benefits.

I will now apply this approach to Class I area protection as defined by the Clean Air Act and EPA's regulations.

VISIBILITY PROTECTION IN THE MANDATORY CLASS I FEDERAL AREAS

Section 169A of the Clean Air Act

How does Section 169A of the Clean Air Act fare against these criteria? This section seeks to protect only visibility in the mandatory Class I federal areas. While it provides for the consideration of all relevant factors in deciding how to retrofit isolated point sources or in formulating long-term strategies, there are no reliable, generic techniques for relating source emissions to the human perception of visibility impairments. It is, therefore, difficult to implement Section 169A.

This may not be a significant practical problem because there may be no need for Section 169A now. This is true because:

1. The primary emissions of particulate matter do not appear to be a problem--at least from power plants. I should add that, if such emissions were not adequately controlled, simple visual observa-

tions and our experience with particulate control equipment would probably be adequate to overcome deficiencies in currently available generic source-receptor relationships and permit the development of particulate matter control strategies for visibility protection.
2. Sulfur oxides do not contribute to plume blight. To the extent they contribute to regional haze, projections of their emissions in the Rocky Mountain area (Arizona, Colorado, Idaho, Montana, Nevada, New Mexico, Utah, and Wyoming) show significant reductions in sulfur dioxide emissions during the next thirty years under current regulatory programs exclusive of visibility protection. These reductions could exceed 50 percent from a 1976 baseline. (See Appendix A and ICF and NERA 1982, Mangeng and Mead 1980, U.S. EPA 1979.)
3. There is no available control technology for significantly reducing nitrogen oxide emissions^3
4. It is uncertain what area or mobile source emissions contribute to visibility impairment in nearby or remote Class I federal areas. As a result, there has been no ground swell of activity to regulate such emissions. I expect that this response is also due in part to the recognition of the practical political realities associated with regulating these sources.

In summary, Section 169A appears to be of little value in light of the current controls on particulate matter, the projected reduction in sulfur dioxide emissions in the geographical areas of principal interest, the absence of efficacious nitrogen oxide control equipment, the uncertain role of area and mobile sources in visibility impairment, and the deficiencies in generic source-receptor relationships.

EPA's Regulations

Current EPA regulations focus only on the narrow issue of single source visibility impairment, usually plume blight. This approach was taken (1) because the lack of reliable, generic source-receptor relationships prompted EPA to postpone regulation of regional haze in mandatory Class I federal areas and (2) because the four items that I have just enumerated suggest that there may be no need to retrofit existing sources to protect visibility. The decision, implicit in EPA's regulations, not to augment controls on existing plants now, therefore, seems to be the correct decision.

One major problem with EPA's regulations for visibility protection in the mandatory Class I federal areas, however, is that they call for the protection of visibility in vistas outside, but "integral" to, the mandatory Class I federal areas. Integral vista protection expands the geographical scope of the visibility protection program in a manner inconsistent with the Act and should be eliminated.

THE PSD PROGRAM AND VISIBILITY

The Act

Let us now turn to the PSD program written into the Clean Air Act. While EEI supports the concept of special protection for the air quality in Class I federal areas, the PSD program does not compare well with the three criteria for sound environmental planning. First, its goals are fuzzy. The Act mentions visibility as an air quality related value but does not embellish further. Second, as noted previously, the generic source-receptor relationships necessary to implement the visibility component of the PSD program do not exist. Third, the statute restricts the factors that can be considered in specifying controls for, and the siting of, new sources once the permitting authority concludes that they will have an adverse impact on the visibility in a Class I federal area. Such a process could foreclose the use of otherwise valuable sites, increasing the costs of energy development and possibly increasing overall environmental effects.

In addition to these general problems, the PSD program suffers from several specific flaws. First, it outlines an elaborate ritual for the licensing of new sources. While this process adds little to the protection of visibility, it does cause costly delays that raise energy prices. Second, the Class I increment system is arbitrary--it does not have a direct correlation with the air quality goals to be attained. As to visibility, the increments are inappropriate surrogates for regional haze and they may result in substantial overkill for plume blight. A recent study done for the National Park Service (Ireson et al. 1981) shows that the "setoff" distances established by the Class I increments and current modeling techniques are greater than necessary to ensure that plume blight from moderately sized power plants is not perceptible from Class I areas. In other words, Class I increments are more stringent than necessary to protect against plume blight. Thus, to the extent that visibility is a value that we want to protect, the increments are artificial and impose energy, economic, social, and environmental costs without providing the requisite benefits. Third, it overlays case-by-case best available control technology (BACT) on the already stringent new source performance standards (NSPS). Neither of these technology-based standards are related directly to our visibility protection. This extra stringency seems inappropriate at a time when emissions of some pollutants are declining.

In short, Congress has enacted an extremely cautious, misguided program for new development. Such a program seems out of place in the west--an area that Americans have selected for population and economic growth. Because a no-growth scenario seems to be politically unpalatable in the west, we should not have an environmental policy that strangles desired growth with programs that cause delay and consume public resources to achieve no discernible benefit.

<u>EPA's Regulations</u>

As already noted, there is little in the way of reliable, generic source-receptor relationships for implementing the visibility portion of EPA's regulations for the PSD program. In addition, these regulations add protection for integral vistas expanding the geographical scope of the program beyond that intended by the Act. Finally, contrary to the scheme contemplated by the Act's PSD program, the EPA regulations expose sources licensed and constructed under the PSD program to the threat of additional federal requirements for visibility protection once they begin operation. This set of regulations, therefore, exposes expensive energy facilities to significant risks even after they have passed the strict requirements of the PSD program.

Such a regulatory approach can only raise energy costs without a guarantee that benefits are being achieved for the public. This can happen either directly by leading to additional expenditures for pollution control equipment or indirectly by forcing development at less efficient sites unnecessarily remote from the Class I areas.

CONCLUSIONS

In conclusion, rational environmental planning requires goal identification, reliable source-receptor relationships, and a political process that considers all relevant factors before decisions are made. Only with such a scheme can we be sure that adequate but not inordinate resources are devoted to environmental goals and that sufficient social benefits are actually achieved for the resources devoted to environmental protection. Our current visibility program does not compare well with this scheme.

NOTES

1. The role of natural forces cannot be ignored. A study of air quality at the airports near Phoenix, Yuma, and Albuquerque shows that meteorological factors alone could explain up to 80 percent of the variability in reported visibility. (Henry et al. 1981).
2. Regrettably, generic models for visibility are not very accurate or reliable. This is so because they must deal with the complex physical and chemical processes of dispersion of primary emissions, their transformation to secondary pollutants and transport over uneven terrain and long distances. Finally, to be of much value, the results have to be reported in frequency distributions describing visibility and translated into terms that can be related to human perception. Because of all of the deficiencies in the science and technology of visibility modeling, we have a significant problem with the implementation of the second component of our visibility planning process.
3. Even if sulfur dioxide emissions were not declining and control technology were available for nitrogen oxides, reliable, generic source-receptor relationships for the visibility impairment caused by the secondary pollutants derived from these emissions do not exist. Without these relationships, the impact of current emissions or the value of emission reductions cannot be determined. Hence, scientifically defensible control strategies cannot be developed.

BIBLIOGRAPHY

Henry, R.C., J. F. Collins, and G.M. Hidy. 1981. Analysis of Historical Visibility Data at Three Airports in the Southwest. Environmental Research & Technology, Inc. for Anaconda Copper Company.

ICF Inc., and NERA. 1982. Summary of Forecasted Emissions of Sulfur Dioxide and Nitrogen Oxides in the United States over the 1980 to 2010 Period. Washington, D.C.

Ireson, R.G., D.R. Souten, D.A. Latimer, and C.D. Johnson. 1981. Perceptibility of Plumes of Facilities Sited at Increment-Limited Distances from Class I Areas. Systems Applications, Inc. for the National Park Service Washington, D.C.

Mangeng, C. and R. Mead. 1980. Sulfur Dioxide Emissions from Primary Nonferrous Smelters in the Western United States. Los Alamos Scientific Laboratory, Los Alamos, NM.

U.S. Environmental Protection Agency. 1979. 1976 National Emissions Report Washington, D.C.

APPENDIX A

Projections of Sulfur Dioxide Emissions for the Rocky Mountain States

Emission estimates for the eight Rocky Mountain states (Idaho, Montana, Wyoming, Nevada, Utah, Colorado, Arizona, and New Mexico) show that, with NSPS for new sources and the present regulatory programs for existing sources, total sulfur dioxide emissions in these states will decline in the future even while new electrical generation capacity is being added.

Two different emission forecasts for this area are shown in Tables 25.1 and 25.2. The emissions estimated for the years 1980 to 2010 in Table 25.1 are based upon results of a study performed for the Utility Air Regulatory Group by ICF and NERA (1982). These results show that total SO_2 emissions for the mountain states will decline by 44 percent between 1980 and 2010. If 1976 emissions are used from the National Emissions Data System (NEDS) data base are used, the decrease is even more dramatic-- emissions in 2010 would be 57 percent of the 1976 emissions.

Table 25.2 uses the same data as in Table 25.1 except that estimates for copper smelter emissions for the 1990 to 2000 period are based upon a study done by the Los Alamos Scientific Laboratory (Mangeng and Mead 1980). Using these numbers, Table 25.2 shows that between 1980 and 2000, SO_2 emissions will decline by 37 percent and by 51 percent between 1976 and 2000. The major cause for the difference in copper smelter emissions between between ICF and NERA (1982) and Mangeng and Mead (1980) is due to the fact that the former's methodology leads to the result that by 2000 most (if not all) smelters would be subject to NSPS, whereas Mangeng and Mead assume that very little of the capacity would be produced by new smelters. Moreover, ICF and NERA assume that NSPS would result in a 99.5 percent clean-up of smelters based on uncontrolled emissions. Thus, even if one were to assume that (a) this degree of clean-up would be too optimistic to expect in practice, and (b) copper smelter emissions could be four times more than calculated by ICF and NERA in 2010, then SO_2 emissions in 2010 would still decline by about 35 percent compared to 1980 levels.

Therefore, even if visibility protection regulations do not influence SO_2 emissions, they are expected to decline dramatically for the next thirty years.

TABLE 25.1
Sulfur Dioxide Emissions (1976 to 2010) in 100,000 Tons for the Mountain States Based on ICF and NERA (1982) Estimates for Copper Smelters

Year	1976*	1980*	1990	2000	2010
Utilities	4.3	4.5	4.9**	5.7	6.1**
Copper Smelters	30.2	22.2	6.2	0.5	0.8
Other			5.1	6.2	7.9
Total	34.5	26.7	16.2	12.4	14.8

* Actual.
** Assuming present NSPS for new sources.

TABLE 25.2
Sulfur Dioxide Emissions (1976 to 2000) in 100,000 Tons for the Mountain States Based on Mangeng and Mead (1980) Estimates for Copper Smelters

Year	1976*	1980*	1990	2000
Utilities	4.3	4.5	4.9**	5.7**
Copper Smelters	30.2	22.2	5.7	4.9
Other			5.1	6.2
Total	34.5	26.7	15.7	16.8

* Actual.
** Assuming present NSPS for new sources.

26. The Importance of Visibility Protection in the National Parks and Wilderness

Robert E. Yuhnke

All too often when we gather to discuss visibility, we lose sight of what visibility is, and what it means to people. Visibility is the ability to see many things. Not just distant vistas, or color, or contrast, but the ability to see the whole of the Grand Canyon from a single vantage, to watch a thunderstorm form over the desert on a clear summer afternoon and feel it advance across the great expanse of time and space, to feel the vastness of the universe while lying under the star-strewn canopy of a desert night, to watch eagles soaring endlessly, to know the beauty of this planet, the joy of consciousness, the great gift of life, and to experience the fullness of creation, and the Creator Himself.

When we destroy visibility, we destroy these experiences. When these experiences are lost, an important part of our vision as a people will be lost with them.

The grand vistas of the American West are a fundamental part of the American heritage. The spirit of countless generations of white man and red has been enriched by the experiences of seeing Navajo Mountain from a hundred miles across the desert, or Mount Rainier from the Olympics, or the Tetons from Yellowstone's peaks. These experiences are part of an American's image of his land, a part of his birthright.

The awe which these special places inspire created the unquestioned consensus among Americans that these places should forever be preserved, not only for ourselves but our posterity. In setting aside Yellowstone, Yosemite, Glacier, the Tetons, the Grand Canyon, and other parks, Congress made clear in the legislative history that one of the goals of those enactments was the preservation of the grand scenic vistas and the inspiration they provide.

That same consensus that led to the preservation of our great national parks in the 19th and early 20th centuries has led to the enactment of wilderness preservation in the 1960s, and visibility protection for parks and wilderness in the 1970s. When the 95th Congress added visibility protection to the Act in 1977, it acted consistent with the same ideals and in response to the same goals that motivated the 52nd Congress to create Yellowstone Park 110 years ago.

Robert Yuhnke is Regional Counsel for the Rocky Mountain Office of the Environmental Defense Fund in Boulder, CO. Assistance with travel funds was provided by the American Petroleum Institute.

The current regulatory scheme for protecting visibility has been the focus of much criticism, both in and out of Congress in recent years. The Western Regional Council, for example, has asked Congress to entirely repeal the visibility program from the Act. Many have cast the issue entirely in terms of economics. I think they have made a serious mistake by failing to take account of the importance which the American people attach to their parks and wilderness. These are not simply places to go play cheaply, they are places where Americans can touch the past and feel a part of the land as God created it and feel communion with the Creator Himself. Such values cannot be measured in the marketplace.

The attachment of the American people to their parks and their wilderness is reflected in the most recent votes in Congress on the visibility program. In March, the Senate Environment and Public Works Committee strengthened the visibility program by incorporating the Parks Service list of proposed integral vistas into the bill. In the House Commerce Committee in April, after a week of votes to nearly dismantle the non-attainment part of the Act, the full committee rejected both the subcommittee language in H.R. 5252 and the functional repeal of the visibility program introduced by the Republic leadership. Instead the Committee adopted a PSD program which would leave the current visibility program vitually intact.

I do not think these votes can be attributed to the clout of the environmental community. Rather, I suggest these votes are a reflection that the broad-based consensus which led to protection of the parks and wilderness in the first place is still abroad in the land. It is also a product of our faith in ourselves as a people: that we have the ingenuity to heat our homes and delight in the grand vistas of the West too. We are not so poor that we must take the Rembrandts from the wall to burn in the fireplace to keep us warm.

So I offer a challenge to the leaders of industry and to all Americans: take a page from the business leaders of the past such as the Rockefellers who helped create the Teton National Park and A.P. Hill whose Great Northern Railroad donated to the park system much of what is Glacier National Park. These men took great pride in our heritage, and were not only leaders of industry, but also in the effort to preserve these unique treasures for future generations. I challenge you to do the same. Let us join together to sincerely figure out how to make the Clean Air Act work. Let us all be able to point with pride a generation from now that we have passed intact to our grandchildren the same opportunities to know and experience the great joys of this Earth as we have received them from our forefathers.

In approaching this challenge, it is important to keep the legal mandate in mind. Section 169A focuses on protecting visibility from all manmade causes. Of course, much attention has been focused in the last two years on plume blight because the EPA regulations only focus on the specific source-related problem. But we must not lose sight of the regional haze problem. In many respects, regional haze is a far more serious problem than plume blight in the Grand Canyon and elsewere, and should be given as much attention in research efforts.

27. Protection of the Visual Experience in the Flat Tops Wilderness

Dennis Haddow
James Blankenship

The 1977 Amendments to the Clean Air Act designated eighty-eight wilderness areas managed by the U.S. Department of Agriculture (USDA) Forest Service as Class I areas. For these and other Class I areas, the federal land manager was given an "affirmative responsibility to protect air quality related values." The only air quality related value specifically identified in the 1977 Amendments to the Clean Air Act is visibility. The federal land manager for the eighty-eight Forest Service Class I areas is the Secretary of Agriculture. However, within the USDA, this responsibility is delegated to the Regional Foresters.

The Flat Tops Wilderness, a Forest Service Class I area in northwest Colorado, lies downwind of a number of proposed synthetic fuel projects and power plants. This paper identifies some of the information needed in Region 2 of the Forest Service to assess visibility impacts in the Flat Tops.

Federal visibility regulations define an adverse impact on visibility to be visibility impairment that interferes with the management, protection, preservation, or enjoyment of the visitor's visual experience of the federal Class I area. This determination must be made on a case-by-case basis taking into account the geographic extent, intensity, duration, frequency, and time of visibility impairments, and how these factors correlate with (1) times of visitor use of the federal Class I area, and (2) the frequency and timing of natural conditions that reduce visibility. However, the end result is the determination of impact on the visitor's visual experience.

Within each Class I wilderness area in Region 2, the Forest Service is selecting one or more views that are representative of the visual attributes of the area. For each identified view, the region should set a maximum acceptable level of impact on visual experience. Any impact beyond this level will be adverse. When selected, this maximum acceptable level of impact on visual experience is the visibility target or standard that should be addressed in prevention of significant deterioration (PSD) permit applications.

Dennis Haddow is an Air Quality Specialist for Region 2 of the USDA Forest Service, Denver, CO. James Blankenship is Manager, Technical Services, USDA, Forest Service in Fort Collins, CO. The views in this paper are those of the authors and not necessarily official USDA Forest Service Policies.

The maximum level of impact should be determined on a case-by-case basis using the factors listed in the Environmental Protection Agency (EPA) Visibility Rule and other factors such as: (1) the physical characteristics of the wilderness, (2) the management objectives of the wilderness (3) visitor expectations in the wilderness.

The Flat Tops Wilderness is located within the White River and Routt National Forests in northwest Colorado. The 235,230 acre wilderness became part of the National Wilderness Preservation System in December 1975 through Public Law 94-146. The dominant feature of the Flat Tops Wilderness is the White River Plateau, a flattened, lava-capped dome, the perimeter of which is sharply defined by sheer volcanic escarpments, below which lie gently rolling benches and deep drainage valleys that shelter scenic lakes, streams, and forests (Forest Service 1978). Elevation in the Flat Tops Wilderness ranges from approximately 8,000 to 12,000 feet. In many instances there is an abrupt elevation change at the wilderness boundary. Precipitation in the area may exceed 60 inches per year, mostly in the form of snow. Snow restricts travel in the wilderness from mid-October through late June.

To the extent that the wilderness resource is not impaired, the Flat Tops Wilderness is managed to provide opportunities for solitude, primitive recreation, physical and mental challenge, freedom from unnatural sights, sounds, and odors, and the chance to experience unmodified ecosystems. Major primitive recreational pursuits include hiking, backpacking, horseback riding, hunting, fishing, photography, and viewing the scenery. Recreation use in 1981 was estimated to be 141,000 recreation visitor-use days (RVDs). This is up from 40,000 RVDs in 1972. The majority of use occurs during two periods: (1) early July through Labor Day and (2) the big game hunting season in October. Winter use of the wilderness is almost nonexistent. The severe topographical changes at much of the wilderness boundary discourage most snowshoe and cross country skiers.

The Region's objective in managing visibility within wilderness is to provide the opportunity for a natural visual experience. The key words are "natural" and "experience." In this context, "natural" is defined as unimpaired by man. This objective is somewhat different than providing the opportunity for visual enjoyment. Enjoyment of any or all parts of a wilderness experience is purely a personal matter for the individual visitor to decide.

The importance of clean air and viewing the scenery to Colorado wilderness users has been documented by a number of studies. In a study of resource attributes, users of one Colorado wilderness identified "clean, fresh air" as the most important of seventy-three physical resource attributes that contributed to their satisfaction (Brown et al. 1977). It is important to realize that wilderness users determine that air is "clean and fresh" by what they see and smell. It is doubtful that any wilderness users carry air quality monitoring equipment in their backpacks. In another study on preservation values, Colorado wilderness users identified "viewing the scenery" as the most important of a list of twenty specific wilderness experiences (Walsh et al. 1982). The same users ranked "protecting air quality" second among thirteen specific reasons for valuing wilderness.

While neither of these studies was designed to research the importance of visibility to Colorado wilderness uses, it is logically easy to project the importance of "clean, fresh air" and "viewing the scenery" to the importance of unimpaired visibility.

EPA, in cooperation with the Forest Service and U.S. Geological Survey, has installed a remotely operated visibility monitor on the southwest bounday of the Flat Tops Wilderness. The monitor consists of three telephotometer and three camera systems. It will provide baseline visibility data for use by the Forest Service, air regulatory agencies, and industry in developing and reviewing PSD permit applications.

Visibility in the Flat Tops Wilderness, as determined by managers and users, is presently considered to be unimpaired. Views from the wilderness include the Unitas, La Sal Mountains, Mt. Zirkel, Gore Range, and Elk Mountains. However, the Region will only consider impacts on the visual experience for views within the wilderness boundary.

Determining the maximum acceptable levels of impact on visual experience in the Flat Tops Wilderness is a management decision for Region 2 of the Forest Service. This decision must fit within the legal frameworks of the Wilderness Act, Clean Air Act, and EPA Visibility Rule. The Region has considerable discretion in setting a maximum acceptable level or standard. As with any standard, there must be measurement methods available to determine if the standard will or is being met. The measurement methods must be both scientifically and legally defendable. Therein lies the problem. EPA and others have developed measurement methods for determining physical visibility impacts (changes in contrast, visual range, coloration). However, there are no scientifically proven and legally defendable measurement methods to determine impacts on visual experience.

The Region has or soon will have the following information for the Flat Tops Wilderness: (1) a list of views which represent the visual resource, (2) Visitor use data detailing numbers and times of visitor use, (3) studies indicating the importance of visibility to wilderness users, (4) baseline physical visibility data.

Unfortunately, without a scientifically and/or legally defensible measurement method, there is not sufficient information for the Region to determine maximum acceptable levels of impact on visual experience.

Until these measurement methods are developed and accepted, it will be difficult for the Region to evaluate impacts on visual experience in the Flat Tops Wilderness. These difficulties are not only those of the Forest Service. Industry needs to know if their permit applications are approvable before the application is submitted. The states have the same needs if they want to do comprehensive planning. Also, the public needs to be able to determine if the states and federal government are meeting their responsibilities in protecting the resource.

The development of measurement methods to determine impact on visual experience should be a high priority for all groups involved in visibility protection.

BIBLIOGRAPHY

Brown, P.G., G. Haas and M. Manfredo. 1977. Identifying Resource Attributes Providing Opportunities for Dispersed Recreation. Final Report to the Rocky Mountain Forest and Range Experiment Station, Fort Collins, CO.

U.S. Department of Agriculture. Forest Service. White River and Routt National Forests. 1978. "Flat Tops Wilderness Management Plan." Denver, CO.

Walsh, R.G., R. Gillman and J. Loomis. 1982. "Wilderness Resource Economics: Recreation Use and Preservation Values." Department of Economics, Colorado State University, Fort Collins, CO.

28. The Lake Tahoe Environmental Thresholds Study

Sheila Brady

INTRODUCTION

The Lake Tahoe Thresholds Study has been of considerable interest in California and Nevada, as well as within the Tahoe Basin itself, partly because no one knew what a threshold was. We all knew it could be a stone that lies under a door, or a beginning point, or even the point at which an effect begins to be produced, but just what it was when applied to Tahoe was another question. We did not know at the beginning of the study, but do know now, that a threshold is something to trip over trying to get from one place to another.

The thresholds study has had the team wrestling with three other major questions in regard to visual values:

1. Can standards be set for visual quality on private lands in the same way they can for air and water quality?
2. If so, how can measurements be made that provide the right kind of data to set standards?
3. Are visual quality standards effective in maintaining or achieving visual quality?

Let me back up and present some facts about Tahoe. Lake Tahoe is a natural resource of rare value. The largest alpine lake in the world, Tahoe's waters are clear enough to see the bottom at great depths. Set in a huge bowl, surrounded by high mountain peaks, Tahoe provides some of the best mountain and lake recreation in the West. The lake basin is one of the most remarkable natural landscapes in this country, and is a major national recreation resource.

About 70 percent of the land within the basin is publicly owned, and most of this is within national forest. Land uses on private lands are primarily residential and commercial, including gaming. Commercial development is concentrated along major roads encircling the lake. The most intensive commercial activity is clustered around the state line areas in Nevada and California at the northern and southern ends of the lake, where the casinos on the Nevada side act as magnets for hotels, motels and related commercial uses.

Sheila Brady is a partner in Wagstaff and Brady, Berkeley, CA.

302

The visual quality of the Lake Tahoe Basin is one of the main attractions of the region. In the early 1900s, the landscape was predominantly natural, and the area was primarily used for summer recreation by a small number of people able to travel long distances and take long vacations. Man's activities were limited, and major visual intrusions on the landscape did not occur. Then, after World War II, when new highways into the Lake Tahoe Basin were constructed, the tourist boom began. Casino development expanded on the Nevada side, and the visual character of the area began to change. Even as late as 1956, however, the area had only 9,500 permanent and summer residents and 36,400 transient tourists. By 1978, the resident population was 73,200 and the total peak day population was estimated at 223,320.

Most major visual deterioration has occurred on privately held lands, which are centered around the perimeter of the lake, in most cases along major roadways. Thus, it is the visual character of the area most frequently seen by visitors that has changed most. The average visitor traveling around the lake now sees urban development of varying types and intensities, and entirely natural landscapes are restricted to publicly owned areas, such as Emerald Bay State Park and the area around Fallen Leaf Lake in the El Dorado National Forest.

Large areas of the basin in public ownership (national forests and state parks) still offer natural landscapes of exceptionally high quality. These areas, however, are accessible only to those who know the area, who are able to hike long distances, or who have advance reservations. In recent years, some areas of publicly held land have undergone visual deterioration as a result of intensive use. Roadside campgrounds are especially susceptible to visual deterioration; large numbers of recreational vehicles and automobiles parked near the roadside are often visible where vegetation screens between campground and roadway have been removed. The visual deterioration of publicly held lands is, however, insignificant compared to the visual degradation caused by private development.

Deterioration in other environmental resources has kept pace with development. Water clarity, for instance, has declined in areas downstream of residential development.

Tahoe has become a headline in the news because its political setting is almost as unique as the natural setting. Divided in half between California and Nevada, the enclosed natural bowl has served well as a political arena where opposing interests are played out year after year and political careers are made. On the Nevada side, the casinos with the only highrise buildings at the lake represent big money interests and high stakes in the development game. The California side, with most of the developed and developable land, resists becoming the provider of housing and services for weekend gamblers and casino employees.

Legislators from both states have long recognized the problem, and Congress in 1959 developed the Bi-State Compact, an agreement that provided the Tahoe Regional Planning Agency (TRPA) with powers to oversee development and conservation on all private lands in the basin. Because of the particular structure of the TRPA Governing Board, and especially the voting procedure by which development applications were automatically approved unless a majority of the Board voted no, Californians were dissatisfied with what they felt was the pro-development stance of the Agency and Board.

In 1980, wide-ranging revisions were made to the Bi-State Compact. Changes included a restructuring of the Governing Board and the voting procedures, and contained provisions that a new regional plan be developed. This new or amended regional plan was to be based on "environmental thresholds" to be set for major natural resources.

The concept behind this provision in the revised Compact was fairly simple: Determine how much change (i.e. growth and development) basin natural resources could sustain without degradation, set threshold levels at that point, and devise a regional plan to implement those levels.

As all of you know who have ever attempted to implement simple concepts though a planning process, the thinking and the doing are vastly different, and subject to endless conjecture.

The actual wording of the Compact was important. As used in the Compact, an "environmental threshold carrying capacity" means "an environmental standard necessary to maintain a significant scenic, recreational, educational, scientific or natural value of the region or to maintain public health and safety within the region. Such standards shall include but not be limited to standards for air quality, water quality, soil conservation, vegetation preservation, and noise."

A federal interagency council developed a work scope to implement Compact legislation entitled "Public Involvement in Establishing Environmental Thresholds". We felt that there were three critical needs for the thresholds study:

1. An accurate updated land-based data system, with information on physical and land use characteristics for individual parcels. Land planning and development regulation in Tahoe has long been hindered by an outdated, fragmented and sometimes inaccurate information system.
2. Models of land use alternatives that would identify implications of various types, levels, and densities of development.
3. An extensive public involvement program from the beginning of the program, so that the results, conclusions, and decisions of the study would have public support.

When the research team presented a detailed work program with these three elements to the TRPA Governing Board, we were surprised by the emphatic reaction of the majority of the Board members; there was to be no public involvement, no land use alternatives, and no analysis of the social and economic implications of the thresholds until after they were adopted as standards. The public involvement program and consideration of land use alternatives and their implications were to be done during Regional Plan development, after the Thresholds Study. The first of many redirections, this response was in fact consistent with the intent of the Compact.

It gradually became clear that the framers of the Compact not only knew what they were saying, they _intended_ to say it, and it was our job to figure out how to do it. It finally also became clear that they had provided us with one extraordinary opportunity, to my knowledge untried before, to set what were in effect visual quality standards, enforceable in the same way as air and water quality standards, on private lands.

We faced a whole set of problems in carrying out this mandate. How were we to set visual thresholds that were (1) _adoptable_ by the strong but divided TRPA Board, (2) _implementable_ by the TRPA staff, and (3) _effective_ in protecting scenic resources.

Our specific problems were the following:

1. There was strong opposition in the basin, well represented on the Board, to adopting visual thresholds if it meant a wholesale shutdown of development. The question of "roll-back" of development to pre-1982 status came up continually.
2. There were genuine conflicts as to what constituted visual quality, especially in the man-made environment.
3. Thresholds had to be set without the benefit of public participation, land use data and/or plans, adequate time for refined visual analyses, or documentation of exising visual resources.

But if thresholds were not set, there would be no mechanisms for addressing visual quality in the regional plan.

These problems prompted several attempts by Board members to throw out visual resources as a suitable thresholds component. Our solutions, or at this stage I should say, responses to these problems have been the following:

1. We developed a "serviceable" values statement to direct the study. (I call it "serviceable because it is general, unsophisticated, but agreeable to the TRPA Board.) Everyone was able to agree that the natural landscape was of primary importance, and that the man-made landscape could stand some improvement, although no one could agree on how, when, or where this improvement should occur.

This value statement was derived from existing goal statements contained in various TRPA, local, state and federal documents relating to the Tahoe Basin such as:

1. Maintain and enhance the dominant natural-appearing landscape for the vast majority of views and lands in the basin.
2. Maintain and/or improve the aesthetic characteristics of the man-made environment to be compatible with the natural environment.
3. Restore, wherever possible, damaged natural landscapes.

The value statement also brought the study in line with the concept of "scenic" with its connotations of "natural" beauty rather than "visual" which would include the man-made landscape; the statement was therefore consistent with the Compact language.

2. The second step was to identify the resource. Just what were we talking about? Where was it? What was it made of?

3. The third step was to attempt to assign evaluative measurements to the resource, so that the thresholds would have some teeth, instead of consisting of vague policy statements, too easily dismissed in the implementation stages.

THE PROCESS

The process included a scenic resource inventory. The inventory was set up in the following way. The study team identified scenic resources within the defined landscape units, as seen from principal travel routes. Individual scenic subcomponents were mapped, photo-documented and also described in narrative text. This inventory provides a baseline of 1982 existing resources so that threshold levels can be tied to measurable degrees of change in resource status that would result from various possible future management decisions.

The following scenic subcomponents in each of forty-six roadway units were identified:

1. Foreground, middleground and background views of natural landscape, seen from the roadway.
2. Views to the lake from roadways.
3. Views of the lake and natural landscape from roadway entry points to the basin.
4. Special landscape features, such as streams, beaches, rock formations, vegetation combinations, distinctive topography, etc.

The scenic quality of each subcomponent was rated based on four criteria of unity, vividness, variety and intactness. Each subcomponent was assigned a number from 0 to 3+. A rating of 3 indicates high visual quality, 2 indicates moderate visual quality, 1 indicates low visual quality, and a rating of 0 indicates an absence of visual quality. A rating of 3+ is assigned to a unit with exceptionally high visual quality. The assignment of numbers to units is intended to express comparative ratings of value rather than absolute numerical measurement.

The quality of each roadway scenic resource unit was ranked based on:

1. the <u>number</u> of major scenic resources within or visible from the unit;
2. the <u>quality</u> of identified major scenic resources--based on unity, vividness, variety and intactness; and
3. the overall scenic quality of the unit, based on unity, vividness, variety and intactness.

A composite number was then assigned for each roadway unit, establishing its relative scenic quality.

The shoreline analysis was completed in a similar fashion. Subcomponents making up visual resources from the lake are:

1. views of predominantly natural shoreline, uninterrupted by man-made features and
2. views with a high degree of natural landscape variety

In addition to the scenic quality ratings, each unit is also assigned a <u>sensitivity to change rating</u>. This rating expresses the degree of vulnerability to change of the resources within a given unit. A rating of 3 indicates a high degree of sensitivity to change, 2 indicates moderate sensitivity, and 1 indicates low sensitivity. The factors considered in determining this rating include the following:

1. Characteristics of the resource subcomponents.
2. Location of the individual resource subcomponents and the likelihood that outside influences, such as land use changes, would alter the quality of the resource.
3. Conditions within the unit, such as erosion, that may be expected to change or impair the scenic resources.

Some scenic resources have especially sensitive elements. For example, a view of a ridgeline or skyline can be easily impaired by changes in land form or land use. The scenic quality of some views may depend upon a compositional arrangement of varied elements; and removal or blocking of any one of these elements could reduce the quality.

The location of the individual resource is an important factor in assessing its sensitivity to change. For example, long-distance views from the Echo Summit area would probably not be dramatically altered by land use change along the south Tahoe shore, since these changes would be viewed at a great distance. On the other hand, views of the lake from Emerald Bay roadway unit would be highly sensitive to changes in land uses in the surrounding area, since the views could be easily blocked or otherwise impaired. The sensitivity to change rating emphasizes changes that would result from man-made activities rather than natural processes, such as vegetative succession or seasonal changes in vegetation color.

RECOMMENDED THRESHOLDS

The threshold number proposed for each unit is a composite of the scenic quality rating (0-3+) and the sensitivity to change rating (1-3). A threshold number between 5 and 6+ would indicate that the units existing scenic resources are of very high quality and also highly sensitive to change. A rating of 3 to 4 would indicate a moderate level of scenic quality and sensitivity to change, or high scenic quality and low sensitivity to change, or low scenic quality and high sensitivity to change. Units identified with threshold numbers between 0 and 2 contain low scenic quality with low sensitivity to change.

Recommended Thresholds Policy

The purpose of scenic resource thresholds is to establish a mechanism for protection of identified resources and a means of monitoring change in these resources. The proposed threshold policy is to maintain or improve the scenic resource threshold number for each unit including the scenic quality rating for each resource identified within that unit. Improvements would occur in the quality of the scenic resources, especially in its degree of intactness. Sensitivity to change of any given resource expresses inherent characteristics that are unlikely to be affected by improvements.

Travel Route Ratings

We developed, in addition to the scenic resource unit rating, a second supplementary and rather interesting measure of scenic quality. This was the so-called travel route rating, and its purpose was solely for monitoring change over time in visual quality in the basin. A Forest Service task group in 1971 evaluated the scenic quality of all the travel routes in the basin. Each of the forty-three roadway units' scenery was evaluated based on man-made features, physical distractions to driving along roadways, roadway characteristics, views of the lake from roadways, general landscape views from roadways and shoreline, and the variety of scenery from roadways and shoreline. Within each of these criteria, a numerical grade was assigned, varying from a low of 1 to a high of 5, as the task group traveled over each roadway unit at sightseeing speed. Each unit could have a score between 6 and 30. To evaluate the thirty-three shoreline units, the same group traveled at slow speed around the lake, approximating the shore distance traveled by sight-seeing boats. Ratings were conducted only on those three criteria relating to lake travel, so scores varied between 3 and 15.

A 1978 update of the roadway portion of this analysis showed a deterioration in visual quality, chiefly resulting from new development and construction, and signing and grading along roadways. The thresholds study team updated this study for 1982 because it provided the only available measure of scenic quality over time.

Although this system has certain limitations (the criteria could be rather easily manipulated by the scorers, and it should be used by observers with some training in visual evaluations), it does provide a measurement of change over time, and should be useful to the TRPA staff as a system for monitoring trends in scenic quality. The 1982 ratings, for instance, were interesting in that they showed improvement in three visual units that correlated with new design review practices and sign ordinances undertaken by jurisdictions in those units.

VISIBILITY THRESHOLD

The threshold for visiblity is being developed as part of the air quality threshold component, and was based on a TRPA/U.S. EPA visibility monitoring study that dove-tailed with the thresholds study.

Traditionally, the purity of the Tahoe Basin alpine air is one of the area's greatest environmental assets. However, recent trends show that

1. both state and federal carbon monoxide air quality standards have been exceeded on numerous occasions,
2. haze, reducing visibility in the basin, is a frequent occurrence,
3. high air pollution levels have caused vegetation damage, and
4. it is believed that air pollution is a source of nutrients that degrades Lake Tahoe water quality.

Natural factors, people-related factors, and background pollutant levels all influence air quality in the basin. Natural factors include large scale weather patterns, local weather patterns and basin topography. The two main sources of air pollution related to human activity are vehicle emissions and wood smoke. They are influenced by population, transportation systems, land use patterns, and land use activities. Background pollutant levels are comprised of pollutants transported into the basin and natural pollutant levels.

Particulate matter is defined as any liquid or solid particle suspended in or falling through the air. Particulate matter not only decreases visibility in the basin, it also causes respiratory problems, aggravation of cardiovascular disease, and cancer in humans. Toxic particles in the air include sulfates, lead, wood smoke, and organic particles formed from hydrocarbons.

The results of the 1981 visibility study show that visual range (measured in the number of kilometers at which an object can be seen) can be related to the concentration of particles in the air. Larger particles settle out of the air more quickly than very small ones, which may stay suspended for as long as six days. The team suggested that a program will need to be established to monitor particulate matter less than 2.5 microns in diameter, as particles of that zone are the most significant cause of visibility degradation in the basin. Setting the regional standard is complicated by that and by the ratio of sulfate to non-sulfate. Sulfates have to be controlled from outside the basin.

Particulate matter in the basin of less than ten microns in diameter is produced chiefly by residential wood combustion, construction activities and vehicular traffic. Basically, concentrations of particulate matter can be reduced and visibility improved by reducing the number of vehicle miles of travel, and limiting the amount of smoke produced from residential wood burning.

The visibility study developed a model to relate visibility range with sources, and developed a recommended regional threshold, to increase contrast or range in all directions. The regional threshold will be expressed as the number of kilometers an object can be seen X percent of the time, not to be exceeded on more than a certain number of days. The sub-regional threshold is at this time still uncertain, and may be visibility range, or may be particle concentration.

CONCLUSIONS

The staff and consultant have recommended adoption of scenic resource thresholds for roadway and shoreline units. Visibility thresholds will be recommended. All environmental thresholds are scheduled for adoption in July of this year. Following adoption, implementation measures will be developed during the regional plan process.

For scenic resources, these measures should include a comprehensive parcel-based design plan or visual resource management system. A design plan could include ordinances, transfer of development rights systems, design standards and criteria, design review procedures for staff, and site-specific development measures. Implementation of scenic resource thresholds would require a new design review function, both at the staff and Board level, perhaps with creation of a design review commission. It would also require increased design capability at the staff level.

In answer to the first two questions posed at the beginning of this paper, I think yes, standards (or thresholds) can be set for visual quality and, yes, measurements can be made in a variety of ways, but they should start with identification of the components of the resource, and they should be derived from comparative evaluations, on the basis of an accepted value or objective. They should also not be taken too seriously, but be used as tools. Numbers are, after all, reflections of values, and have validity only if the implementation measures can be carried out through the political and regulatory system.

The third question is more problematic, and, at least in Tahoe, the answer is unknown. I sincerely hope that standards for visual quality can be effective on private lands.

At this point, at risk of one more pun, I will say that the thresholds study is "a foot in the door" for visual quality. Opening that door will depend on commitment of the community to invest the time and money and to accept changes in the level and type of allowable development.

PART VII
Conclusions and Future Directions

Understanding of the kinds of visibility and visual changes that are perceptable and thought undesirable by human observers is improving, as is understanding of the importance of scenic resources in the context of recreation activities. Methodologies are being developed to help in defining undesirable changes in visibility and visual resources and in quantifying the importance or value of preventing such changes. Much has yet to be accomplished, however, before we will have adequate tools to address the resource management decisions that are being made today and that will have to be made in the future. Many of the tools examined here are specific to the requirements of the Clean Air Act regulations, but these same kinds of tools for will be necessary for resource management decisions under any legislation that strives to attain the best possible allocation of our natural resources taking into consideration the importance of environmental aesthetic values to the quality of life.

At the end of the Visual Values Workshop several of the participants offered comments and suggestions as to what the current problems are and the appropriate future research directions to solve these problems. These discussions, as well as written comments submitted after the workshop, are the basis for the summary that follows.

Among the most important deficiencies in our current research and management of visibility and visual resources is in determining an accepted standardized definition and measurement of what constitutes an adverse change. Resource management decisions concerning what levels of degradation are tolerable are having to be made. While substantial technical information can now be obtained, it is still often insufficient to effectively meet the requirements of the Clean Air Act regulations, especially those that require a determination of whether a change is considered "adverse." This was a concern expressed by most of the industry representatives and land managers at the workshop. There is agreement and recognition that visibility and visual values are important and deserve protection, but the lack of a clear cut procedure for identifying and measuring whether the change is adverse, or measuring the degree of adverseness, is a problem for those who must make specific resource management decisions.

In short, we do not yet have an adequate and acceptable visual value assessment procedure with which to quantify the importance of visibility changes to the observer's recreation experience. A related concern is that we still do not have thoroughly reliable tools for determining the link between emissions and the visibility and visual changes that would result, but this was not the focus of this research reported in this volume.

Land managers were also concerned that research be conducted with an understanding that management decisions are currently being made and that defensible tools are needed now. All participants agreed that more interaction between land managers, industry, and researchers would be beneficial, perhaps through a formal working group. One overall suggestion was that uncertainty be incorporated more explicitly into research results. It is apparent that there is sometimes a large margin of error in the identification and quantification of impacts that may result from increased emissions or other visual changes and that the usefulness of such results for policy decisions would be enhanced if the error bounds were included. Another point that was emphasized by all groups at the workshop is that the context of concern for visibility and visual quality must be kept in mind. It is the importance of the scenic resources in the context of the entire recreation experience that is of concern. Due to the complexity of this context, each research effort tends to focus on only one aspect. The plea here is for researchers to be aware of and respond to the needs of the land managers and to keep in mind the original reasons why the questions they are addressing are being asked.

Studies addressing human perceptions of visibility changes have been ongoing for some time, yet more work is needed to improve the usefulness of these results for future adverseness determinations. These studies attempt to determine whether one visibility circumstance, or set of circumstances, is perceived as different and preferable to another. This aids in defining probability thresholds of perceptibility and provides insight into what changes in which characteristics will likely lead to adverse impacts on visitor enjoyment. It is important to keep in mind that these studies do not, however, actually determine how important these changes are to visitor enjoyment and, therefore, are not able to answer whether a particular change, while perhaps less preferred, is adverse to visitor enjoyment. Debate continues over whether or not an absolute threshold of a just perceivable change can be quantified or is the appropriate focus of the perception research. Concern was also expressed that the approach used in most of this work, asking a volunteer respondent to rate or in some way evaluate photographic representations of various views under different visibility conditions, is subject to problems of preconditioning of the respondent. Evidence indicates that the ratings can be influenced by the way which the question is phrased. These studies have not, in general, addressed the effects of what people expect to see as opposed to what they do see. Certain preference tendencies have emerged, but is still not clear upon exactly what elements in the view respondents' ratings are based on. More work on the human perception process, rather than simply its empirical reflections, is needed, as indicated by Henry (Chapter 7) and Middleton, Stewart, Dennis and Ely (Chapter 6).

Visual resource management tools have been widely applied in the fields of landscape architecture and are being used for resource management decisions in the Forest Service and the Bureau of Land Management, where multiple land uses coexist. Visual resource management has typically focused on specific landscape elements including form, color, texture, line, contrast, and content of fairly close range scenery. Changes in visibility and the types of long distance views that tend to be affected by air quality have not been addressed with these tools. Visual resource management techniques provide some insights into how the scenic content of an affected view can be systematically addressed in visibility management decisions. Two examples highlight this point. Malm, MacFarland, Molenar and Daniel (Chapter 4) found that layered hazes in the sky are less detrimental than layered hazes blocking parts of the landscape. Latimer, Hogo, Hern and Daniel (Chapter 5) found visibility impacts to be most significant when prominent, distant landscape features are obstructed. Many feel visual resource management systems should be more thoroughly integrated into visibility perceptions and preference studies to more adequately address these landscape effects.

It is in the social and psychological approaches to valuing changes in (or measuring adverseness of) visibility and visual resources where the most questions remain. These types of studies are beginning to address what kinds of behaviors are affected by scenic resources and some methodologies have been applied to determine the attitudes and responses of park visitors to changes in scenic resources. The work is, however, in early stages of development and is lacking adequate theoretical understanding of how subjective and environmental attributes interact in perception processes and influence judgments. There are many inputs into these judgments concerning scenic quality that are not well understood and have not been specifically defined. Much theoretical and methodological development, as well as validation of methodologies already in use, is needed in this area. The literature needs considerable more work in defining operational measures of adverseness that can be used in decision making and that are related to visitor attitudes.

Somewhat contradictory concerns arose from the discussions about the importance of changes in actual behavior. In one sense the land managers are concerned about when a deterioration in visibility or visual resources would cause people to not visit a park or when the use for which the park was intended would be adversely affected; however, it is not clear that small changes in behavior tied to changes in visual resources imply the impact is adverse. Concern was also expressed that baseline air quality conditions may be widely deteriorating causing visitor's expectations of parks to diminish while visitation does not decline due to lack of alternatives and other factors. It was suggested that researchers could directly query visitors as to whether a change was adverse and relate their response to behavior and the characteristics of the visibility change. It is also of concern that the value of the park to the visitors derives from the visitor's recreation experience, which is an outcome of recreation behavior. The experience and the behavior are not equivalent. It seems as though more use of observed behavior would be useful for validation purposes and to suggest the kinds of reactions visitors may have to changes in visibility, but how these changes affect the visitor's experience is what must be ultimately addressed.

Several methodological questions about the appropriate way to identify and quantify values related to visibility were raised in the papers as well as in the discussions. Stewart pointed out that the survey approaches that have been used to date presume that respondents have a clear set of values established in their own minds concerning environmental quality which we need only lead them to reveal. Others voiced similar concerns about using survey results for policy decisions when respondents have not been told the uses to which their responses could be put. Such concerns apply to economic as well as social and psychological methodologies.

The economic approaches attempt to quantify in dollars the importance of protecting visibility and visual resources to park users and those who derive satisfaction from the knowledge that these resources are being protected. Several techniques are being developed and have been applied to elicit values for park users, as presented in this volume, although validation of results across different techniques is still necessary as the accuracy of any one approach is still in question. Initial studies have indicated that preservation values may overwhelm user values, but the theoretical understanding of preservation values is still in developmental stages and methodologies for quantifying such values are currently fraught with difficulties. Economics studies must also strive to respond to the needs of resource managers who may not be interested in how much additional a visitor would pay to prevent a degradation, but may be interested in how visitor behavior and attitudes would change if a degradation occured irrespective of whether visitors could or would pay to prevent these changes. On the other hand, the likelihood is small that individuals will be willing to pay to prevent visibility degradation if they do not consider these changes as adverse, so that these approaches present one quantifiable measure of adverseness.

In summary, more work is still needed in all of the above areas, but it must be integrated and coordinated in a consistent framework to be most useful to those who must implement the findings to make resource management decisions. This suggests that an ongoing working group must try to establish assessment procedures that are within legislative requirements and that are acceptable to land managers, agencies, and industry. The process of developing and preparing to implement such a procedure will require an evaluation of the usefulness of current research and its accuracy. It will also provide a natural mechanism to prioritize future research directions to effectively make resource management decisions.